SUSTAINABLE USE OF WASTEWATER IN AGRICULTURE

Applications, Methods, and Technology

First edition published 2025

Apple Academic Press Inc.
1265 Goldenrod Circle, NE,
Palm Bay, FL 32905 USA

760 Laurentian Drive, Unit 19,
Burlington, ON L7N 0A4, CANADA

CRC Press
2385 NW Executive Center Drive,
Suite 320, Boca Raton FL 33431

4 Park Square, Milton Park,
Abingdon, Oxon, OX14 4RN UK

Library and Archives Canada Cataloguing in Publication

CIP data on file with Canada Library and Archives

Library of Congress Cataloging-in-Publication Data

CIP data on file with US Library of Congress

ISBN: 978-1-77491-908-8 (hbk)
ISBN: 978-1-77491-909-5 (pbk)
ISBN: 978-1-00361-069-4 (ebk)

DEDICATION

To our esteemed mentor

Professor Megh R. Goyal

Senior Technical Editor-in-Chief (Agricultural and Biological Engineering)
Apple Academic Press Inc. (Copartner CRC Press/Taylor and Francis Group)

For your leadership, motivation, and guidance have been instrumental in shaping this book volume.

For your invaluable contributions to the field of science and technology research.

For being a guiding light, providing us with profound impact both personally and academically, throughout our professional careers.

For your mentorship at every stage of our professional and personal development is truly priceless.

For your unwavering support and dedication.

Global Educational Community

ABOUT THE BOOK SERIES: INNOVATIONS IN AGRICULTURAL AND BIOLOGICAL ENGINEERING

Under this book series, Apple Academic Press Inc. is publishing book volumes over a span of 8–10 years in the specialty areas defined by the American Society of Agricultural and Biological Engineers (asabe.org). Apple Academic Press Inc. is a principal source of books in agricultural and biological engineering. We welcome book proposals from readers in areas of their expertise.

The mission of this series is to provide knowledge and techniques for agricultural and biological engineers (ABEs). The book series offers high-quality reference and academic content on agricultural and biological engineering (ABE) that is accessible to academicians, researchers, scientists, university faculty and university-level students, and professionals around the world.

Agricultural and biological engineers ensure that the world has the necessities of life, including safe and plentiful food, clean air and water, renewable fuel and energy, safe working conditions, and a healthy environment by employing knowledge and expertise of the sciences, both pure and applied, and engineering principles. Biological engineering applies engineering practices to problems and opportunities presented by living things and the natural environment in agriculture.

ABE embraces a variety of the following specialty areas (<asabe.org>): aquaculture engineering, biological engineering, energy, farm machinery and power engineering, food and process engineering, forest engineering, information and electrical technologies, soil and water conservation engineering, natural resources engineering, nursery and greenhouse engineering, safety and health, and structures and environment.

For this book series, we welcome chapters on the following specialty areas (but not limited to):

- Academia-to-industry to end-user loop in agricultural engineering
- Agricultural mechanization
- Aquaculture engineering
- Biological engineering in agriculture

- Biotechnology applications in agricultural engineering
- Energy source engineering
- Food and bioprocess engineering
- Forest engineering
- Hill land agriculture
- Human factors in engineering
- Information and electrical technologies
- Irrigation and drainage engineering
- Nanotechnology applications in agricultural engineering
- Natural resources engineering
- Nursery and greenhouse engineering
- Potential of phytochemicals from agricultural and wild plants for human health
- Power systems and machinery design
- GPS and remote sensing potential in agricultural engineering
- Robot engineering and drones in agriculture
- Simulation and computer modeling
- Smart engineering applications in agriculture
- Soil and water engineering
- Micro irrigation engineering
- Structures and environment engineering
- Waste management and recycling
- Rural electrification.
- Sanitary engineering
- Farm to fork technologies in agriculture
- Impact of global warming and climatic change on agriculture economy
- Any other focus areas.

For more information on this series, readers may contact:

Megh R. Goyal, PhD, PE
Book Series Senior Editor-in-Chief: Innovations in Agricultural and Biological Engineering

OTHER BOOKS ON AGRICULTURAL AND BIOLOGICAL ENGINEERING BY APPLE ACADEMIC PRESS, INC.

Management of Drip/Trickle or Micro Irrigation
Megh R. Goyal, PhD, PE, Senior Editor-in-Chief
Evapotranspiration: Principles and Applications for Water Management
Megh R. Goyal, PhD, PE and Eric W. Harmsen, PhD Editors

Book Series: Research Advances in Sustainable Micro Irrigation
Senior Editor-in-Chief: Megh R. Goyal, PhD, PE

- **Volume 1:** Sustainable Micro Irrigation: Principles and Practices
- **Volume 2:** Sustainable Practices in Surface and Subsurface Micro Irrigation
- **Volume 3:** Sustainable Micro Irrigation Management for Trees and Vines
- **Volume 4:** Management, Performance, and Applications of Micro Irrigation Systems
- **Volume 5:** Applications of Furrow and Micro Irrigation in Arid and Semi-Arid Regions
- **Volume 6:** Best Management Practices for Drip Irrigated Crops
- **Volume 7:** Closed Circuit Micro Irrigation Design: Theory and Applications
- **Volume 8:** Wastewater Management for Irrigation: Principles and Practices
- **Volume 9:** Water and Fertigation Management in Micro Irrigation
- **Volume 10:** Innovation in Micro Irrigation Technology

Book Series: Innovations and Challenges in Micro Irrigation
Senior Editor-in-Chief: Megh R. Goyal, PhD, PE

- Engineering Interventions in Sustainable Trickle Irrigation: Water Requirements, Uniformity, Fertigation, and Crop Performance
- Management Strategies for Water Use Efficiency and Micro Irrigated Crops: Principles, Practices, and Performance

- Nanotechnology Applications in Agricultural and Bioprocess Engineering: Farm to Table
- Nanotechnology Applications in Dairy Science: Packaging, Processing, and Preservation
- Nanotechnology Horizons in Food Process Engineering, 3-volume set:
  - Volume 1: Food Preservation, Food Packaging and Sustainable Agriculture
  - Volume 2: Scope, Biomaterials, and Human Health
  - Volume 3: Trends, Nanomaterials, and Food Delivery
- Nonthermal Plasma Technology for Food and Food Products
- Nonthermal Pulsed Electric Field Technology for the Food Industry
- Novel and Alternative Methods in Food Processing: Biotechnological, Physicochemical, and Mathematical Approaches
- Novel Dairy Processing Technologies: Techniques, Management, and Energy Conservation
- Novel Processing Methods for Plant-Based Health Foods: Extraction, Encapsulation and Health Benefits of Bioactive Compounds
- Novel Strategies to Improve Shelf-Life and Quality of Foods: Quality, Safety, and Health Aspects
- Ohmic Heating Technology for Processing of Foods and Food Products: Design, Food Quality, and Applications
- Phytochemicals and Medicinal Plants in Food Design: Strategies and Technologies for Improved Healthcare
- Processing of Fruits and Vegetables: From Farm to Fork
- Processing Technologies for Milk and Milk Products: Methods, Applications, and Energy Usage
- Quality Control in Fruit and Vegetable Processing: Methods and Strategies
- Scientific and Technical Terms in Bioengineering and Biological Engineering
- Soil and Water Engineering: Principles and Applications of Modeling
- Soil Salinity Management in Agriculture: Technological Advances and Applications
- State-of-the-Art Technologies in Food Science: Human Health, Emerging Issues and Specialty Topics
- Sustainable and Functional Foods from Plants: Health Impact, Bioactive Compounds, and Production Technologies

- Sustainable Biological Systems for Agriculture: Emerging Issues in Nanotechnology, Biofertilizers, Wastewater, and Farm Machines
- Sustainable Green Nanomaterials: Synthesis, Characterization, and Engineering Applications
- Sustainable Nanomaterials for Biomedical Engineering: Impacts, Challenges, and Future Prospects
- Sustainable Nanomaterials for Biosystems Engineering: Trends in Renewable Energy, Environment, and Agriculture
- Sustainable Use of Wastewater in Agriculture: Applications, Methods, and Technology
- Technological Interventions in Dairy Science: Innovative Approaches in Processing, Preservation, and Analysis of Milk Products
- Technological Interventions in Management of Irrigated Agriculture
- Technological Interventions in the Processing of Fruits and Vegetables
- Technological Processes for Marine Foods, From Water to Fork: Bioactive Compounds, Industrial Applications, and Genomics
- The Chemistry of Milk and Milk Products: Physicochemical Properties, Therapeutic Characteristics, and Processing Methods

ABOUT THE EDITORS

Megh R. Goyal, PhD, PE

Retired Professor in Agricultural and Biomedical Engineering, University of Puerto Rico, Mayaguez Campus; Senior Acquisitions Editor, Biomedical Engineering and Agricultural Science, Apple Academic Press, Inc.

Megh R. Goyal, PhD, PE, is currently a Retired Professor of Agricultural and Biomedical Engineering from the General Engineering Department at the College of Engineering at the University of Puerto Rico–Mayaguez Campus (UPRM); and Senior Acquisitions Editor and Senior Technical Editor-in-Chief for Agricultural and Biomedical Engineering for Apple Academic Press Inc. During his professional career, he has worked as a Soil Conservation Inspector; Research Assistant at Haryana Agricultural University and Ohio State University; Research Agricultural Engineer/Professor at the Department of Agricultural Engineering of UPRM; and Professor of Agricultural and Biomedical Engineering in the General Engineering Department of UPRM. He spent a one-year sabbatical leave in 2002–2003 at the Biomedical Engineering Department of Florida International University, Miami, USA.

Dr. Goyal was the first agricultural engineer to receive the professional license in agricultural engineering from the College of Engineers and Surveyors of Puerto Rico. In 2005, he was proclaimed the "Father of Irrigation Engineering in Puerto Rico for the Twentieth Century" by the American Society of Agricultural and Biological Engineers, Puerto Rico Section, for his pioneering work on micro irrigation, evapotranspiration, agroclimatology, and soil and water engineering.

During his professional career of 52 years, he has received many awards, including Scientist of the Year, Membership Grand Prize for the American Society of Agricultural Engineers Campaign, Felix Castro Rodriguez Academic Excellence Award, Man of Drip Irrigation by the Mayor of Municipalities of Mayaguez/Caguas/Ponce and Senate/Secretary of Agriculture of ELA, Puerto Rico, and many others. He has been recognized as one of the experts "who rendered meritorious service for the development of the irrigation sector in India" by the Water Technology Centre of Tamil Nadu Agricultural University in Coimbatore, India, and ASABE (American Society of Agricultural and Biological Engineers), which was bestowed on him the 2018 Netafim Microirrigation Award.

Dr. Goyal has authored more than 200 journal articles and edited more than 100 books. AAP has published many of his books, including *Management of Drip/Trickle or Micro Irrigation; Evapotranspiration: Principles and Applications for Water Management;* ten-volume set on *Research Advances in Sustainable Micro Irrigation.* He has also authored the textbooks *Elements of Agroclimatology* (Spanish) by UNISARC, Colombia, and two *Bibliographies on Drip Irrigation.* Dr. Goyal has also developed several book series with AAP, including Innovations in Agricultural & Biological Engineering (with over 60 titles in the series to date), Innovations and Challenges in Micro Irrigation; and Innovations in Plant Science for Better Health: From Soil to Fork.

Dr. Goyal received his BSc degree in Engineering from Punjab Agricultural University, Ludhiana, India, and his MSc and PhD degrees from the Ohio State University, Columbus, Ohio, USA. He also earned a Master of Divinity degree from the Puerto Rico Evangelical Seminary, Hato Rey, Puerto Rico, USA.

Gaurav Saini, PhD

Gaurav Saini, PhD, is working as an Associate Professor & Head of the Department of Civil Engineering at Netaji Subhas University of Technology, Delhi, India. His current areas of research include sustainability, natural materials, pollution control, pollution and human health, and, broadly, environmental engineering. He has guided 48 theses at the master's level and five at the doctoral level (one awarded), in addition to numerous BTech-level projects. His research work has resulted in several research papers in journals of repute, conference presentations at the international level, and 20 patent filings (fourteen awarded). He serves as a resource person for various governmental bodies and universities for academic, research, and consultancy-related endeavors. He also serves as an ad-hoc reviewer for more than 25 international journals. He has edited one book and guest-edited a special issue of a Springer journal. He has more than a decade's experience in teaching at undergraduate and graduate levels and has been actively engaged in research for the last two decades. He is an alumnus of Delhi College of Engineering, India, where he earned his Bachelor of Engineering degree. He is also an alumnus of the Oregon State University, United States, where he earned his MS and PhD degrees in Environmental Engineering. He also completed a postdoctoral at the University of Delaware, USA.

CONTENTS

CONTRIBUTORS

Shehu Abdulmumin
Transmission Company of Nigeria, Nigeria

Muhammad Isa Abdurrahman
School of Environmental, Civil, Agricultural, and Mechanical Engineering, University of Georgia, Athens, Georgia, USA

Priya Agarwal
Department of Civil Engineering, Sharda University, Greater Noida, Uttar Pradesh, India

Muhammad Makarfi Ahmad
Department of Agricultural Economics and Extension, Bayero University Kano (BUK), Kano, Nigeria

Usman Usman Aliyu
Department of Civil Engineering, Sharda University Greater Noida, Delhi-NCR, India
Department of Civil Engineering, Federal University, Dutsin MA, Katsina State, Nigeria

Thamer Alomayri
Department of Physics, Faculty of Applied Science, Umm Al-Qura University, Makkah, Saudi Arabia

Rumaisa Ashraf
Division of Entomology, Sher-e-Kashmir University of Agricultural Sciences and Technology-Kashmir, Srinagar, Jammu and Kashmir, India

Yusuf Bashir
Department of Civil Engineering, Kano University of Science and Technology, Wudil, Kano, Nigeria

Javid Ahmad Bhat
Division of Soil Science, Sher-e-Kashmir University of Agricultural Sciences and Technology- Kashmir, Wadura Sopore, Jammu and Kashmir, India

M. H. Chesti
Division of Soil Science, Sher-e-Kashmir University of Agricultural Sciences and Technology- Kashmir, Wadura Sopore, Jammu and Kashmir, India

Abdul Gani
Department of Civil Engineering, Netaji Subhas University of Technology West Campus, Jaffarpur, New Delhi, India

Lasmar Garba
Department of Civil Engineering, Lovely Professional University, Punjab, India
Department of Works, Jigawa State Polytechnics, Dutse, Nigeria

Akash Garg
Department of Civil Engineering, Gautam Buddha University, Gautam Budh Nagar, Greater Noida, Uttar Pradesh, India

Mustapha Nuhu Garko
Department of Civil Engineering, Kano University of Science and Technology, Wudil, Kano, Nigeria

ABBREVIATIONS

Å	Angstrom
ADI	acceptable daily intake
ADP	Agricultural Development Projects
AI	artificial intelligence
ANOVA	analysis of variance
ARGs	antibiotic resistance genes
AWM	agricultural water management
B	boron
BIS	Bureau of Indian Standards
BOD	biochemical oxygen demand
BRIC	Brazil, Russia, India, and China
Ca	calcium
CCME	Canadian Council of Ministers for the Environment
Cd	cadmium
CEC	cation exchange capacity
CEP	Center for Economic Performance
CGIAR	Consultative Group for International Agricultural Research
CNT	carbon nanotube
COD	chemical oxygen demand
CRF	capital recovery factor
Cu	copper
D	difference of m and s
DIM	daily intake of metal
DO	dissolved oxygen
E	energy
EC	electrical conductivity
EC	emerging contaminant
E_g	band gap energy
EI	evaporation isograms
ENM	engineered nanomaterial
EPA	Environmental Protection Agency
EU	European Union
FF	free flow
FAO	Food and Agriculture Organization

TOC	toxic organic compound
TWW	treated wastewater
UV	ultraviolet radiations
V F	vertical flow
v	frequency
V	volume
γ_{AL}	correlation between operators' age and the quantity of water lifted
γ_{AW}	correlation between age and weight
γ_{WL}	correlation between operators' weight and the quantity of water lifted high
WA	weighted arithmetic
WHO	World Health Organization
WQI	water quality index
WRECA	Water Resources & Engineering Construction Agency
WSN	wireless sensor network
WUA	Water User Associations
WWI	wastewater irrigation
WWTP	wastewater treatment plant
Zn	zinc
β	line broadening at the half-maximum intensity
Θ	Bragg angle
λ	wavelength
τ	domain size or particle size
φ	work function

PREFACE

Agriculture is the mainstay for a significant portion of the populations of developing and underdeveloped countries. It not only provides employment but also serves as a key economic activity contributing a dominant share in the gross domestic product (GDP) of these countries. According to the world bank data, the agriculture (along with forestry and fishing) sector contributed to 4.3% of the world's GDP in 2022. This value was 25% for countries classified as "low-income" compared to 15.4% for countries classified as "lower-middle-income." Similarly, for South Asia, the contribution to GDP was 16.5 and 17.3% for sub-Saharan Africa. These values show the critical importance of agriculture in the lives of millions (and perhaps billions) of people. Unfortunately, these regions are also plagued by a scarcity of resources and a severe dependence on nature for irrigation requirements. This coupled with climate change and variabilities (in both geographical and duration/intensity matrices) of precipitation has resulted in extreme hardships for the farmers, who face the price inflation.

To minimize such hardships while avoiding overexploitation of limited water resources, sustainable management of sources of water and wastewater for agriculture production are needed and have been studied throughout the world. One such promising source is the wastewater generated by municipalities and industries, which can potentially serve as a source of irrigation water, with or without prior treatment. Rainwater and farm recycled water are also good potential resources for irrigation. This, however, is not without its own drawbacks, which include the uptake of heavy metals and emerging contaminants by the resulting crops, especially if the irrigation water was derived from industrial sources without appropriate treatment. In addition, several low-income countries may not even have the appropriate resources or technologies for implementation of such schemes at commercial scales.

This edited volume is a timely and much-needed addition to knowledge on the new sources of water for agriculture with a special focus on low- and middle-income regions of the world. The themes explored in this book include the application of wastewater for irrigation, concerns associated with it, innovative methods of its application, current state of affairs, and application of latest technology in this domain. The overarching theme

CHAPTER 1

TREATED WASTEWATER FOR AGRICULTURAL USAGE: A REVIEW

MANJEETA PRIYADARSHI, ATHAR HUSSAIN, and GAURAV SAINI

ABSTRACT

With inexorable increase in global populace and rapid industrialization, demand of water has increased at an alarming rate. This has increased an enormous breach between the demand and supply of water such that it is increasing a risk to human existence. However, the researchers around the world are looking for other options and best alternative practices of conserving the water. Also, large metropolitan cities have limited available land area with limited space for wastewater treatment and are hence forced to discard wastewater, which is a foremost issue the metropolitan cities and government bodies face. However, this wastewater is a prominent resource that contains an ample amount of nutrients and minerals and therefore can be best utilized for agricultural, aquaculture, and other usages after proper treatment. This makes the agriculture production cost effective when close to such treatment facilities as it requires the considerations of: (i) reuse safety/guidelines, (ii) present reuse status, (iii) production, (iv) future availability, (v) storage facilities, (vi) governance/policies/ legislation/monitoring and compliance, and (vii) long-distance transportation networks and pumping requirements. In developed and developing countries the treated wastewater can be utilized in various areas such as irrigation of fiber, feed, and seed yields, landscaping (wildlife habitat improvement, playgrounds, golf courses, schoolyards, and parks), industry, freeways, groundwater recharge, construction, dust control, and aquaculture. However, the reuse of wastewater possesses some negative effects on environment, ecology, and health of human beings and therefore needs to be assessed and identified. Therefore, the water to be used

Sustainable Use of Wastewater in Agriculture: Applications, Methods, and Technology. Megh R. Goyal & Gaurav Saini (Eds.)

for agricultural irrigation should be treated thoroughly as per the standard quality requirements that include the reduction of concentration of chemicals in it. Therefore, there is a need of mix-up of measures by chemicals reduction at source and utilizing good agricultural practices with supplementary preventive measures that will confirm safe irrigation practices in the future.

1.1 INTRODUCTION

The unaccounted and haphazard increase in population has resulted in rapid depleting of domestic and other water usage has elevated the level of freshwater demand across the globe. However, the reuse of wastewater through reclamation has gained popularity and is now considered an essential requirement with respect to the current scenario. The worldwide total consumption of water from underground water sources and river water bodies for agriculture accounts 92%, of which irrigation usage accounts 70% of freshwater storage.[26] The wastewater reuse in agriculture is considered to be an ultimate support that replaces the freshwater usage and saves enormous quantity of water that can be utilized in other demanding areas.

The significant use of treated wastewater along with domestic water is because of the presence of different types of nutrients including nitrogen, phosphates, and sulfates as major ones. These nutrients are important stimulants required for the plant cell that helps in growth and reproduction of plants.[17] The concentration of nitrogen and phosphorous as major nutrients in wastewater attracts its usage for irrigation purposes as these nutrients are available in assimilative form and are absorbed by plants easily. This high accessibility of such nutrients in treated wastewater thus lessens the usage of enrichers with improved soil fertility along with increases in crop productivity. This ultimately reduces the cost incurred in production of crops.

In developed countries the usage of treated wastewater for agriculture and other purposes falls under strict rules, regulations, and standards being developed by these countries. However, in developing countries such rules, regulations, and standards exist; still the untreated wastewater is used directly and enforcing any sound regulatory policies is not properly enforced. This ultimately ends up with environmental threats and public health issues. Therefore, considering the above-mentioned facts in view an attempt has been made to review the works and studies being carried out by different researchers and investigators on the issues associated with usage of wastewater for agriculture purposes. In the present review an attempt has been made to discuss the hazard related to the use of untreated wastewater and its exposures. Also, possibilities of reuse of the treated wastewater for agricultural usage and advanced water treatment techniques have also been discussed.

In spite of being an ancient practice, the usage of wastewater for agriculture has not been properly managed and even used without following any norms and standards. Accordingly, the history of mankind indicates that the knowledge pertaining to wastewater use is not new. Wastewater reuse in agriculture comprises further usage of "treated" wastewater for crop irrigation and is recognized as an efficient tool for managing water resources. This ultimately reduces the usage of freshwater supplies that ultimately reduces shortages of water during dry weather seasons or the unbalanced accessibility sources of water from other means for the irrigation of crops all through the hydrological year. However, the United Nations body known as Food and Agriculture Organization (FAO), an international organization, has also developed various guiding principles for the appropriate usage of wastewater to be used for agriculture purpose.[12]

The continuous application of fertilizers and agrochemicals along with wastewater irrigation contaminates the agricultural land with heavy metals as this wastewater from urban areas comes directly to the surface water bodies or it may be discharged from other point sources. The crops production ultimately decreases. This also leads to an additional problem of soil erosion as it is evident that under such conditions, the bacteria, microorganisms, and fungi present in the soil begin to decrease.[8] The use of industrial effluents for irrigation also deteriorates the soil as such effluents contain high concentration of heavy metals and other toxic chemicals. These surface contaminants percolate through the soil and deteriorate it quality. The pollution on the ground surface and surface water bodies is the chief cause of soil pollution.[9]

In this chapter, an attempt has been made to review the community health problems linked with environmental aspects with the effective usage of treated wastewater for farming purposes. Various concepts to handle the problems pertaining to wastewater usage for agriculture purpose with the fact that the ultimate requirement for overall approaches for cultivation has been discussed and projected.

1.2 SOURCES OF WASTEWATER

The wastewater discharged from households, institutions, and commercial buildings, storm water, industries and by groundwater seepage termed as sewage generally enters the municipal sewerage network system (Figures 1.1

and 1.2). The wastewater discharged from industries known as industrial effluent is discharged by various industrial units and plants located around a city or an urban area.

Around the globe especially in developing countries the industrial wastewater from various activities in or around the city is discharged directly to the municipal wastewater sewerage system. In developed countries the cities consist of separate systems, while the sewerage network carries the storm water too in developing countries like India.[10]

(A) (B) (C)

FIGURE 1.1 Various sources of wastewater: (A) domestic sources, (B) agricultural sources, and (C) industrial sources.

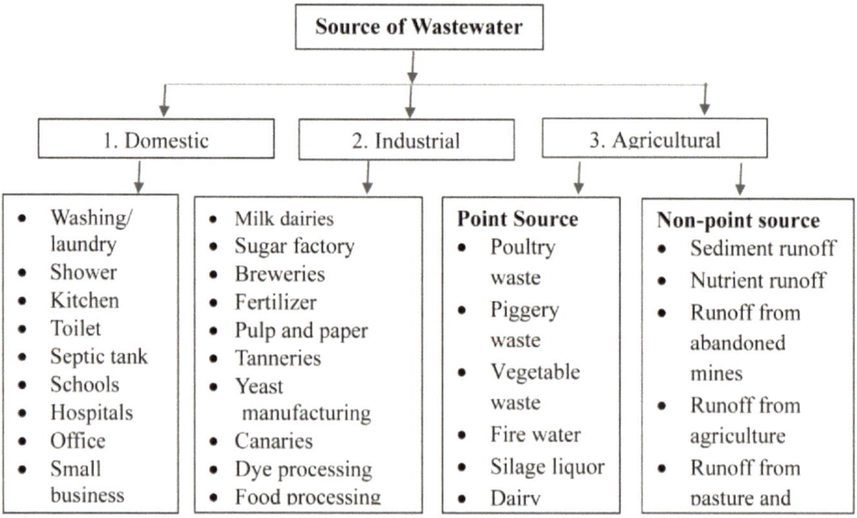

FIGURE 1.2 Classification of water based on sources.

1.3 CHARACTERISTICS OF TREATED WASTEWATER FOR AGRICULTURAL USE

The production of greenhouse crops is another critical aspect that depends upon the irrigation water quality and depends on many factors. The presence of alkalinity and soluble salts with required pH values are important one. The occurrence of hardness, salts of magnesium and calcium, and heavy metals are other important factors that are also considered. This may be due to the presence of such parameters in water can clog irrigation systems or considered toxic ions. Therefore, prior to usage, the characteristics of such water should be determined when used for agricultural purpose.

However, the poor quality of water when used for agriculture can affect quality and growth of the crops. It results in slow growth and ends up with poor aesthetic quality and sometimes the plant even dies eventually. The existence of soluble salts in higher concentration in irrigation water injures the plant root directly and creates interference in nutrient uptake of plants. Such salts in water enter the plant and accumulate in the leaves, thus causing the burning of plant leaf margins at the edges. The high pH of irrigation water results in high alkalinity and the growing medium significantly. This ultimately results in interference of nutrient uptake in plants, thus ending with nutrient deficiencies resulting in poor plant health. The quality and characteristics of the treated wastewater can be categorized as physical, chemical, and biological.

1.3.1 PHYSICOCHEMICAL CHARACTERISTICS OF WASTEWATER USED FOR IRRIGATION

The wastewater (treated or untreated) should satisfy the appropriate quality criteria for it to be used for irrigation purposes. These standards are summarized in Tables 1.1 and 1.2. Individual water quality parameter-wise details are described as follows:

- **Suspended solids:** The presence of suspended solids and dissolved solids is an indicator of organic matter in the wastewater or industrial effluents. When such untreated wastewater containing suspended solids containing a part of organic fraction is discharged into streams or surface water bodies, it ultimately ends up with high sludge deposits and subsequently leads to anaerobic/septic conditions. This

NH_4-N in water is corrosive and high phosphorus concentration in water leads to scale formation. The water quality criteria for water to be used for irrigation purpose: the range of nitrogen and phosphorus in secondary and tertiary effluents should be 10–30 mgN/L and 0.1–30 mgP/L respectively, as well as for recycled water it should be <10–15 mgN/L and <0.1–2 mgP/L, respectively.

- **Sodium and chloride:** Surface and ground water bodies such as streams, lakes, rivers, ponds, lagoons, and private tube wells are considered to be major sources of water to be used for agriculture purposes. There is a possibility of presence of excess sodium (Na) and chloride (Cl) in the water. Wastewater generally contains minimum level of Na and Cl, maybe due to the low applications of road salt on the nearby locations of reservoirs and public wells. However, the water from wells and ponds used for irrigation may contain high Na and Cl concentration. This is because of the excessive runoff from snowmelt or during summer seasons when water level becomes low especially during drought period. According to FAO standards, prescribed limit for Na and Cl for irrigation water lies in the range of 0–40 and 0–30 meq/L, respectively.[6]
- **Potassium and sulfate:** The presence of nutrients in many plants generally exists at lower concentrations. When water to be used for irrigation contains these nutrients at higher concentrations, this indicates the pollution due to the contaminants or fertilizers. In fertilizers, an essential element sulfur is not included which is useful for plant growth. It is analyzed especially in the irrigation water to provide a knowledge of available deficiency problems. Thus, sulfate as a supplemental may be added for the proper growth of plant if concentration is lower than 50 ppm. The prescribed limit for sulfate for water to be used for irrigation purpose as per FAO standards for irrigation water lies in the range of 0–20 meq/L.[6]
- **Bacterial parameter (fecal coliform):** In sewage, the dissolved suspended solids present as an organic matter is the very useful parameter in the sewage. While considering the human health aspect, the focus is to be upon the major bacterial parameters viz. Fecal Coliform (FC), which shows the fecal pollution. Adaptability of bio-fouling for sprinklers and nozzles in irrigation system may reduce the risk of infection which can cause due to pathogens. The water quality criteria for water to be used for irrigation purpose are that the range of suspended solids in secondary and tertiary effluents should

be <10–10^7cfu/100 mL, as well as for recycled water it should be <1–200 cfu/10 mL.

- **Biological oxygen demand (BOD) and chemical oxygen demand (COD):** BOD indicates the quantity of oxygen essential by the microorganisms to degrade the available active organic matter completely in aerobic environment, whereas COD is the total oxygen required to oxidize the available organic matter completely using potassium dichromate as an oxidizer. However, when organic matter is present in a distribution system it can lead to bacterial regrowth and microbial fouling. The range of BOD and COD as per water quality criteria for irrigation in secondary and tertiary effluents should be 10–30 mg/L and 50–150 mg/L, as well as for recycled water it should be <5–45 mg BOD/L and <20–200 mg COD/L, respectively.
- **Electrical conductivity** is a parameter generally used to measure the potential of a solution or a substance to pass an electric current. Electrical conductivity indicates the presence of dissolved and dissociated substances. Conductivity can be impacted by different parameters such as ionic valences, temperature, and ionic mobility. Also, to check the presence of sea water in the groundwater, and to monitor desalination in plants, electrical conductivity can be obtained. The prescribed limit for sulfate for water to be used for irrigation purposes as per FAO standards for irrigation water lies in the range of 0–3 dS/m.[6]
- **Heavy metals:** Any element that possesses a silvery lustre with good conduction of heat and electricity is termed as metal. Metals can be classified by their known essentiality to living organisms as essential nonliving elements. Some dissolved mineral salts are termed as nutrients such that they support the plant growth and are considered stimulants. However, some of the specific elements such as Chromium (Cr), Arsenic (As), Cadmium (Cd), Lead (Pb), Mercury (Hg), Zinc (Zn), and Nickle (Ni) are hazardous to vegetations beyond certain limits, and therefore highest concentration limits for such elements in water to be used for agriculture purpose have been fixed. The criteria of quality of water to be used for irrigation purpose are that the range of suspended solids for recycled water should be 0.001 mgHg/L, 0.01 mgCd/L, and 0.02–0.1 mgNi/L.
- **Chlorine residual:** The addition of chlorine helps to restrict the growth of various types of bacteria and microbes in water and also helps in destroying the trace amount of remaining organic matter in treated wastewater. Also, some sensitive crops can be damaged when the free

chlorine excessive concentration is greater than 0.05 mg/L. The water quality criteria for water to be used for irrigation purpose are that the range of suspended solids for recycled water should be 0.5–5 mgCl/L.

1.4 WASTEWATER TREATMENT

Wastewater treatment depends upon the characteristics of effluent emanating from domestic, industrial, and commercial activities and concentration of impurities and contaminants present in it. The conventional wastewater treatment system comprises combination of various units and the process can be bifurcated as unit operations and unit processes. Both operation and processes are designed with an objective to reduce the concentration of different contaminants present in wastewater within acceptable limits. The systems can be categorized as (i) primary treatment, (ii) secondary treatment, and (iii) tertiary treatment (Figure 1.3).

1.4.1 PRIMARY TREATMENT

The first step of wastewater treatment is utilized with an objective to eliminate the large floating debris and solids materials present in the wastewater. The first unit of primary treatment process is screens that are generally intended to remove debris floating material existing in wastewater. However, grinding devices are generally provided before the screens so that the size of the bigger floating materials can be reduced considerably. After screens, Grit Chamber is another unit being provided in the treatment method in order to remove the inorganic solids present in the wastewater through sedimentation. Thereafter, oil and grease traps are provided to eliminate the oil and greasy material present in the wastewater.

The removal of oil and grease at the initial step will not hamper the efficiency of biological systems in secondary treatment. Thereafter, sedimentation process is applied to remove the organic solids that can settle out by mere action of gravity while providing suitable hydraulic retention time to the wastewater. This separation of solids and liquid is a physical process where the solids are separated by using a mechanical process. However, in order to enhance the removal efficiency of sedimentation process some chemicals such as Alum and Ferric salts are also used. During the entire process of primary treatment around 20–30% of the BOD along with 50–65% of total suspended solids can be reduced from the wastewater.

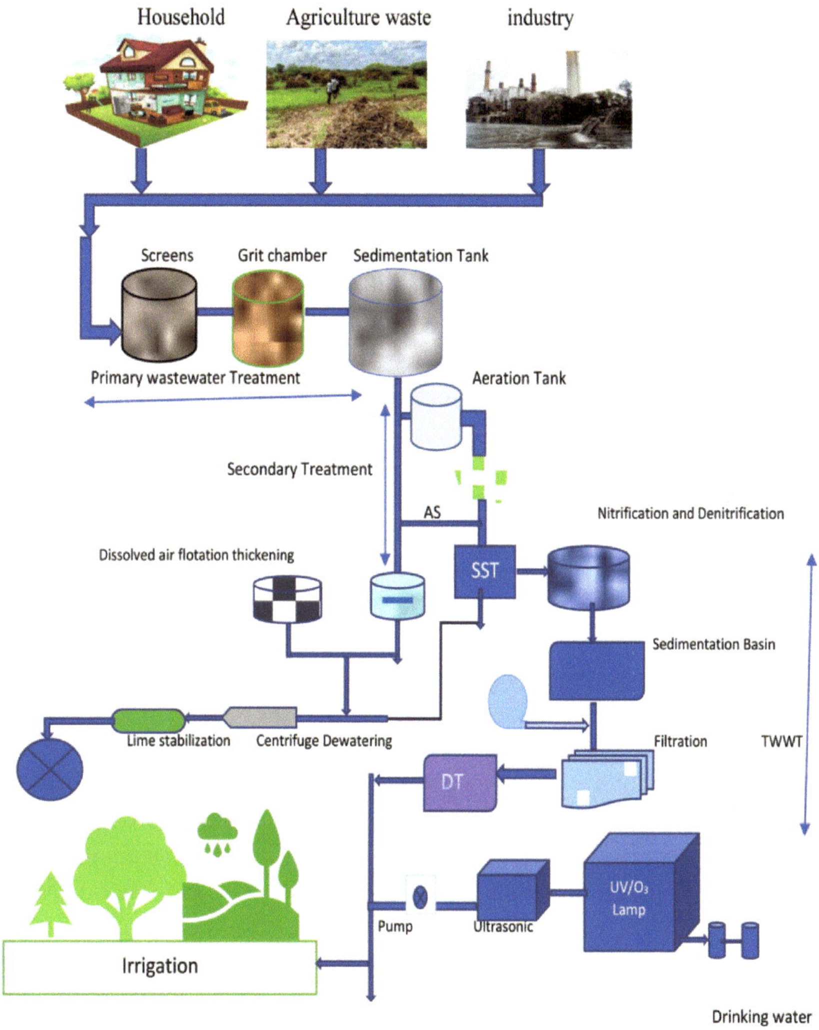

FIGURE 1.3 Various methods for wastewater treatments for irrigation use.

1.4.2 SECONDARY TREATMENT

The effluent from primary sedimentation tank still contains colloidal and dissolved organic matter that contains high BOD concentration. Therefore, it is important to remove the remaining colloidal and dissolved organic matter through the use of microorganisms that can utilize this organic matter as a substrate and some portion of the dissolved and colloidal organics is converted into biomass. The excessive biomass can be removed subsequently by sedimentation. This entire process of solids removal biologically is termed as secondary treatment. The bacteria utilize the available organic matter as food, and the end products are water, carbon dioxide, and biomass as new cells in the form of sludge. The entire process of biological treatment followed by secondary sedimentation removes around 85% BOD from the system in secondary treatment.

Various conventional treatment schemes consisting of aerobic and anaerobic processes are generally utilized to treat the various types of wastewater. Activated sludge process is such a popular aerobic process generally used worldwide. In this system, supply of required oxygen to the microorganism is being provided through aeration that provides the intimate contact between these microorganisms and organic matter present in wastewater through turbulence. These microorganisms thus utilize the oxygen and metabolize the organic matter present in the wastewater and ultimately reduction in BOD takes place in the system. However, recirculation of biomass portion from the secondary clarifier is carried out in order to maintain an optimum substrate to microorganisms ratio in the system. This ultimately optimizes the BOD/COD removal efficiency in the activated sludge process. The trickling filter system is another example of aerobic process that has gained wide popularity around the globe in terms of treatment of wastewater in large scale for megacities and metropolitan cities.

In trickling filter system wastewater is trickled and passed through a bed of coarse aggregates defined as media over which microorganisms are allowed to grow in the form of biofilm. When wastewater containing organic matter comes in contact with the biofilm attached on the solid media, the residing bacterium in the biofilm metabolizes the organic matter before it could reach to effluent. In a conventional trickling filter system coarse aggregates or stones are used as media. However, in high-rate trickling filters plastic saddles or rings are used as media as it contains high surface area and therefore the area of biofilm growth is higher. This makes the system possible to handle high organic loading rates and thus large quantum of wastewater can be treated with lesser area.

1.4.3 TERTIARY TREATMENT

The conventional systems of wastewater treatment have some limitations as it can remove the organic matter efficiently, but the wastewater still contains some other contaminants in the form of nutrients. These nutrients in the form of nitrogen and phosphorus should be restricted at the effluent end so that when reclaimed wastewater is disposed in surface water bodies it should not harm the ecology of the surface water system. The nitrogen and phosphorus are major nutrients that are generally present in all types of wastewaters and at high concentrations they can led to the excessive enrichment of algal proliferation that results in eutrophication problems in surface water bodies.

The tertiary treatment process includes nitrification and denitrification process that eliminates the nitrogen from the wastewater. Another important method is removal of phosphorus aerobically and anaerobically. The tertiary treatment processes also include chlorination that kills all harmful microorganisms and remaining organic matter is eliminated, filtration through activated carbon or biochar and chemical precipitation for elimination of many heavy metals. The wastewater can also be utilized for drinking purpose after thorough treatment using ultrafiltration, UV disinfection, ozonation, etc.[21]

1.5 SOIL QUALITY PARAMETERS

The quality of any type of soil can be measured by using some soil quality parameters. It is a simple attribute that assesses the quality with respect to its function. The selection of these attributes is an important task that is helpful in the determination of complexity of the soil and remarkably many soil parameters can be determined. The factors that are affecting and influencing the quality of a soil can be bifurcated in main three categories (physical, chemical, and biological) as listed below:

Physical Factors

- Aeration, drainage, structure, texture, density: Physical environment
- Soil depth and water-holding capacity
- Water and erosion of wind: soil erosion

Chemical Factors

- Ion exchange phenomena
- Nutrient availability: capacity and intensity factor
- Presence of toxic elements
- Soil reaction: acidic, saline, sodic soil

Biological Factors

- Activity of enzymes in soil

gets influenced by sodium, calcium, magnesium, and bicarbonate contents in the soil. Permeability index (PI) of the groundwater samples can be determined[3] as follows:

$$PI = \frac{Na^{+} + (HCO_3^{-})}{(Ca^{2+} + Mg^{2+} + Na^{+})} \times 100 \tag{1.6}$$

The presence of calcium, magnesium, sodium, and carbonate in water for agriculture for long duration affects the soil quality on which crops are grown. The permeability of soil is greatly affected due to long exposure with the use of such water. It is recommended to have a PI value <80 for agriculture purpose.

However, all these ionic concentrations are measured in terms of meq/L (milli equivalent per liter). The hydro-geo-chemical parameters should be compared with all international and national standards for addressing the suitability of groundwater for agriculture purpose.[21] Table 1.3 summarizes the applicable values of all these parameters for ascertaining their suitability for irrigation purpose.

TABLE 1.3 Different water quality parameters for Agriculture Use

S. No.	Parameters	Range	Class
1.	SSP	<200	Suitable
		>200	Unsuitable
2.	SAR	<20	Excellent
		20–40	Good
		40–60	Permissible
		60–80	Doubtful
		>80	Unsafe
3.	RSC	<1.25	Safe
		1.25–2.50	Marginally suitable
		>2.50	Unsuitable
4.	MAR	<50	Suitable
		>50	Unsuitable
5.	KR	<1	Suitable
		>1	Unsuitable
6.	PI	<80	Good
		80–100	Moderate
		100–120	Poor

1.7 WASTEWATER USAGE IN AGRICULTURE AND ITS IMPACTS

1.7.1 COMMUNITY HEALTH

The untreated wastewater when used for irrigation possesses many health hazards to human beings of various age groups due to presence of harmful microorganisms and toxicants in it. This health risk may vary from one age group to the other. The untreated wastewater when used for irrigation can have a tendency of having high frequency of hookworm and Ascariasis infections among lower age groups like children. Several findings in the literature indicate the fact that the community health hazards evaluation is generally linked with wastewater used for agriculture purpose. The public health risk is a parameter used to be a decision variable in wastewater to be used for irrigation. The potential exposure group that are generally affected are adult population as well as children. The second important criteria are such that the whole populace residing outside and within a wastewater zone of irrigation, are possible exposed groups for economic evaluation purposes.

1.7.2 CROPS

It is reported that most of the crops when irrigated with wastewater provides very high yield that ultimately proves to be economical as the use of chemical fertilizers reduces considerably. It is also observed that nitrogen dose has a significant effect on the production of crops. When such crops are irrigated with wastewater the high concentration of nitrogen proves to be stimulating in terms of vegetative growth. However, it can delay in ripening and maturity of such crops and ultimately it can result in yield losses.

1.7.3 SOIL RESOURCES

The wastewater may contain high salts that can accumulate in the root zone and ultimately results in harmful impacts on crop yields and health of soil health. These salts when leaches through the soil underneath the root zone cause pollution of soil and groundwater. The crops when irrigated with wastewater also affect flora and fauna of the soil as the wastewater percolation results in heavy metals transportation to the soil and can cause deterioration of crops. It is reported that the occurrence of high concentration of nutrient such as nitrogen and phosphorus, excessive total dissolved solids, diarrhea disease, of which 88% are attributable to contaminated water supply, sanitation, and hygiene.[13]

1.8 BENEFITS OF WASTEWATER USE IN AGRICULTURE

Farmers in several countries utilize the wastewater for irrigation purposes. A summary of the same is provided in Table 1.4. The wastewater application in agriculture has its own pros and cons and the same are listed in the following sections:

1.8.1 BENEFITS FOR AGRICULTURE

- Food security contribution with high urban agricultural production
- High nutrient content in wastewater results in increased crop yields, as compared with freshwater.
- In urban areas increase in employment and income
- Large share of women in poor subsistence farmers results in improved livelihoods for urban agriculturalists.
- Low cost and reliable, irrigation water supply system

1.8.2 ENVIRONMENTAL BENEFITS

- It also helps to reduce the surface water pollution as the wastewater is not discharged into surface water bodies such as rivers or lakes; this ultimately reduces the environmental pollution problems such as eutrophication, foaming, dissolved oxygen depletion, and fish kills.
- The conservation of soil because of build-up humus and therefore the prevention of land erosion takes place.
- The irrigation and fertilization of tree belts helps in desertification control and reclamation of desert.[18]
- The lesser requirement and reduced usage of chemical fertilizers results in reduction of contaminants and the cost involved in their elimination from wastewater.
- This results in conservation of freshwater resources where its usage is more rational especially in arid and semiarid areas.
- Various environmental benefits for schemes related to wastewater management are utilized properly.

TABLE 1.4 Comparison of Treated Wastewater and Freshwater Consumption Status in Various Countries.[17]

Country	Water-utilizing sectors	Status of water reuse (major sectors reusing water	Utilization (%)	Refs.
Europe	Agriculture	Landscape irrigation	44	[5]
Greece	Irrigation	Agricultural irrigation	83	[7]
India	Agriculture	Agricultural irrigation	87	[14]
South Africa	Agriculture	Landscape and sports field irrigation	60	[1]
USA	Freshwater thermoelectric plants	Agricultural irrigation	41	[16]

1.9 UNTREATED WASTEWATER AND ECOLOGICAL TOXICITY

The untreated wastewater when disposed directly in surface water bodies or used for crop irrigation results in bacteriological pollution and heavy metals along with development of pathogens in crops and soil. Also, when soils are irrigated with untreated wastewater continuously for long durations it has determinantal effects like soil salinity. It is also important to mention that the inappropriate use of wastewater for agriculture purpose makes it risky and environmentally aggressive. The crops irrigated with untreated effluents such as sewage, sewage sludge, industrial effluents, and agricultural wastewaters ultimately results in heavy metals accumulation in crops and soil of the area and it enters into the food chain resulting in several health issues and problems among the communities.

1.9.1 WASTEWATER IMPACT ON AGRICULTURE

Around the globe it is observed that the farming sector is the chief consumer of water, and it is assessed that the total water usage around the globe is approximately 70% of water being used for agriculture purposes.[25] However, around the world wastewater is used for irrigation of approximately 20 million hectares of land, thus indicating the wastewater to be the major source for irrigation of which maximum wastewater is untreated.[4] In many developing countries the wastewater used for irrigation purposes is either partially treated or untreated.[22] This nontreated wastewater mostly comprises huge concentration of chemicals and toxic pollutants, toxic wastes from industrial discharges, food-processing wastes, pesticides, heavy metals, chemical fertilizers, textile, leather, paper, and waste from sewage. The WHO has also advised not to use the untreated wastewater for irrigation as this direct use has substantial health implications.[24] The health issues exposure pathway of crops irrigated with wastewater is shown in Figure 1.4.

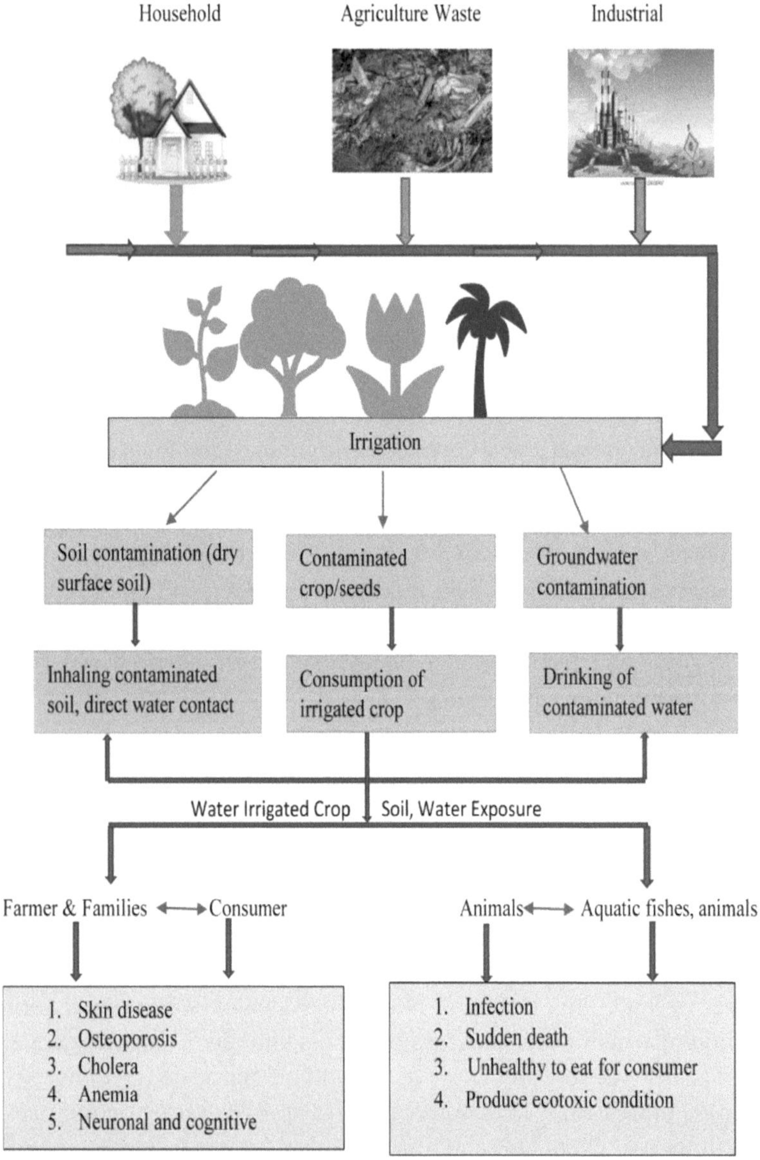

FIGURE 1.4 Health issues exposure pathway of crops irrigated with wastewater.

1.10 REUSE, RECLAMATION, AND RECYCLING OF WASTEWATER

1.10.1 WATER REUSE

The treated wastewater for industry, agriculture, and numerous other direct and indirect usages has certain advantages. The wastewater when used indirectly or is reused exits within the context of natural water systems such as rivers, aquifers, and canals. It can also be defined in terms of indirect potable reuse of wastewater. The global hydrologic cycle also reflects the ultimate indirect reclaim wastewater. However, the direct reclaim of treated wastewater is its beneficial reuse for agriculture, industry, and other applications. The reclaimed water for potable uses is an example of direct potable reuse of treated wastewater.

1.10.1.1 ADVANTAGES OF REUSE IN AGRICULTURE

- The implementation of training and awareness programs for the staff members also maximizes the organic carbon and nutrients.
- The prevention of wastewater runoff or excessive infiltration minimizes the risk of polluting the ground water and surface water resources.
- The suitable separation distances for irrigation help in preventing the ecological annoyance by classifying the wastewater for its pretreatment necessities.
- The use of salt elimination techniques by choosing appropriate land, feasible and lenient crops, and appropriate cropping exercises also proves to be helpful.
- The utilization of sustainable irrigation applications also minimizes soil degradation and damage that can be caused to crops.
- Using suitable irrigation equipment thus prevents public and animal health impacts.

1.10.2 RECLAIMED WASTEWATER: MAJOR AREAS OF REUSE

Reclamation involves all processes used to treat wastewater so that it can be beneficially reused. The major areas of usage are as:

- Artificial lakes and for flushing of toilet
- Concrete mixing during construction activities along with dust control
- Dust suppression, general-surface, and floor washings
- Feasible processes: industry

- Flushing: domestic
- Gardens and parks, irrigation, crops
- Golf course irrigation and community parks
- Landscape and agriculture
- Meting out water requirements for mills and other such plants
- Oil refineries and cooling water requirements for thermal power plants

1.10.2.1 ADVANTAGES: USE OF RECLAIMED WATER IN AGRICULTURE

- The use of treated wastewater is more prevalent due to environmental constraints, deterioration of water quality, and shortages of water.
- More and more stringent rules on water quality requirements and economy related to achieving this quality.
- Moderate requirements in wastewater treatment attract the agricultural reuse and make it advantageous because of the occurrence of nutrients in wastewater and for soil amendments.
- The treatment plants should be located near the agricultural areas and cash crops grown on such soils adds economy value.

1.10.3 WASTEWATER RECYCLING: ADVANTAGES

- Reuse of wastewater back in the same cycle where it is generated is termed as recycling. It helps in compensating the effects of:
- Depleting water sources and water scarcity
- Increasing per capita water demand—rapid industrialization and development, shift in life style—more consumerism
- Population increases
- Rising cost of fresh water
- Water conservation for future generations

1.10.3.1 PROBLEMS: USE OF RECYCLED WATER IN AGRICULTURE

- Pathogens can be removed efficiently by using the disinfection process.
- Percolation of nitrate to groundwater, pathogenic hazards to farm workers, and retaining various heavy metals in soil are the major contamination problems.

- The biological process is considered to be the best method that can be utilized to eliminate organic compounds from the effluents, while heavy metals removal can be carried out by using deep-bed sand filtration.

1.11 SUMMARY

Because water resources are under threat due to haphazard growth of population, the reuse potential of the treated wastewater has also been discussed. The increased production of wastewater emanating from industrial, municipal, and farming processes in developing countries including India has proficiency to aid as a substitute of freshwater resource for its use in paddy farming. The wastewater should be treated thoroughly and appropriately, and distribution of such types of wastewater should be carried out with suitable techniques.

The untreated wastewater when disposed directly in surface water bodies or used for crop irrigation results in heavy metals and bacteriological pollution along with contact of pathogens in crops and soil. Therefore, an attempt has been made to discuss the suitable techniques for proper usage of wastewater to be used for irrigation. Conventional wastewater treatment methods have also been discussed, thus giving a big possibility of better water quality for irrigation purpose. Various combined methods concept based wastewater treatment techniques and suitability of the treatment model on various types of wastewater effluents have also been discussed. The model recommends the appropriateness of wastewater treatment for irrigation and also for the better usage of produced solid wastes as fertilizers. The tertiary treatment helps in removing the toxic contaminants such as pesticides and heavy metals from the wastewater to be used for irrigation purpose. This can have two-fold advantage as the toxicants can be removed from the food chain, thus keeping the overall environment safe and sound as well as the risk of public health issues can be reduced considerably.

KEYWORDS

- **chemicals**
- **ecology**
- **irrigation**
- **nutrients**
- **wastewater**

CHAPTER 2

USE OF WASTEWATER IN AGRICULTURE

SUMYRAH MUKHTAR, JAVID AHMAD BHAT, M. H. CHESTI, AANISA MANZOOR, RUMAISA ASHRAF, and AABIRO RASHID

ABSTRACT

Environmental degradation on account of anthropogenic activities poses a hefty threat to sustainable, sparkling, and green environment. Irrational expansion of global population and industrialization has resulted in the production of an enormous number of environmental contaminants at the expense of environmental and human health. The pressing issue of water scarcity looms large, with alarming projections suggesting that billions will be deprived of clean water by 2025. Furthermore, the escalating global population demands a 70% increase in agricultural production by 2050 leading to a significant surge in water consumption. Water, the lifeblood of our planet, becomes the linchpin for maintaining food security and a thriving environment for all living beings. In this landscape of challenges, a transformation is necessary—a shift toward a sustainable, closed-loop urban wastewater management system that sustains crop production with limited resources. Reusing treated wastewater for irrigation of ready-to-eat produce emerges as a safe and viable solution. However, the risks associated with this practice cannot be ignored. It is critical to strike a balance between using the vital nutrients found in wastewater and reducing the amount of hazardous elements like heavy metals. There is a rising need to investigate the immense potential of treated wastewater for agricultural uses as freshwater supplies become more and more scarce. We can prepare the path for a future where the demand for reuse water can be satisfied without compromising the health and well-being of both persons and ecosystems by examining the current research and various risk reduction strategies.

Sustainable Use of Wastewater in Agriculture: Applications, Methods, and Technology. Megh R. Goyal & Gaurav Saini (Eds.)

2.1 INTRODUCTION

The global population growth has led to a significant disparity between the availability and demand for water, reaching critical levels in some regions that pose a substantial threat to human existence. The adverse effects of climate change exacerbate the situation, impacting water quality, availability, food security, and human health worldwide. The shortage of freshwater resources is swayed by multiple factors, including population growth, urbanization, individual consumption patterns, water pollution, and the effects of climate change.[49] Water scarcity is a significant factor impacting public health and contributing to poverty, particularly in rural regions characterized by high population density.[18] Recent estimations suggest that around 1.2 billion individuals reside in river basins experiencing physical water scarcity, while an additional 1.6 billion people reside in areas with insufficient water resources and lack affordable access to water delivery systems.[9] Researchers worldwide are actively involved in developing innovative approaches to water conservation. At this time, it is appropriate to draw attention to one specific water-recycling technique that includes reusing urban wastewater for a variety of uses, including irrigation. By using this strategy, it will be possible to provide clean water for other industries that need it while also supplying water for industries that can use treated wastewater for things like irrigation and other ecosystem services.

The rate of groundwater depletion has experienced a significant increase globally, reaching a double of the previous rate. Projections indicate that by the year 2050, the global water demand will continue to rise, estimated to be 20–30% higher than the current level. As an outcome, 40% of the global inhabitants are anticipated to experience severe water stress.[30] This situation will particularly affect regions such as South Asia, Middle East, North Africa, and parts of China. Measures need to be implemented to address this issue and prevent further deterioration of the global water supply. Water scarcity is an escalating concern due to the combination of increasing water demand and the amplified impact of climate change.[29] Presently, around 20% of the world's aquifers are being excessively exploited, resulting in reduced availability and deteriorating quality of water resources.[12] These unsustainable practices have far-reaching consequences, affecting various sectors including agriculture, industry, domestic water supplies, power generation, navigation, fishing, recreational activities, and most critically, the environment. To address this pressing issue and eliminate plagiarism entirely, it is imperative to prioritize originality and cite the relevant sources accurately.[25]

The excessive depletion of freshwater resources is primarily attributed to the expansion of agricultural needs, encompassing irrigation, livestock farming, and aquaculture. Agriculture stands as the foremost water consumer, accounting for 70% of global annual water withdrawals. Its primary purposes include food and fiber production, as well as the processing of agricultural goods.[51] Conversely, the industrial sector, encompassing power generation, comprises 19% of water usage, while households account for 12%.[6] As a result, agriculture will continue to be the largest overall consumers for many years to come even though its proportion of consumption of water will be decreasing relative to other industries.

The escalation of intensive agriculture significantly contributes to global water consumption; it becomes imperative to seek sustainable solutions. The broadening of international efforts for collecting information and hydrological surveillance is necessary to meet this problem. Strengthening global monitoring networks is vital in this regard, while also exploring the possibilities offered by emerging technologies to enhance data collection capabilities.[47] In prospect of this, the utilization of wastewater for agricultural purposes is gaining traction as a viable solution, particularly in water-scarce regions.[26]

In addition, it has been put forth that over 10% of the global population relies on agricultural products cultivated through wastewater irrigation.[28] Wastewater encompasses liquid waste produced through the daily activities, usage, and production processes of households, industries, and commercial sectors. This type of waste can contain valuable raw materials, including nutrients and specific metals, particularly in the case of industrial wastewater.[50] A study was conducted demonstrating that these resources found in wastewater have the potential to serve as alternative sources of fertilizers, thereby mitigating the need for energy-intensive extraction methods. The utilization of these raw materials from wastewater not only reduces reliance on conventional extraction processes but also contributes to the promotion of sustainable resource management practices. The growing demand for water, particularly in agriculture, coupled with the diminishing availability of clean water, has led to a rise in the utilization of alternative water sources, including urban effluents. Nonetheless, implementing reuse programs in the current national context entails certain considerations, such as public acceptance and the feasibility in terms of technical and economic factors. It is crucial to address these issues to ensure the successful implementation of such schemes in order to sustain the agricultural production with limited resource base and without compromising the ecosystem health too!

An approach that is frequently employed in industrialized and developing countries for fertilization of land is the use of wastewater from municipalities, whether it has been treated or not. In advanced countries with strict environmental regulations, a sizable quantity of wastewater undergoes treatment properly before being used to irrigate particular crops including feed, fiber, and seed crops. Additionally, although to a lesser degree, it is also utilized for irrigating orchards, vineyards, and various other crop types. Wastewater has diverse significant uses, including replenishing groundwater, landscaping areas like freeways, golf courses, playgrounds, and parks, supporting industrial processes, aiding construction activities, controlling dust, improving wildlife habitats, and facilitating aquaculture.[23] Nevertheless, standards have been determined; it is of the utmost importance to understand that they are not always rigorously adhered to in developing nations. Untreated wastewater is frequently utilized for aquaculture and agricultural purposes; this is a customary practice in countries like India, China, and Mexico.

Wastewater possesses both potential benefits and challenges, serving as both a valuable resource and a pressing issue. Its reuse has the capacity to bring positive advantages to the agricultural sector, communities, and local governments. However, the practice of wastewater reuse also carries negative impacts on human health and ecological systems, which necessitate careful identification and evaluation. Explanatory economic research should be conducted before utilizing wastewater irrigation as a method of enhancing water supplies for agriculture. This investigation must extensively figure out the costs and advantages involved with the reuse of wastewater. As a result, using treated water and secure irrigation techniques, it is possible to strategically execute the use of these wastewater effluents in agriculture. Since utilizing untreated wastewater directly (undiluted) or indirectly (diluted wastewater) might provide dangers to both farmers and consumers, this technique should be undertaken with caution and proper information. Therefore, proper planning and precautions are necessary to ensure the safety of all stakeholders involved.[36]

2.2 BENEFITS OF WASTEWATER USE IN AGRICULTURE

Wastewater that is discharged down the drain contains valuable raw materials that can be utilized effectively. This resource has the potential to be utilized to replenish surface water and groundwater, as well as to irrigate gardens and agricultural areas. Water recycling presents a sustainable solution that not only helps in conserving our precious water resources but also offers environmental and economic advantages. By adopting careful planning, effective implementation, and efficient management, the practice of wastewater irrigation in agriculture can yield numerous benefits for the environment, agriculture, and water resource management. Additionally, using treated wastewater in agriculture encourages breakthroughs in the economics, environment, and health of people. These benefits can be summarized as follows:

Benefits for agriculture

- Decreased Pressure on Freshwater Sources: Wastewater use in agriculture reduces the demand for freshwater, thereby relieving the strain on limited freshwater resources. This is especially crucial considering that agriculture accounts for 70% of global water consumption.[39] Thus, by serving as an alternative source of irrigation, it decreases the pressure on freshwater sources.[53]
- High nutrient content: Wastewater naturally contains significant amounts of nutrients, which can serve as an alternative to expensive chemical fertilizers. This nutrient-rich wastewater can supply essential macro- and micronutrients required for plant growth and reduce agricultural costs, particularly benefiting communities with limited financial resources. Wastewater is a good source of both macro-nutrients (N, P, and K) as well as micro-nutrients (Cu, B, Mg, Zn, and Fe).[3, 33]
- Economical means of irrigating crops: A straightforward and effective approach to conserve water while lowering waste is to utilize wastewater for irrigation. It can also eliminate the need for extracting groundwater, leading to cost savings.
- Moreover, the treated wastewater discharged for human consumption can be valued, and serves as an implicit benefit of using wastewater in agriculture. Apart from this it helps in reducing the financial burden on municipalities to search for more expensive water sources.
- Increased Crop Production: By providing a consistent water supply throughout the year, irrigation using wastewater enables farmers to extend the growing season, increase crop yields, and cultivate in areas that would otherwise be too dry. This serves as an insurance policy against drought and seasonal variability.
- Improved Crop Quality: Irrigation with wastewater eliminates water stress, resulting in better crop quality and enhanced productivity. For agricultural commodities and pastures to perform well and succeed at their best, adequate water availability is essential. Increased Property Value: Irrigated land has the potential to support a higher yield of

crops and animal production, making it more valuable. The presence of reliable irrigation infrastructure contributes to the overall value of the property.

- Food Safety: Reusing wastewater in agriculture stimulates output from agriculture, especially in areas with water restrictions, helping guarantee the safety and security of food.[15]
- Income Generation and Employment Opportunities: Reusing wastewater in urban areas may provide income and job possibilities, which is especially advantageous for urban farmers, belong to subsistence farming community, encompassing a sizable number of women.

Benefits from an environmental perspective

When successfully carried out, wastewater management plans might provide an assortment of environmental upsides, including:[35]

- **Water pollution escaping:** As contrasted to being consumed once more, the recycled use of wastewater in agriculture helps prevent the water degradation that could come from its discharge into rivers and lakes. This method has a number of benefits and reduces environmental problems including fish casualties, eutrophication, foaming, and dissolved oxygen depletion. A reduction of wastewater discharge into water bodies improves their water quality.[16,48]
- **Reduced dependency on artificial fertilizers:** Wastewater's nutrient content reduces the reliance on synthetic fertilizers, leading to a reduction in energy consumption and industrial pollution associated with fertilizer production.
- **Soil conservation:** By fostering humus production and minimizing erosion, the application of wastewater in agriculture promotes soil conservation and environmentally friendly land management.
- **Combating the effects of desertification:** Wastewater irrigation can aid in desertification control and reclamation efforts by providing water and fertilization for tree belts.
- **Preserving groundwater reservoirs:** using wastewater spares enough good quality water for recharging groundwater reservoirs.

Benefits for water resources management

- **Improved treatment facilities:** The expanding use of wastewater in agriculture may have an impact on the building and development of treatment facilities. These facilities can produce good quality effluent that can be employed for irrigation. This not only enhances the effectiveness of wastewater reuse but also represents an economic benefit for sanitation projects.

- **Additional drought-proof water supply:** Wastewater reuse provides an additional source of water that is resistant to drought conditions. This alternative water supply is often more cost-effective compared to other methods such as expanding water storage, implementing water transfers, or resorting to desalination.
- **Local sourcing of water:** Incorporating wastewater into water resources management enables a shift toward more localized water sourcing. Instead of relying solely on distant water sources, wastewater can be utilized within the same region, reducing the dependence on external water supplies.
- **Inclusion of wastewater in water management:** Integrating wastewater into the broader context of water resources management ensures its proper consideration and utilization. This holistic approach recognizes wastewater as a valuable resource and maximizes its potential within the overall water management framework.
- **Enhanced urban water resources management:** Wastewater management in agriculture promotes more effectively integrated management of urban water resources. By utilizing treated wastewater for irrigation, urban areas can achieve a more sustainable and efficient water cycle, reducing the strain on freshwater sources and improving overall water management practices.

Risks affiliated with wastewater use in agriculture

- **Microbial threats:** Pathogens that are present in wastewater pose a significant risk of excreta-related diseases, including bacteria, viruses, helminths, and protozoa. These diseases can be endemic, continuously spreading within a community, or become epidemics when introduced to susceptible communities by outsiders. The utilization of wastewater in agriculture may contribute to the transmission of excreta-related diseases, especially when individuals consume raw wastewater-irrigated foods like salad crops and specific vegetables (refer to Table 2.1). Consuming food that was accidentally irrigated using wastewater is simply one possible method of transmission, and the effects on local public health may differ. To effectively address local public health concerns, thorough consideration must be given to all possible routes of transmission.[45]

TABLE 2.1 Environmental Classification of Illnesses Associated with Excrement that are Instrumental in Wastewater Irrigation in Agriculture.

Category	Aspects of environmental/horizontal transmission[a]	Foremost occurrences	Visible populations and comparative infection risks[b, c]
Oral illnesses that are not caused by bacteria	High Infectivity, Non-Latent, Reduced to No Perseverance, No Propagation	Hepatitis A, E, and F, Amebiasis, Cryptosporidiosis, Cyclospora cayetanensis diarrhea, Giardiasis, and Isopora belli diarrhea, Enterocytozoon bienusi diarrhea, are all protozoan diseases, are viral infections. Rotavirus, norovirus, and adenovirus-related diarrhea are caused by viruses.	Consumers: +++, Fieldworkers: +
Bacterial oral and fecal illnesses	Nonlatent, moderate to low infectiousness, high to medium perseverance aware of multiplying	Escherichia coli infections, Campylobacteriosis, Almonellosis, Pathogenic Cholera, shigellosis	Fieldworkers: + Consumers: +++
Geohelminthiases	High contagiousnes, latent, extraordinarily persistent, and deficient of multiplication	Hookworm illness, trichuriasis, and ascariasis	Consumers: +++ Fieldworkers: +++

[a]*The minimal amount of time a pathogen can exist without a human host is known as persistence, whereas the time period necessary for a virus to become spreadable outside of a human host is recognized as latency.*

[b]*+++ = high risk, + = low risk. The aforementioned risks are brought on by the use of wastewater that has not been treated for agricultural irrigation/sprinkle purposes; they can be decreased by treatment of effluent/wastewater and the implementation of subsequent treatment health-protection management methods.*

[c]*Bearing in mind that fieldworkers normally act like consumers.*

Microbiological disease transmission by wastewater

The existing rigorous epidemiological research on the health concerns associated with irrigation with treated wastewater were analyzed.[44] The key findings from their study were:

- Helminthic infections transmitted through the soil are the greatest health concern for those who operate in fields that are irrigated with untreated wastewater and consume uncooked vegetables that have been irrigated with wastewater. However, when treated wastewater is utilized for crop irrigation, the danger is drastically decreased to less than 1%.

- Those who consume salad crops and raw vegetables irrigated with wastewater may be exposed to bacterial feco-oral illnesses like diarrhea and cholera. However, the evidence for viral and protozoan disease transmission was less convincing. These findings were based on the epidemiological data obtained from the investigation.[44] Blumenthal[7] examined the epidemiological findings provided following the investigation in reference.[44]

- **Unrestricted irrigation:** Consuming vegetables that have been irrigated with untreated wastewater has been connected to an increase in helminth infections, which includes bacterial infections like typhoid, cholera, Ascaris lumbricoides infections, diarrheal sickness, as well as *Helicobacter pylori* infection symptoms. When some raw vegetables are irrigated with water having less than 105 fecal coliforms per 100 mL, evidence indicates that the likelihood of bacterial and viral enteric diseases remains elevated.
- **Restricted irrigation:** The risk of infection worsened when spray irrigation exposed the general public to wastewater droplets with a quality of 106 total coliforms per 100 mL, corresponding to investigations on dangers of bacterial and viral enteric disorders associated with the utilization of reclaimed water. When the discharge water quality was between 103 and 104 feces-borne coliforms per 100 mL, there was no heightened risk of infection. When the wastewater quality exceeded >104 fecal coliforms per 100 mL for furrow or flood irrigation, research revealed a heightened likelihood of sickness in kids. Additionally, studies have linked personal contact with treated wastewater while working on farms to risks of enteric virus infections and symptomatic diarrheal illness in both adults and children. The threshold for symptomatic diarrhea in adults was 105 fecal coliforms per 100 mL, but even in areas where the virus spread by alternative channels, dissemination of a Mexican strain of norovirus occurred at 104 fecal coliforms per 100 mL.

An increased prevalence and severity of ascariasis and hookworm disease in fieldworkers was found,[44,45] when untreated wastewater was utilized for irrigation compared to a control group (Figure 2.1). Blumenthal[7] validated these findings for bacterial and geohelminth disorders.

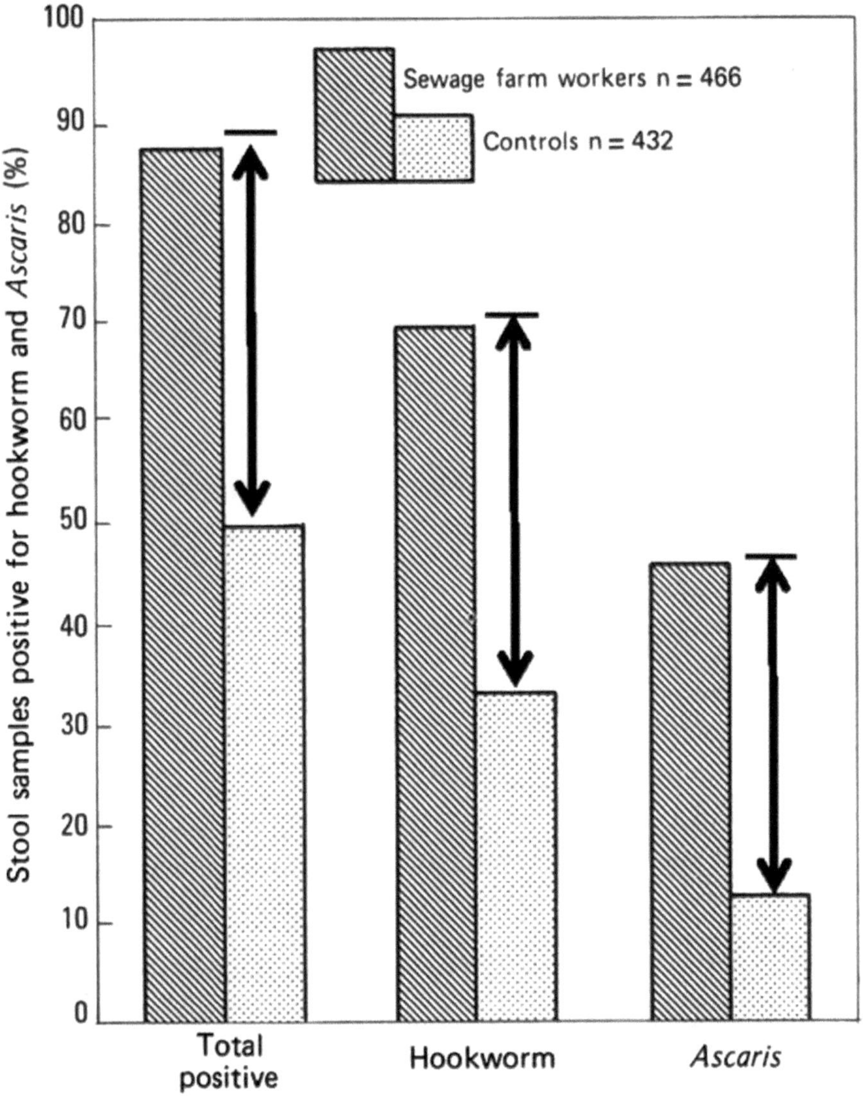

FIGURE 2.1 Compared to a control group, the incidence of ascariasis and hookworm illness among Indian "sewage farm" workforce who use wastewater that is untreated for irrigation.

2.2.1 CHEMICAL HAZARDS TO HUMAN HEALTH

Various health hazards, such as heavy metals (cadmium, lead, and mercury) and organic substances like pesticides, are associated with the use of industrial wastewaters, which find their way into municipal wastewaters. Anthropogenic chemicals have long-lasting effects on human health, such as those of drugs, hormones, endocrine disruptors, antimicrobials, and antibiotics are not fully understood.[38] In a study,[13] the major chemical dangers linked to consuming wastewater-irrigated foods were examined. The findings indicated the following:

- For over a hundred years, land application has been an established method of managing municipal wastewater and sewage sludge. The widespread use of untreated urban and industrial wastewaters for crop irrigation, particularly in China, has raised concerns owing to the improper control of potentially toxic component discharge, which may affect both crops and people.
- The level of dangerous pollutants in municipal wastewater and sewage sludge differs depending on the location and is impacted by industrial point-source discharges. Inorganic contaminants are detected more frequently (50–100%), whereas organic contaminants are found at a lower frequency (5–10%), with modest concentrations. Implementation of industrial wastewater pretreatment regulations at the community level has helped lower pollutant concentrations in wastewater and sewage sludge.

To mitigate chemical threats to human health, the following principles have been proposed:[13]

- Preventing pollutant buildup in waste-receiving soils to ensure that the soil retains its ecological and chemical integrity, and future land uses are supported. This requires implementing guidelines with upper limits for contaminants, necessitating advanced wastewater treatment technology to reduce pollutant levels.
- Utilizing the soil's ability to absorb, diminish, and detoxify pollutants while considering the most effective approaches for wastewater and sewage sludge management. The approach establishes maximum allowable pollutant loading and promotes the appropriate utilization of recycled wastewater for irrigation, given the water quality. It meets standards and water volume does not exceed crop requirements. However, extra care is necessary in addressing chemical concerns in underdeveloped nations, where comprehensive industrial wastewater pretreatment programs are not common, and household chemicals may also contribute to graywater contamination.

TABLE 2.2 Maximum Estimated Acceptable Equates of a Particular Organic and Inorganic Contaminant on Soils Irrigated Via Wastewater.

Inorganic substances (Maximum allowable estimations (mg/kg soil))					
Barium (Ba)	302	Arsenic (As)	8	Antimony (Sb)	36
Cadmium (Cd)	4	Boron (B)	1.7	Beryllium (Be)	0.2
Mercury (Hg)	7	Lead (Pb)	84	Fluorine (F)	635
Selenium (Se)	6	Nickel (Ni)	107	Molybdenum (Mo)	0.6
Vanadium (V)	47	Thallium (Tl)	0.3	Silver (Ag)	3
Organic substances (Maximum allowable estimations (mg/kg soil))					
Pentachlorophenol	14	PAH[a]	16	Benzene	0.14
Hexachlorobenzene	1.4	Chlorobenzene	211	Dieldrin	0.17
2,4,5–T	3.82	DDT	1.57	Aldrin	0.48
Styrene	0.68	Heptachlor	0.18	Dioxins	1.2×10^{-4}
PCBs	0.89	Methoylchlor	4.27	Chlorodane	3
Chloroform	0.47	Pyrene	41	Dichlorobenzene	15
Lindane	12	Toxophene	0.0013	Toluene	12
2,4–D	0.25	Phthalate	13,733	Trichloroethane	0.68
Tetrachloroethane	1.25				

[a]*As benzo(a)pyrene.*

2.2.2 RISKS FROM CHEMICALS TO PLANT HEALTH

Crop yields may be adversely affected by irrigation wastewater's inappropriate physical and chemical properties, such as high salinity or concentrations of hazardous chemicals including boron, heavy metals, nitrogen, and salt. The FAO's Water Quality for Agriculture[1] serves as the primary source on the subject, along with later publications like Utilization of Brackish Waters for Crop Cultivation,[42] Supervision of Wastewater Quality for Agricultural Irrigation,[52] and Salinity Management for Sustainable Irrigation.[22] Treating residential wastewaters can be manageable; however, industrial effluents need attention as they may lead to potential issues, sodium adsorption ratio (SAR), electrical conductivity (EC), boron concentrations (B), total nitrogen (N) concentrations, and pH should all be constantly tracked during an irrigation season. In certain hydrogeological circumstances, microbiological contaminants can travel long distances in groundwater, potentially affecting water supplies.

2.3 ENVIRONMENTAL HAZARDS

One renowned downside of wastewater irrigation in agriculture is the contamination of soil and groundwater. While slow sand filtration usually retains pathogens in the topsoil layers, specific hydrogeological conditions can lead to longer-range microbiological contaminant transmission. Additionally, chemical contaminants, especially nitrates, can travel longer distances, affecting drinking water supplies in areas near wastewater irrigation. Soil contamination may arise from improper leaching and draining practices, leading to salinization, and waterlogging. Appropriate irrigation practices and suitable crop choices are necessary to minimize these negative consequences.[1,22,42,46] To address concerns like odor, vector growth, and inadvertent hazardous material discharges, properly treated wastewater should be used.

2.3.1 MITIGATION OF WASTEWATER USE IN AGRICULTURE

Improved water treatment is essential for safe water reuse, especially to eliminate harmful chemicals. Combining different treatment methods can reduce risks associated with organic micro-contaminants and pathogens. Natural processes like natural attenuation may complement commercial treatment solutions for soil and groundwater remediation, depending on specific conditions.[37]

2.3.2 IRRIGATION TECHNIQUES

Irrigation techniques, such as drip irrigation and subsurface watering, can limit exposure to pathogens and pollutants, as well as enhance water efficiency. Restricted access and control measures can further reduce potential interactions between reclaimed water and crops.[4] However, stakeholders should consider the broader environmental risks and factors like groundwater sensitivity, soil conditions, and climate in their risk management approaches.

2.3.3 MONITORING OF ORGANIC MICROCONTAMINANTS

The results presented emphasize the significance of environmental monitoring to identify potentially harmful consequences associated with organic micro-contaminants, such as their translocation and accumulation. At the EU level, recent initiatives seek to find chemical and toxicological indicators for managing and lowering hazards related to these pollutants. According to Lazarova,[32] monitoring based on the assessment of effects can also tackle the potential hazards caused by unidentifiable micro-contaminants, offering new information that will aid in the development of water reuse procedures. Given that some substances might affect soil populations of microbes and processes, techniques based on effects should be modified to account for the soil matrix and ecosystems.[4]

Monitoring certain molecules as indicator substances could be a promising strategy, and inorganic compounds like cations (positive ions) and anions (negative ions) might reflect the influence of wastewater on ecological resources, assuming that there are no other ion sources like seawater incursion into groundwater. However, selecting relevant indicators for monitoring requires considering the specific persistence and transport behaviors of the compounds. For example, while caffeine has been proposed as an indicator for anthropogenic wastewater pollution due to its presence in treated wastewater, its rapid degradation in the environment and complete elimination through soil passage make it less suitable for monitoring groundwater impacts in reclaimed water-irrigated areas.[10,19] It should be noted that California's Recycled Water Policy includes observing necessities for micro-contaminants in water reuse.

However, while the Science Advisory Panel's recommendations[43] resulted in changes to the Water Quality standards, an all-encompassing structure is needed to establish a purposeful monitoring system for organic micro-contaminants. This will help validate water treatment methods and evaluate the possible build-up in soils, groundwater, and plants, which could have significant implications for ecology and human well-being.

2.4 SUMMARY

Optimum wastewater use as a significant resource in agricultural output depends on advances in research in science, growth and development, and innovation. We can successfully address the problems caused by water shortages across the world and raise the standard of life in vulnerable and underprivileged areas by supporting technical solutions that improve the effective utilization of reclaimed water. Wastewater may partly obstruct the water requirements for irrigation of agricultural crops in the face of rising worldwide temperatures and diminishing freshwater resources. While recycling wastewater has many benefits, including lowering the total quantity of wastewater dumped into waterways, boosting crop yields and less need on fertilizers that contain chemicals by using nutrients found in wastewater, it is important to identify the potential health menaces linked to untreated effluent (wastewater). It is obvious that utilizing only treated wastewater of the proper quality that are appropriate for irrigation purposes in agriculture is the best course of action given the expanding acceptance of wastewater reclamation and the necessity to protect the well-being of humans and the natural world. The dangers to the environment and human health will be substantially decreased by putting into practice current, required legislation for secure disposal and reuse of wastewater. Nevertheless, for these attempts to succeed, further development and creativity are required. Gathering information and evidence can result in new knowledge that helps with an improved comprehension of the use of wastewater proficiently for agrarian purpose. This option is possible if its potential risks have been thoroughly evaluated and the necessary precautions for environmental and health protection are put in place.

KEYWORDS

- **agriculture**
- **climate change**
- **environment**
- **hazards**
- **irrigation**
- **organic compounds**
- **inorganic compounds**
- **wastewater**

REFERENCES

1. Ayers, R. S.; Westcot, D. W. Water Quality for Agriculture. FAO Irrigation and Drainage Paper No. 29, Rev. 1. Food and Agriculture Organization (FAO): Rome, 1989.
2. Banzhaf, S.; Nödler, K.; Licha, T.; Krein, A.; Scheytt, T. Redox-sensitivity and Mobility of Selected Pharmaceutical Compounds in a Low Flow Column Experiment. *Sci. Total Environ.* **2012,** *438,* 113–121.
3. Barreto, A. N.; do Nascimento, J. J. V.; de Medeiros, E. P.; da Nóbrega, J. A.; Bezerra, J. R. Changes in Chemical Attributes of a Fluvent-Cultivated with Castor Bean and Irrigated

with Wastewater. *Soil, Water Plant Manag.(Rev. Bras. Eng. Agríc. Ambient)* **2013,** *17* (5). https://doi.org/10.1590/S1415-43662013000500003; (accessed Aug 12, 2023).
4. Becerra-Castro, C.; Lopes, A. R.; Vaz-Moreira, I.; Silva, E. F.; Manaia, C. M.; Nunes, O. C. Wastewater Reuse in Irrigation: A Microbiological Perspective on Implications in Soil Fertility and Human and Environmental Health. *Environ. Int.* **2015,** *75,* 117–135.
5. BGS. Guidelines for Assessing the Risk to Groundwater from On-site Sanitation [Online] 2001. https://www.susana.org/en/knowledge-hub/resources-and-publications/library/details/1926; (accessed Aug 12, 2023).
6. Bijl, D. L.; Bogaart, P. W.; Kram, T.; de Vries, B. J.; van Vuuren, D. P. Long-term Water Demand for Electricity, Industry, and Households. *Environ. Sci. Policy* **2016,** *55,* 75–86.
7. Blumenthal, U. J.; Peasey, A. *Critical Review of Epidemiological Evidence of Health Effects of Wastewater and Excreta in Agriculture*; Background paper for the WHO Guidelines for the Safe use of Wastewater and Excreta in Agriculture. World Health Organization (WHO): Geneva, 2002.
8. Brack, W.; Aissa, S. A.; Backhaus, T.; Dulio, V.; Escher, B. I.; Faust, M. The European Collaborative Project SOLUTIONS Recommends Integrating Effect-based Methods for Diagnosis and Monitoring of Water Quality. *Environ. Sci. Eur.* **2019,** *31,* 10–15.
9. Brar, B.; Sonu, N.; Rawat, J. Wastewater Reuse for Irrigation of a Vegetable Crops and its Impacts. *Pharma Innov. J.* **2022,** *11* (SP-7), 111–117
10. Buerge, I. J.; Poiger, T.; Müller, M. D.; Buser, H. R. Caffeine, an Anthropogenic Marker for Wastewater Contamination of Surface Waters. *Environ. Sci. Technol.* **2003,** *37,* 691–700.
11. California Water Board (CWD). Water Quality Control Policy for Recycled Water. State Water Resources Control Board—California Environmental Protection Agency. 2019.
12. Carrard, N.; Foster, T.; Willetts, J. Groundwater as a Source of Drinking Water in Southeast Asia and the Pacific: A Multi-country Review of Current Reliance and Resource Concerns. *Water* **2019,** *11* (8), 1605–1610.
13. Chang, A. C.; Pan, G.; Page, A. L.; Asano, T. Developing Human Health-Related Chemical Guidelines for Reclaimed Wastewater and Sewage Sludge Applications in Agriculture. Background Paper for the WHO Guidelines for the Safe use of Wastewater and Excreta in Agriculture: World Health Organization: Geneva, 2002.
14. Christou, A.; Papadavid, G.; Dalias, P.; Fotopoulos, V.; Michael, C.; Bayona, J. M.; Pina, B. Ranking of Crop Plants According to Their Potential to Uptake and Accumulate Contaminants of Emerging Concern. *Environ. Res.* **2019,** *170,* 422–432.
15. Corcoran, E., Ed. *Sick Water? the Central Role of Wastewater Management in Sustainable Development: A Rapid Response Assessment*; UNEP/ Earthprint, 2010.
16. Davide, B. I. X. I. O. *Water Reuse System Management Manual: Aquarec*; European Commission. 2007.
17. DWA. *Non-Potable Water Reuse: Development, Technologies and International Framework for Agricultural, Urban and Industrial Uses.* In *DWA Deutsche Vereinigung für Wasserwirtschaft* (DWA German Association for Water Management); Abfall, E. V., Ed.; 2019.
18. Food and Agriculture Organization of the United Nations (FAO). Coping with Water Scarcity: An Action Framework for Agriculture and Food Security. FAO Water Reports #38, FAO: Rome - Italy, 2012.
19. Fries, E.; Mahjoub, O.; Mahjoub, B.; Berrehouc, A.; Lions, J.; Bahadir, M. Occurrence of Contaminants of Emerging Concern (CEC) in Conventional and Non-Conventional Water Resources in Tunisia. *Fresh Environ. Bull.* **2016,** *25,* 3317–3339

20. González O, Bayarri B, Aceña J, Pérez S, Barceló D Treatment Technologies for Wastewater Reuse: Fate of Contaminants of Emerging Concern. In *Advanced Treatment Technologies for Urban Wastewater Reuse. The Handbook of Environmental Chemistry*; Fatta-Kassinos, D., Dionysiou, D., Kümmerer, K., Eds.; Springer: Cham, 2016; Vol 45, pp 5–37.
21. Harwood JJ Molecular Markers for Identifying Municipal, Domestic and Agricultural Sources of Organic Matter in Natural Waters. *Chemosphere* **2014,** *95,* 3–8.
22. Hillel, D. *Salinity Management for Sustainable Irrigation: Integrating Science, Environment, and Economics*; World Bank: Washington, DC, 2000.
23. Hussain, I.; Raschid, L.; Hanjra, M. A.; Marikar, F.; Van Der Hoek, W. Wastewater use in Agriculture: Review of Impacts and Methodological Issues in Valuing Impacts. Working Paper 37, International Water Management Institute: Sri Lanka, 2002.
24. ISO. *Guidelines for Treated Wastewater use for Irrigation Projects,* Part 1-3, ISO 16075-1-3:2015. Beuth Verlag: Berlin, 2015.
25. Jago-on, K. A. B.; Siringan, F. P.; Balangue-Tarriela, R.; Taniguchi, M.; Reyes, Y. K.; Lloren, R.;... Bagalihog, E. Hot Spring Resort Development in Laguna Province, Philippines: Challenges in water use Regulation. *J. Hydrol.: Regional Stud.* **2017,** *11,* 96–106.
26. Jaramillo, M. F.; Restrepo, I. Wastewater Reuse in Agriculture: A Review About its Limitations and Benefits. *Sustainability* **2017,** *9* (10), 1734–1740.
27. Jekel, M.; Dott, W. Guide: Polar Organic Trace Substances as Indicators in the Anthropogenically Influenced Water Cycle: Results of the Cross-cutting Theme "Indicator substance" [Online]. 2013. https://worldwidescience.org/topicpages/o/organic+substances+labelled.html; (accessed Aug 13, 2023).
28. Jeong, H.; Kim, H.; Jang, T. Irrigation Water Quality Standards for Indirect Wastewater Reuse in Agriculture: A Contribution Toward Sustainable Wastewater Reuse in South Korea. *Water* **2016,** *8* (4), 169–174.
29. Khalid, S.; Shahid, M.; Natasha.; Bibi, I.; Sarwar, T.; Shah, A. H.; Niazi, N. K. A Review of Environmental Contamination and Health Risk Assessment of Wastewater use for Crop Irrigation with a Focus on Low and High-income Countries. *Int. J. Environ. Res. Public Health* **2018,** *15* (5), 895–904.
30. Kölbel, J.; Strong, C.; Noe, C.; Reig, P. Mapping Public Water Management by Harmonizing and Sharing Corporate Water Risk Information. *World Resour. Inst.* **2018,** *2018,* 1–20.
31. Krzeminski, P.; Tomei, M. C.; Karaolia, P.; Langenhoff, A.; Almeida, C. M. R.; Gritten, F.; Andersen, H. R.; Fernandes, T. Performance of Secondary Wastewater Treatment Methods for the Removal of Contaminants of Emerging Concern Implicated in Crop Uptake and Antibiotic Resistance Spread: A Review. *Sci. Total Environ.* **2019,** *648,* 1052–1081.
32. Lazarova, V. Global Milestones in Water Reuse: Keys to Success and Trends in Development. *Water* **2013,** *21,* 12–22.
33. Liu, Y. Y.; Haynes, R. J. Origin, Nature, and Treatment of Effluents from Dairy and Meat Processing Factories and the Effects of Their Irrigation on the Quality of Agricultural Soils. *Crit. Rev. Environ. Sci. Technol.* **2011,** *41* (17), 1531–1599.
34. Luo, Y.; Guo, W.; Ngo, H. H.; Nghiem, L. D.; Ibney, Hai F.; Zhang, J.; Liang, S.; Wang, X. C. Review on the Occurrence of Micropollutants in the Aquatic Environment and Their Fate and Removal During Wastewater Treatment. *Sci. Total Environ.* **2014,** *473,* 619–641.

35. Mara, D. D.; Cairncross, S.; WHO. *Guidelines for the Safe use of Wastewater and Excreta in Agriculture and Aquaculture: Measures for Public Health Protection*; World Health Organization: Geneva, 1989.
36. Mateo-Sagasta, J.; Raschid-Sally, L.; Thebo, A. Global Wastewater and Sludge Production, Treatment, and Use. In *Wastewater: Economic Asset in an Urbanizing World*; 20151 pp 15–38.
37. Mulligan, C. N.; Yong, R. N. Natural Attenuation of Contaminated Soils. *Environ. Int.* **2004,** *30* (4), 587–601.
38. Orr, S.; Cartwright, A.; Tickner, D. Understanding Water Risks: a Primer on the Consequences of Water Scarcity for Government and Business [Online]. 2009; https://awsassets.panda.org/downloads/understanding_water_risk_iv.pdf (accessed on Aug 13, 2023).
39. Pimentel, D.; Pimentel, M. H., Eds. *Food, Energy, and Society*; CRC Press, 2007.
40. Pina, B.; Bayona, J. M.; Christou, A. On the Contribution of Reclaimed Wastewater Irrigation to the Potential Exposure of Humans to Antibiotics, Antibiotic Resistant Bacteria, and Antibiotic Resistance Genes: NEREUS COST Action ES1403 Position Paper. *J. Environ. Chem. Eng.* **2018,** *8* (1), 102131.
41. Pinnekamp, J. (Editor), In 52nd ESSEN Conference on Water Management (Water and Health) from March 20–22, *Aachen. Society for the Promotion of Urban Water Management at RWTH Aachen University*, Aachen, 2019.
42. Rhoades, J. D.; Kandiah, A.; Mashali, A. M. The use of Saline Waters for Crop Production. Irrigation and Drainage Paper no. 48. Food and Agriculture Organisation: Rome, 1992.
43. SCCWRP *Monitoring Strategies for Constituents of Emerging Concern (CECs) in Recycled Water: Recommendations of a Science Advisory Panel.* Drewes, J. E., Anderson, P., Denslow, N., Jakubowski, W., Olivieri A., Schlenk, D., Snyder, S. Eds.; SCCWRP Technical Report 1032, 2018.
44. Shuval, H. I.; Adin, A.; Fattal, B.; Rawitz, E.; Yekutiel, P. Wastewater Irrigation in Developing Countries: Health Effects and Technical Solutions. Technical Paper No. 51. World Bank: Washington, DC, 1986.
45. Tahereh, M.; Bagheri, A.; Mohammadjavad, A. In *Benefits and Risks of Wastewater use in Agriculture*, The First International and the Fourth National Congress on Recycling of Organic Waste in Agriculture, 26–27 April, in Isphan, Iran, 2021.
46. Tanji, K. K.; Kielen, N. C. *Agricultural Drainage Water Management in Arid and Semiarid Areas*; Irrigation and Drainage Paper No. 61. Food and Agriculture Organization: Rome, 2002.
47. Tauro, F.; Selker, J.; Van de Giesen, N.; Abrate, T.; Uijlenhoet, R.; Porfiri, M.; Grimaldi, S. Measurements y Observations in the XXI Century (MOXXI): Innovation and Multi-disciplinarity to Sense the Hydrological Cycle. *Rev. Cienc. Hidrológicas* **2018,** *63* (2), 169–196.
48. Toze, S. Reuse of Effluent Water—Benefits and Risks. *Agric. Water Manag.* **2006,** *80* (1–3), 147–159.
49. Turral, H.; Burke, J.; Faurès, J. M. *Climate Change, Water and Food Security (No. 36)*; Food and agriculture Organization of the United Nations (FAO), 2011.
50. Wang, X.; Daigger, G.; Lee, D. J.; Liu, J.; Ren, N. Q.; Qu, J.; Butler, D. Evolving Wastewater Infrastructure Paradigm to Enhance Harmony with Nature. *Sci. Adv.* **2018,** *4* (8), EAAQ0210.

51. Weinzettel, J.; Pfister, S. International Trade of Global Scarce Water use in Agriculture: Modeling on Watershed Level with Monthly Resolution. *Ecol. Econ.* **2019,** *159,* 301–311.
52. Westcot, D. W. *Quality Control of Wastewater for Irrigated Crop Production. Water Reports No. 10*; Food and Agriculture Organization: Rome, 1997.
53. Winpenny, J.; Heinz, I.; Koo-Oshima, S.; Salgot, M.; Collado, J.; Hérnandez, F.; Torricelli, R. Reutilización del agua en la Agricultura: ¿Beneficios Para Todos (Water Reuse in Agriculture: Is it Beneficial for All?). Reports Related to Water Topics. Food & Agriculture Organization: Rome, 2013.

CHAPTER 3

NANOMATERIALS FOR WASTEWATER TREATMENT

MUHAMMAD MUNIR SAJID, ALI RAZA ISHAQ, THAMER ALOMAYRI, HAIFA ZHAI, and SADAF BASHIR KHAN

ABSTRACT

Developing low-cost and simple photocatalysts is a primary objective for those active in the energy conversion sector. TiO_2, ZnO, Fe_2O_3, WO_3, and bismuth vanadate are few examples of nanostructured materials for eradication of organic contaminants. Insufficient valence or conduction band edge energy leads to an increased charge carrier recombination rate, which reduces the application's potential even more. As a result, designing a photocatalyst that is both effective and affordable and has stable and continuous band edges is essential. Our expanding energy needs can be met by harnessing the sun's power. Research on high-performance light-dependent nano-photocatalysts can benefit from a thorough understanding of the fundamental variables influencing material photocatalytic activity. Despite a great deal of effort and success, there are still many challenges to overcome in nano photocatalysis. There has been an enormous focus on developing new methods for altering semiconductor nanomaterials, such as morphology and structural governable syntheses.

3.1 INTRODUCTION

It has long been realized that sufficiently scaling down the count of particles in a structure gives dramatic significance to its attributes, behavior, and future practical application.[1–3] When the size of microcrystalline formations

Sustainable Use of Wastewater in Agriculture: Applications, Methods, and Technology. Megh R. Goyal & Gaurav Saini (Eds.)

is reduced, the unique features of particular atoms, generally useless in bulk, begin to gain relevance.[4,5] The recent size reduction of device elements, particularly electronic components, has sparked intense interest in achieving the exceptional behavior of current small-scale structures.[6,7] Size effects become prominent when at least one crystal dimension has hundreds of atoms on the length scales of nanometers.[8–10] This has fueled the rapid expansion of the field of "Nanotechnology." Nanotechnology is the discipline and handling of materials on a microscopical scale. Nanotechnology commonly refers to structures with dimensions ranging from 1 to 100 nanometers in at least one dimension.[11,12]

The initial goal of the idea of nanotechnology was historically represented and named by Richard Feyman (physicist) in his speech titled "*There's Plenty of Room at the Bottom,*"[13,14] Richard Phillips Feynman advocated for more potent synthetic chemical research than was at the time possible by directly manipulating individual particles. Professor Norio Taniguchi came up with the term "Nanotechnology" in the following passage in 1974:[15,16] "*The basic component of nanotechnology is the treating, detachment, consolidation, and contortion of materials from one atom or single molecule.*" Dr. K Eric Drexler generalized the term "Nanotechnology." The Engines of Creation was released in 1986 and is arguably the most important report on the matter ever written.[1,17,18]

Nanotechnology gained acceptance in the 1980s, owing primarily to the advent of assembly discipline and the innovation of STM. This followed the breakthrough and use of fullerenes in the middle of the 1980s. Its progress has aided the growth of semiconducting material nanocrystals.[1,19] The characteristic bulk materials are very well understood, while the characteristic of particles or atoms at the nanoscale level results in different attributes when the size reaches the 100 nm range. The properties of elements are generally linked to space availability for electron motion. So, any changes in physiochemical are observed in shape and size changes. There is a bright future for "Nanotechnology" in terms of applications; it is predicted that in the next 10–20 years, the nanotech industry could be worth up to $1 trillion.[20–22] The significance of nanomaterials for water remediation is displayed in Figure 3.1.

This chapter provided an in-depth review on the development of photocatalysis and nanomaterials for wastewater treatment.

3.2 NANOMATERIALS

Nanomaterials proceed materials in at least one dimension 1–100 nm range.[21,23] Nanomaterials are building blocks for the field of nanotechnology. Based on the dimensions, nanomaterials are classified into four classes[24]: (1) zero-dimensional materials, specified as nanoparticles and pulverizes; (2) unidimensional nanomaterials, for instance, nanowires, nanotubes, and so on; (3) two-dimensional nanomaterials, e.g., nanosheets, quantum wells; and (4) three-dimensional such as nanospheres and quantum dots.

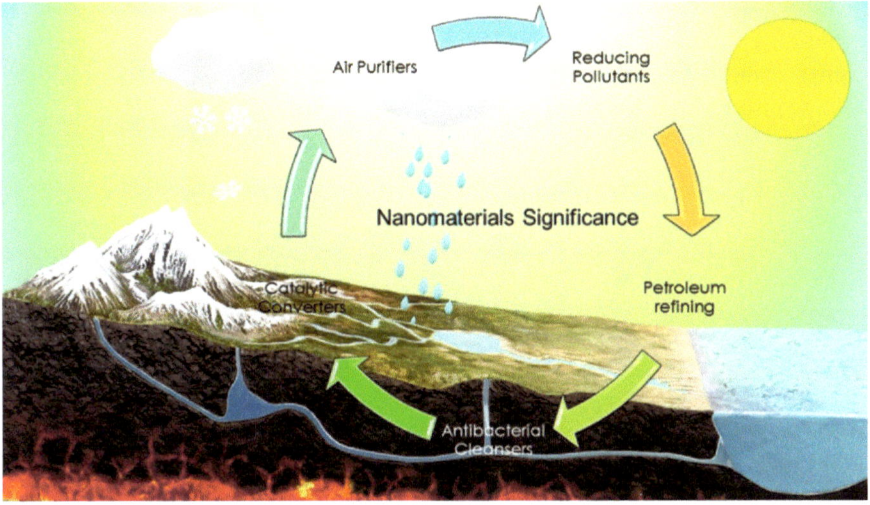

FIGURE 3.1 Significance of nanomaterials in photocatalysis for the recycling of wastewater.

The originality of nanomaterials in bulk form may be classified into two broad categories: (1) surface-related phenomena and (2) quantum confinement effects. The surface expanse to volume ratio increments as speck size decrements, implying that the surface area of the decreased particle will be more significant for the same volume, increasing reactivity.[25] The ratio of surface particles to bulk materials is insignificant in bulk materials; therefore, activities that occur at the surface of material typically have minimal influence on the substance's overall behavior. However, the surface-to-volume ratio in nanoscale fabrication is significantly superior, which cannot be overlooked.[26,27]

Figure 3.2 describes the consequence of quantum confinement on the compactness of energy states and energy bands. The energy differential between energy levels and the bandgap is increased by quantum confinement. Phenomena are closely connected to the optical and electrical characteristics of materials.

Quantum confinement effects arise when an electron's wavelength in a material is of the same order as the substance's dimension.[28,29] This restricts the travel of an electron in the material, causing it to become quantized in that restricting dimension. The number of dimensions in which electrons are quantized therefore influences the compactness of electron states.[18,30] Mechanical, electrical, and optical qualities alter at the nanoscale compared to the macroscopic level.[31] Quantum confinement contributes to the increase in the energy differential between the energy level and the bandgap. Due to its huge dimension compared to the wavelength, a particle operates freely; when the scale reaches a nano-level, the continuous energy state transforms into a discrete energy state, resulting in a shift in the optical energy bandgap. Consequently, the bandgap becomes size-dependent.[32–34]

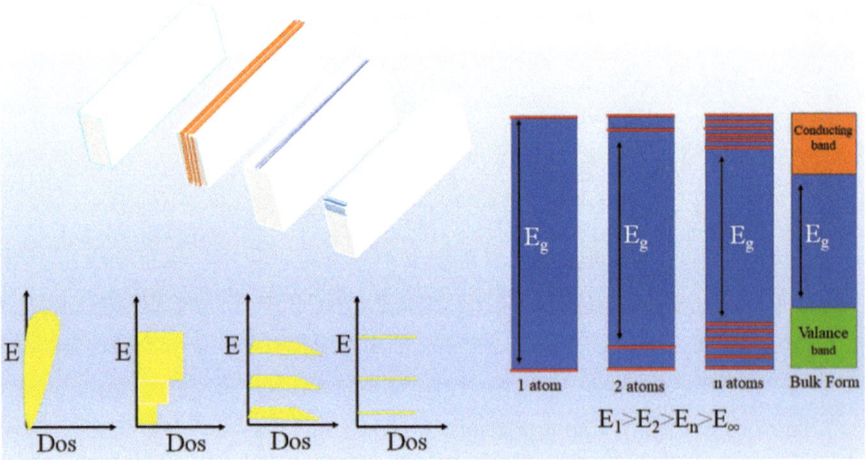

FIGURE 3.2 Schematic depiction demonstrates the changes in the electrical density of states when the dimensionality is changed from 3D to 0D.

3.2.1 TECHNIQUES FOR SYNTHESIS OF NANOMATERIALS

In nanotechnology, two competing ways of producing nanostructures are used (Figure 3.3): "top-down" and "bottom-up."[18,35] The Foresight Institution, founded by Dr. K. Eric Drexler in 1989, is credited with being the first organization to apply these phrases to the field of nanotechnology to recognize the systematic differences between molecular and formal fabrication.

The top-down strategy employs traditional small production processes such as dethronement, engraving, and ionic beam milling on two-dimensional substrates to decrease the adjacent dimensions of the films into nanometers. Extremely organized nanowires can result from the top-down method.

Nonetheless, this technique does not meet the industrial requirement of creating inexpensive and vast quantities of twists. Moreover, the one-dimensional nanostructures generated by these processes are not always exclusively crystallized. Bottom-up approaches involve the fabrication of molecular establishing interferences or chemical preparation from evaporation phase transport and solvent-supported procedures. The bottom-up technique allows for low-priced manufacturing, albeit it may be problematic to have it considerably ordered as well as modeled. Furthermore, dominance and perceptiveness in the growing process are required to successfully consolidate functioning order gadgets.[35–37] The different preparation methods, including top-down and bottom-up, are represented in Figure 3.3.

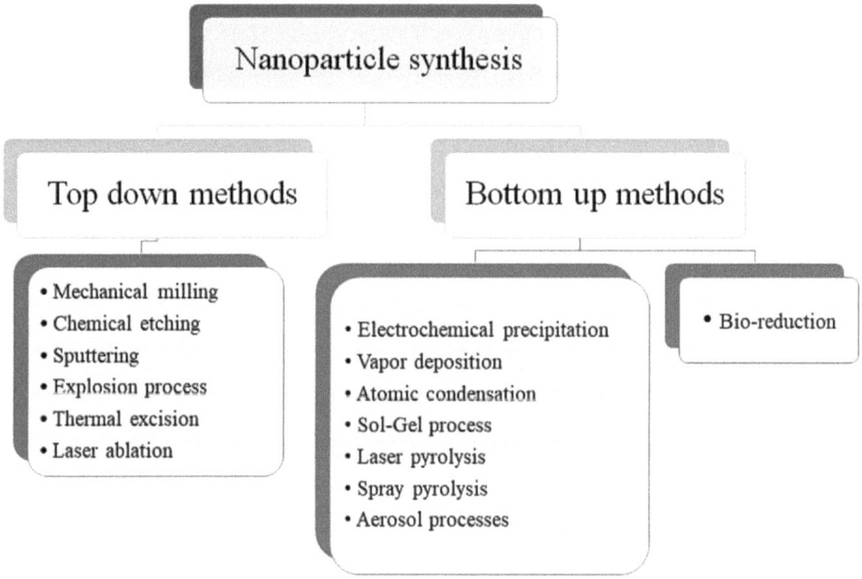

FIGURE 3.3 Major approaches for formulating nanostructures.

Krystek et al.[38] explained how current advancements in nanotechnology need rigorous characterization of nanoparticles owing to ease of manufacturing, product manufacture, and safety research. Hubert et al.[39] described major purpose of fundamental characterization to clarify the 3-D array of pieces in an a-solid. The main goal of structural characterization in the workplace is to evaluate the measurements and locations of the unit cell. For the characterization of semiconductors, numerous classification approaches have been used. Scanning electron microscope (SEM), XRD, photoluminescence,

FTIR, and Raman spectroscopy are among the most operating techniques. These methods are primarily employed to capture images of the superficial and, as a result, characterize the surface shape and characteristics. Mittal et al. described that the majority of nanoparticle applications require these qualities to be homogenous.[40]

3.3 SCOPE OF NANOMATERIALS IN WASTEWATER PURIFICATION

Sánchez et al. reported nanotechnology has emerged as a critical participant in the growth of research and engineering in environmental remediation. It is suitable for both in situ and ex situ environmental contamination and cleanup.[41] Nanoscale material remediation (also called nano-remediation) possesses numerous advantages over conventional techniques of removing pollutants. For instance, it may minimize toxin absorption to near nil, reducing cleaning time and expenses. Nanoparticles bump off the contaminants from the surroundings by surface assimilation, oxidation-reduction processes, or photocatalytic chemical reaction. Qu described that adsorption is a step forward in wastewater treatment for lowering organic and inorganic pollutants. As they significantly improve the surface area, choosiness, and adsorption kinetics, nano adsorbents outperform standard adsorbents in terms of adsorption efficiency.[42] Zhu et al. reported that magnetic nanocomposites might be recycled and utilized for heavy removal (Cr, As, Hg, and Pd).[43]

3.3.1 PHOTOCATALYTIC DEGRADATION OF CONTAMINANTS

Due to their good chemical exposure stability, simplicity of manufacture, and low cost, metallic oxide semiconductors frequently become redundant in photocatalytic applications. Since Fujishima and Honda discovered TiO_2 as an effectual catalyst for photoinduced water splitting in 1972, numerous semiconductor photocatalysts have been actively researched.[44–46] Although TiO_2 is still the most affordable photocatalyst for the breakdown of many carbon-based compounds, TiO_2's broad band gap (3.3 eV) makes it unsuitable for solar photocatalysis in UV light irradiation, which restricts the usefulness of TiO_2 in sunlight.[47] Visible radiation (400–800 nm) contributes considerably to the solar spectrum; there has been a lot of interest in developing novel visible light-dependent photocatalysts. The creation of a high-level, efficient, and effective semiconductor photocatalyst is therefore one of the current key concerns in photocatalysis research.[48–53]

Bokare et al. demonstrate that Fe/Ni bimetallic nanoparticles are used to remove the toxic organic color Orange G from an aqueous solution.[54] Fe/Pd nanoparticles with bimetallic dechlorinate 5 mg/L of lindane limited through first-order kinetics in 5 min, according to research investigating the degradation of Lindane-120 in a water-based solution by nanoscale Fe/Pd particles. Mohamed et al. described photocatalysis processes into two types.[55]

Figure 3.4 indicates the catalyzed photoreactions, and sensitized photoreactions. On the catalyst, the initial photoexcitation happens continuously. The electron is then transferred into a ground-state molecule via a photoexcited catalyst. Early light excitation happens in engageable molecules (dye molecules) associated with the ground-state catalyst in the case of sensitized photoexcitation. Figure 3.4 depicts the suggested graphic illustration of photocatalytic processes.

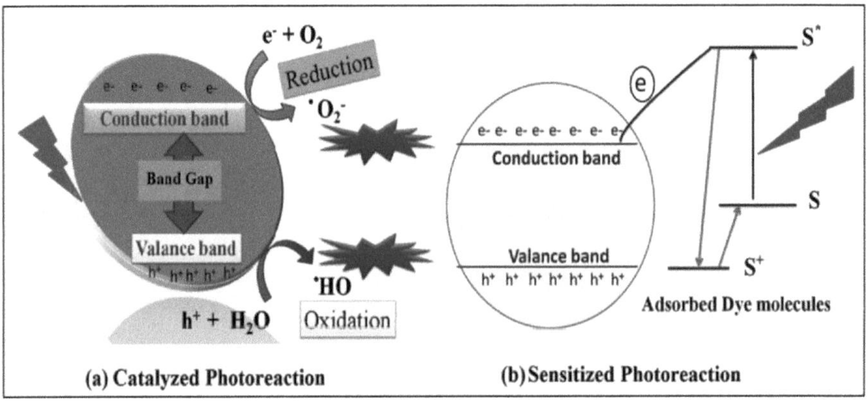

FIGURE 3.4 Schematic representation of photocatalytic reaction.

Qu[42] stated that photocatalysis has significant promise because of its low cost, environmental friendliness, and supportability. The use of nanoparticles in this therapeutic procedure, in particular, is a significant problem. Among the most often used photocatalytic nanoparticles in wastewater treatment are TiO_2, ZnO, CdS, and WO_3. TiO_2 is the most desirable substance because of its low toxicity, chemical strength, low cost, and global availability. However, a number of obstacles have prevented photocatalysts like TiO_2 from being effectively used. For instance, the need for aqueous solutions limits its widespread practical application because photocatalyst recycling and residue separation are challenging. Another difficulty represents optimizing the photocatalysts to take advantage of visible radiation of solar irradiation. Fortunately, the last problem may have been fixed by adding metals to nanostructures of nano-TiO_2 with dye sensitizers or by combining nano-TiO_2 with a material that constricts the bandgap to create interbred nanocomposites.

3.3.2 FUNDAMENTAL CONCEPTS OF PHOTOCATALYTIC OXIDIZATION

Ahmed et al. described that photocatalysis represents an advanced oxidization process (AOP) for the evacuation of harmful pollutants in semiconducting material photocatalysis, a light source, and an oxidizing intermediator, once a conductive catalyst is sufficiently illuminated with photon energy adequate to or higher than the band-gap energy difference, atoms transfer from the band of valence to the conduction band.[56] A hole (h), also known as an electron vacancy, forms simultaneously in the valence band. A potential design of the photocatalysis concept is shown in Figure 3.5.

FIGURE 3.5 Graphic diagrams illustrating the principle of photocatalysis.

Mohamed investigated that photo-produced positive holes oxidize with H_2O to form $OH^{\bullet}$ radicals, whereas electrons make superoxide radicals via the oxygen reaction.[57] These significant products, notably (OH•), are long-lasting oxidizing kinds that destroy organic contaminants. The rate of oxygen breakdown and contaminant oxidation must stay constant in the photocatalytic removal of carbon-based contaminants.[56] Otherwise, it stimulates electron collection in the CB (typically due to a lack of oxygen atoms in the solvent), maximizing the rank of ehp recombination and retarding the photocatalytic oxidization process. Nonetheless, photocatalytic actions strongly rely on internal catalyst features such as crystal structure, surface binder, molecule size, optical band gap, and superficial hydroxyl radical concentration.

Yao et al. and Li research groups reported that bismuth vanadate ($BiVO_4$) is a photocatalyst that may exist in three distinct crystalline phases: scheelite, monoclinic scheelite, and tetragonal-Zircon structures.[58–60] Monoclinic $BiVO_4$ (m-$BiVO_4$) was shown to have exceptional light absorption properties, with a bandgap of 2.4 eV. $BiVO_4$'s photocatalytic properties are strongly controlled by its crystal structure; hence, m-$BiVO_4$ (2.4 eV) performs better than tetragonal $BiVO_4$ (3.1 eV). A simple and uncomplicated hydrothermal process was used to generate zinc vanadate ($Zn_2V_2O_7$) nanorods.[61]

The $Zn_2V_2O_7$ nanorods were characterized using XRD, SEM, TEM, and FTIR. Additionally, it is UV-visible. According to them, sodium dodecyl sulfonate (SDS) is required to synthesize $Zn_2V_2O_7$ nanorods. The photocatalytic activity of zinc vanadate nanorods was evaluated using photocatalytic deprivation of methylene blue organic dye under solar illumination. They proposed that the structure, size, shape, pH, hydrothermal temperature, duration time, and SDS concentration of $Zn_2V_2O_7$ nanorods rendered them suitable for the photocatalytic degradation of organic pollutants when exposed to solar light.

According to Hu et al., nanoparticles of ferric vanadate ($FeVO_4$) can act as visible light photocatalysts.[58] Utilizing a scanning electron microscope and X-ray diffraction, they looked at the material's physical and chemical characteristics. The photocatalytic reactivity of the compounds as they were synthesized was examined by observing the degradation of the MB organic hazardous dye under visible light irradiation. They discovered that various parameters, including solution pH, H_2O_2, and catalyst amount, as well as $FeVO_4$ heat treatment temperature, all impacted the photocatalytic removal of MB. Their findings showed that under visible-light irradiation, 99% photodegradation was obtained in 40 min at pH levels between 1.5 and 2.0. The importance of nanomaterials in different areas of life is given in Figure 3.6. Table 3.1 shows the photocatalytic activity of different photocatalysts that have been previously reported for the breakdown of various organic contaminants in industrial effluent.

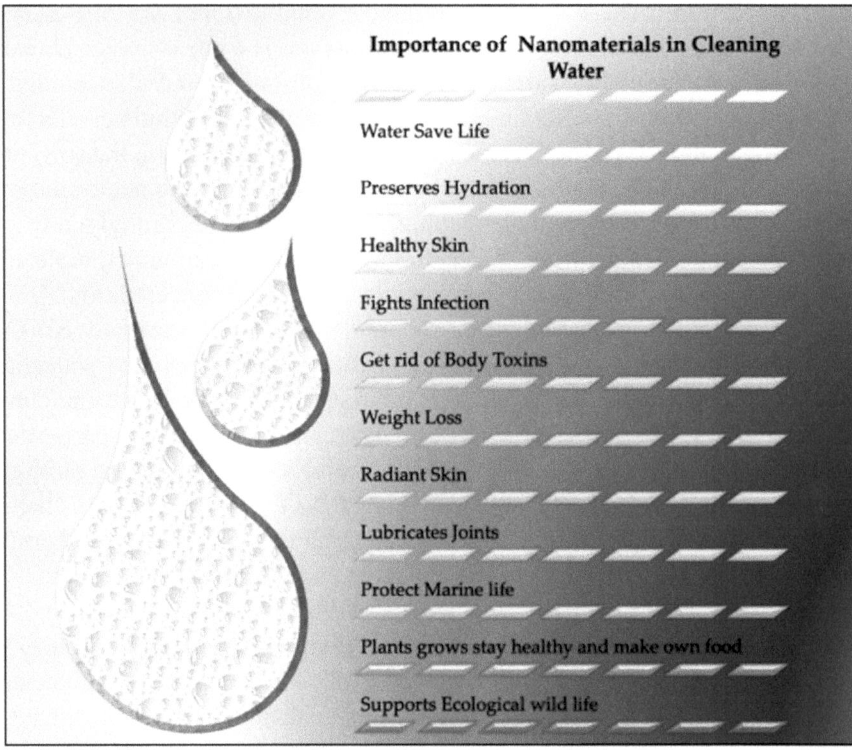

FIGURE 3.6 Graphic diagrams illustrating nanomaterial's significance.

3.4 SUMMARY

The limitations of existing and conventional polluted water treatment can be overcome using nanomaterials, which are both promising and environmentally beneficial. Technology advancements can be made without damaging the Earth's natural ecology. Thus, the main aspects are summarized as follows:

- Water filtration with nanomaterials is both economical and effective, resulting in lower overall expenses. By concurrently removing dangerous and undesirable contaminants, this is achieved.
- Nanomaterials' nanoscale dimensions bring unexpected benefits to cleaning and poison elimination. These benefits include atomic resolution information, improved low signal levels, accelerated response times, and continuous change regulation.

- Research and development of novel photocatalysts with desirable features, such as oxidizing organic substrates or reducing pollutants when exposed to visible light, should be prioritized.

TABLE 3.1 Some Previously Reported Photocatalysts' Photocatalytic Activity for Degrading Organic Pollutants.

Photocatalyst	Efficacy %	Organic Dye	Time	References
Ag_2Se-G-TiO_2	85.2	RhB	1 h 20 min	[62]
$Bi_{0.90}Gd_{0.05}Sm_{0.05}FeO_3$	95	CR	40min	[63]
$Bi_{0.90}Gd_{0.10}Fe_{0.95}Sn_{0.05}O_3$	94.99	CR	3 h	[64]
$Bi_7O_9I_3$/GO	96	CV	2 h	[65]
$BiFeO_3$–grapheme	70.99	CR	2 h	[66]
$BiFeO_3$/TiO_2 (1 : 1)	69.99	CR	2 h	[67]
BiOI/GR	80	RhB	1 h 20 min	[65]
$BiVO_4$ nanoparticles	82	RhB	1 h 20 min	[68]
$BiVO_4$ nanoparticles	30	RhB	30	[69]
$BiVO_4$ nanoparticles	74	RhB	2 h	[70]
BWO/1MG	77.31	RhB	2 h	[71]
Cd-Al/C	52	CV	1 h	[72]
CdS/$CdTiO_3$–TiO_2	92.59	CV	2 h	[73]
Chitosan/nano-CdS	86.01	CR	3 h	[74]
Cu_2O/RGO	69.75	MO	3 h	[75]
CZ-400	64.99	CR	2 h	[76]
Fe_3O_4@SiO_2@TiO_2@Ho	78.4	MO	3 h	[77]
m-alginate/Fe_2O_3/ CdS composite	91.6	CR	5 h	[78]
m-lginate/Fe_2O_3/ CdS composite	91.5	CR	5 h	[78]
TiO_2	89.95	MO	3 h	[79]
TiO_2-SWNT-P-21	81.99	CR	1 h 20 min	[80]
V_2O_5	99.55	CR	3 h	[81]
V_2O_5	99	MO	3 h	[81]
$Zn2SiO_4$	42	MO	1 h	[82]
ZnO	92	MO	3 h	[83]
ZnO/UV-A	86.3	CR	1 h	[84]

- The complexity of the wastewater structure is the primary cause of the problems currently experienced in wastewater management. This intricacy causes difficulties in treatment processes by calling for various equipment and advanced technologies, which in turn causes handling times to be stretched out and higher operational costs.

KEYWORDS

- **degradation efficiency**
- **byproduct**
- **organic hazardous dye**
- **nanocomposites**
- **nanotechnology**
- **photocatalytic reaction**

REFERENCES

1. Abid, N.; Khan, A. M.; Shujait, S.; Chaudhary, K.; Ikram, M.; Imran, M.; Haider, J.; Khan, M.; Khan, Q.; Maqbool, M. Synthesis of Nanomaterials Using Various Top-Down and Bottom-Up Approaches, Influencing Factors, Advantages, and Disadvantages: A Review. *Adv. Colloid Interface Sci.* **2021,** *2021*, 102597.
2. Ahmed, S.; Mofijur, M.; Nuzhat, S.; Chowdhury, A. T.; Rafa, N.; Uddin, M. A.; Inayat, A.; Mahlia, T.; Ong, H. C.; Chia, W. Y. Recent Developments in Physical, Biological, Chemical, and Hybrid Treatment Techniques for Removing Emerging Contaminants from Wastewater. *J. Hazard. Mater.* **2021,** *416*, 125912.
3. Ahmed, S.; Rasul, M.; Brown, R.; Hashib, M. Influence of Parameters on the Heterogeneous Photocatalytic Degradation of Pesticides and Phenolic Contaminants in Wastewater: A Short Review. *J. Environ. Manag.* **2011,** *92* (3), 311–330.
4. Ali, A.; Oh, W.-C. Preparation of Ag_2Se-Graphene-TiO_2 Nanocomposite and Its Photocatalytic Degradation (Rh-B). *J. Korean Ceram. Soc.* **2017,** *54* (5), 388–394.
5. Anwar, K.; Naqvi, F. K.; Beg, S. Synthesis of Tetragonally Stabilized Lanthanum Doped Bismuth Vanadium Oxide Nanoparticles and Its Enhanced Visible Light Induced Photocatalytic Performance. *Phase Transit.* **2022,** *95* (1), 64–79.
6. Baig, N.; Kammakakam, I.; Falath, W. Nanomaterials: A Review of Synthesis Methods, Properties, Recent Progress, and Challenges. *Mater. Adv.* **2021,** *2* (6), 1821–1871.
7. Bayda, S.; Adeel, M.; Tuccinardi, T.; Cordani, M.; Rizzolio, F. The History of Nanoscience and Nanotechnology: From Chemical–Physical Applications to Nanomedicine. *Molecules* **2019,** *25* (1), 112–116.

8. Biondo, F. *Design and Development of Novel Biocompatible Nanosystems for Drug Delivery. Curriculum Scienze Chimiche E Scienze Farmaceutiche Ciclo XXXII-Anno Accademico 2018-2019, Dipartimento Di Scienze Pure E Applicate*; (Curriculum Chemical Sciences & Pharmaceutical Sciences Cycle XXXII-Academic Year 2018–2019, Department of Pure And Applied Sciences), University of Urbino Carlo Bo, 2020; p 132.
9. Bokare, A. D.; Chikate, R. C.; Rode, C. V.; Paknikar, K. M. Iron-Nickel Bimetallic Nanoparticles for Reductive Degradation of Azo Dye Orange G in Aqueous Solution. *Appl. Catal. B: Environ.* **2008,** *79* (3), 270–278.
10. Broderick, D. *The Time Machine Hypothesis*; Springer, 2019.
11. Bruno, F.; Sciortino, A.; Buscarino, G.; Soriano, M. L.; Ríos, Á.; Cannas, M.; Gelardi, F.; Messina, F.; Agnello, S. A Comparative Study of Top-Down and Bottom-Up Carbon Nanodots and Their Interaction with Mercury Ions. *Nanomaterials* **2021,** *11* (5), 1265–1270.
12. Cai, J.; Liu, W.; Li, Z. One-Pot Self-Assembly of Cu_2O/RGO Composite Aerogel for Aqueous Photocatalysis. *Appl. Surface Sci.* **2015,** *358,* 146–151.
13. Chirayil, C. J.; Abraham, J.; Mishra, R. K.; George, S. C.; Thomas, S. *Thermal and Rheological Measurement Techniques for Nanomaterials Characterization*; Elsevier, 2017; pp 1–36.
14. Chou, S.-Y.; Chung, W.-H.; Chen, L.-W.; Dai, Y.-M.; Lin, W.-Y.; Lin, J.-H.; Chen, C.-C. A Series of BiO x I y/GO Photocatalysts: Synthesis, Characterization, Activity, and Mechanism. *RSC Adv.* **2016,** *6* (86), 82743–82758.
15. Danish, M. S. S.; Estrella, L. L.; Alemaida, I. M. A.; Lisin, A.; Moiseev, N.; Ahmadi, M.; Nazari, M.; Wali, M.; Zaheb, H.; Senjyu, T. Photocatalytic Applications of Metal Oxides for Sustainable Environmental Remediation. *Metals* **2021,** *11* (1), 80–87.
16. Davies, J. C. *Oversight of Next Generation Nano Technology*. Citeseer, 2017.
17. Edvinsson, T. Optical Quantum Confinement and Photocatalytic Properties in Two-, One-and Zero-Dimensional Nanostructures. *R. Soc. Open Sci.* **2018,** *5* (9), 180387.
18. Fujishima, A.; Honda, K. Electrochemical Photolysis of Water at a Semiconductor Electrode. *Nature* **1972,** *238* (5358), 37–38.
19. Ghorbanpour, M.; Manika, K.; Varma, A. *Nanoscience and Plant-Soil Systems*; Springer, 2017.
20. Giwa, A.; Yusuf, A.; Balogun, H. A.; Sambudi, N. S.; Bilad, M. R.; Adeyemi, I.; Chakraborty, S.; Curcio, S. Recent Advances in Advanced Oxidation Processes for Removal of Contaminants from Water: A Comprehensive Review. *Proc. Saf. Environ. Protect.* **2021,** *146,* 220–256.
21. Goyal, M.; Singh, M. Size and Shape Dependence of Optical Properties of Nanostructures. *Appl. Phys. A* **2020,** *126* (3), 1–8.
22. Hofmann, M.; Rainer, M.; Schulze, S.; Hietschold, M.; Mehring, M. Nonaqueous Synthesis of a Bismuth Vanadate Photocatalyst by Using Microwave Heating: Photooxidation Versus Photosensitized Decomposition in Visible-Light-Driven Photocatalysis. *ChemCatChem* **2015,** *7* (8), 1357–1365.
23. Hu, L.; Pan, H.; Zhou, Y.; Zhang, M. Methods to Improve Lignin's Reactivity as a Phenol Substitute and as Replacement for Other Phenolic Compounds: A Brief Review. *BioResources* **2011,** *6* (3), 3515–3525.
24. Hubert, M.; Martin, S.; Yacobi, B. G.; Burgmann, T. A.; Passy, P.W. Phototherapeutic Treatment Method Using a Passive Host Medium Containing Nanoparticles. *Google Patents* 2011.

25. Irfan, S.; Rizwan, S.; Shen, Y.; Li, L.; Butt, S.; Nan, C.-W. The Gadolinium (Gd^{3+}) and Tin (Sn^{4+}) Co-Doped $BiFeO_3$ Nanoparticles as New Solar Light Active Photocatalyst. *Sci. Rep.* **2017,** *7,* 42493.
26. Jafry, H. R.; Liga, M. V.; Li, Q.; Barron, A. R. Single-Walled Carbon Nanotubes (SWNTs) as Templates for the Growth of TiO_2: the Effect of Silicon in Coverage and the Positive and Negative Synergies for the Photocatalytic Degradation of Congo Red Dye. *N. J. Chem.* **2011,** *35* (2), 400–406.
27. Jatoi, A. S.; Akhter, F.; Mazari, S. A.; Sabzoi, N.; Aziz, S.; Soomro, S. A.; Mubarak, N. M.; Baloch, H.; Memon, A. Q.; Ahmed, S. Advanced Microbial Fuel Cell for Waste Water Treatment—A Review. *Environ. Sci. Pollut. Res. Int.* **2021,** *28* (5), 5005–5019.
28. Jiang, R.; Yao, J.; Zhu, H.; Fu, Y.; Guan, Y.; Xiao, L.; Zeng, G. Effective Decolorization of Congo Red in Aqueous Solution by Adsorption and Photocatalysis Using Novel Magnetic Alginate/γ-Fe_2O_3/CdS Nanocomposite. *Desalinat. Water Treatment* **2014,** *52* (1–3), 238–247.
29. Joung, S.-K.; Amemiya, T.; Murabayashi, M.; Itoh, K. Relation Between Photocatalytic Activity and Preparation Conditions for Nitrogen-Doped Visible Light-Driven TiO_2 Photocatalysts. *Appl. Catal. A: Gen.* **2006,** *312,* 20–26.
30. Khan, S. A.; Khan, S. B.; Asiri, A. M. Layered Double Hydroxide of Cd-Al/C for the Mineralization and De-Coloration of Dyes in Solar and Visible Light Exposure. *Sci. Rep.* **2016,** *6,* 35107.
31. Khan, S. B.; Lee, S. L. Nanomaterials Significance; Contaminants Degradation for Environmental Applications. *Nano Express* **2021,** *2* (2), 022002.
32. Krystek, P.; Ulrich, A.; Garcia, C. C.; Manohar, S.; Ritsema, R. Application of Plasma Spectrometry for the Analysis of Engineered Nanoparticles in Suspensions and Products. *J. Anal. Atom. Spectr.* **2011,** *26* (9), 1701–1721.
33. Kumar, D. S.; Kumar, B. J.; Mahesh, H. Quantum Nanostructures (QDs): An Overview. *Synthes. Inorg. Nanomater.* **2018,** *2018,* 59–88.
34. Kuraj, N. *REACH and the Environmental Regulation of Nanotechnology: Preventing and Reducing the Environmental Impacts of Nanomaterials*; London: Routledge, 2019.
35. Lan, S.; Liu, L.; Li, R.; Leng, Z.; Gan, S. Hierarchical Hollow Structure ZnO: Synthesis, Characterization, and Highly Efficient Adsorption/Photocatalysis Toward Congo Red. *Ind. Eng. Chem. Res.* **2014,** *53* (8), 3131–3139.
36. Li, J.; Zhao, W.; Guo, Y.; Wei, Z.; Han, M.; He, H.; Yang, S.; Sun, C. Facile Synthesis and High Activity of Novel $BiVO_4$/$FeVO_4$ Heterojunction Photocatalyst for Degradation of Metronidazole. *Appl. Surface Sci.* **2015,** *351,* 270–279.
37. Li, S.; Lin, Y.-H.; Zhang, B.-P.; Li, J.-F.; Nan, C.-W. $BiFeO_3$/TiO_2 Core-Shell Structured Nanocomposites as Visible-Active Photocatalysts and Their Optical Response Mechanism. *J. Appl. Phys.* **2009,** *105* (5), 054310.
38. Li, Y.; Zhang, W.; Li, L.; Yi, C.; Lv, H.; Song, Q. Litchi-Like CdS/$CdTiO_3$–TiO_2 Composite: Synthesis and Enhanced Photocatalytic Performance for Crystal Violet Degradation and Hydrogen Production. *RSC Adv.* **2016,** *6* (56), 51374–51386.
39. Li, Z.; Shen, Y.; Guan, Y.; Hu, Y.; Lin, Y.; Nan, C.-W. Bandgap Engineering and Enhanced Interface Coupling of Graphene–$BiFeO_3$ Nanocomposites as Efficient Photocatalysts Under Visible Light. *J. Mater. Chem. A* **2014,** *2* (6), 1967–1973.
40. Linsebigler, A. L.; Lu, G.; Yates Jr., J. T. Photocatalysis on TiO_2 Surfaces: Principles, Mechanisms, and Selected Results. *Chem. Rev.* **1995,** *95* (3), 735–758.

41. Liu, C. J.; Xu, Y. H. *Synthesis, Characterization and Photocatalytic Activities of Bismuth Vanadate by Facile Co-Precipitation Method*; Trans Tech Publ., 2011; pp 1469–1472.
42. Liu, H.; Han, W. Wüstite-Based Catalyst for Ammonia Synthesis: Structure, Property and Performance. *Catal. Today* **2017,** *297,* 276–291.
43. Liu, M.; Xue, X.; Yu, S.; Wang, X.; Hu, X.; Tian, H.; Chen, H.; Zheng, W. Improving Photocatalytic Performance from Bi_2WO_6@ MoS_2/Graphene Hybrids Via Gradual Charge Transferred Pathway. *Sci. Rep.* **2017,** *7* (1), 3637.
44. Madhusudan, P.; Yu, J.; Wang, W.; Cheng, B.; Liu, G. Facile Synthesis of Novel Hierarchical Graphene–Bi 2 O2 CO3 Composites with Enhanced Photocatalytic Performance Under Visible Light. *Dalton Trans.* **2012,** *41* (47), 14345–14353.
45. Manolis, G. D.; Dineva, P. S.; Rangelov, T.; Sfyris, D. Mechanical Models and Numerical Simulations in Nanomechanics: A Review Across the Scales. *Eng. Anal. Boundary Elements* **2021,** *128,* 149–170.
46. Masjedi-Arani, M.; Salavati-Niasari, M. A Simple Sonochemical Approach for Synthesis and Characterization of Zn_2SiO_4 Nanostructures. *Ultrason. Sonochem.* **2016,** *29,* 226–235.
47. Mezzetti, A. Nanostructured WO_3 Films for Photoelectrochemical Water-Splitting; 2017 [Online]. https://www.politesi.polimi.it/handle/10589/131174 (accessed July 31, 2023).
48. Mittal, A. K.; Chisti, Y.; Banerjee, U. C. Synthesis of Metallic Nanoparticles Using Plant Extracts. *Biotechnol. Adv.* **2013,** *31* (2), 346–356.
49. Mohamed, H. H.; Bahnemann, D. W. The Role of Electron Transfer in Photocatalysis: Facts and Fictions. *Appl. Catal. B: Environ.* **2012,** *128,* 91–104.
50. Mohamed, R.; McKinney, D.; Sigmund, W. Enhanced Nanocatalysts. *Mater. Sci. Eng. Rep.* **2012,** *73* (1), 1–13.
51. Mohammad, A.; Kapoor, K.; Mobin, S. M. Improved Photocatalytic Degradation of Organic Dyes by ZnO-Nanoflowers. *Chem. Select* **2016,** *1* (13), 3483–3490.
52. Mortazavi-Derazkola, S.; Salavati-Niasari, M.; Amiri, O.; Abbasi, A. Fabrication and Characterization of Fe_3O_4@ SiO_2@ TiO_2@ Ho Nanostructures as a Novel and Highly Efficient Photocatalyst for Degradation of Organic Pollution. *J. Energy Chem.* **2017,** *26* (1), 17–23.
53. Moscow, S.; Jothivenkatachalam, K.; Jaganathan, K. Facile Fabrication, Characterization of Bismuth Vanadate Nanoparticles Via Hydrothermal Method and Its Photocatalytic Properties. *Struct. Opt. Study Titanium Dioxide Thin Films Prepared Sol–Gel Tech.* **2012,** *24* (3), 46.
54. Mourdikoudis, S.; Pallares, R. M.; Thanh, N. T. Characterization Techniques for Nanoparticles: Comparison and Complementarity Upon Studying Nanoparticle Properties. *Nanoscale* **2018,** *10* (27), 12871–12934.
55. Pankajakshan, A.; Sinha, M.; Ojha, A. A.; Mandal, S. Water-Stable Nanoscale Zirconium-Based Metal–Organic Frameworks for the Effective Removal of Glyphosate from Aqueous Media. *ACS Omega* **2018,** *3* (7), 7832–7839.
56. Paolini, A.; Kollmannsberger, S.; Rank, E. Additive Manufacturing in Construction: A Review on Processes, Applications, and Digital Planning Methods. *Addit. Manuf.* **2019,** *30,* 100894.
57. Pei, L.; Lin, N.; Wei, T.; Liu, H.; Yu, H. Zinc Vanadate Nanorods and Their Visible Light Photocatalytic Activity. *J. Alloys Compd.* **2015,** *631,* 90–98.
58. Prakash, J.; Khan, S.; Chauhan, S.; Biradar, A. Metal Oxide-Nanoparticles and Liquid Crystal Composites: A Review of Recent Progress. *J. Mol. Liq.* **2020,** *297,* 112052.

59. Qiu, L.; Zhu, N.; Feng, Y.; Michaelides, E. E.; Żyła, G.; Jing, D.; Zhang, X.; Norris, P. M.; Markides, C. N.; Mahian, O. A Review of Recent Advances in Thermophysical Properties at the Nanoscale: From Solid State to Colloids. *Phys. Rep.* **2020,** *843,* 1–81.
60. Qu, X.; Alvarez, P. J.; Li, Q. Applications of Nanotechnology in Water and Wastewater Treatment. *Water Res.* **2013,** *47* (12), 3931–3946.
61. Quhe, R.; Xu, L.; Liu, S.; Yang, C.; Wang, Y.; Li, H.; Yang, J.; Li, Q.; Shi, B.; Li, Y. Sub-10 nm Two-Dimensional Transistors: Theory and Experiment. *Phys. Rep.* **2021,** *938,* 1–72.
62. Sajid, M. M.; Shad, N. A.; Javed, Y.; Khan, S. B.; Zhang, Z.; Amin, N.; Zhai, H. Preparation and Characterization of Vanadium Pentoxide (V_2O_5) for Photocatalytic Degradation of Monoazo and Diazo Dyes. *Surf. Interfaces* **2020,** *2020,* 100502.
63. Saleh, T. A. Nanomaterials: Classification, Properties, and Environmental Toxicities. *Environ. Technol. Innovat.* **2020,** *20,* 101067.
64. Samsudin, M. F. R.; Sufian, S.; Hameed, B. Epigrammatic Progress and Perspective on the Photocatalytic Properties of $BiVO_4$-Based Photocatalyst in Photocatalytic Water Treatment Technology: A Review. *J. Mol. Liquids* **2018,** *268,* 438–459.
65. Sánchez, A.; Recillas, S.; Font, X.; Casals, E.; González, E.; Puntes, V. Ecotoxicity and Remediation with Engineered Inorganic Nanoparticles in the Environment. *TrAC Trends Anal. Chem.* 2018, **2011,** *30* (3), 507–516.
66. Shang, Y.; Hasan, M.; Ahammed, G. J.; Li, M.; Yin, H.; Zhou, J. Applications of Nanotechnology in Plant Growth and Crop Protection: A Review. *Molecules* **2019,** *24* (14), 2558.
67. Shen, Y.; Dierking, I. Perspectives in Liquid-Crystal-Aided Nanotechnology and Nanoscience. *Appl. Sci.* **2019,** *9* (12), 2512.
68. Shi, Q.; Murcia-López, S. N.; Tang, P.; Flox, C.; Morante, J. R.; Bian, Z.; Wang, H.; Andreu, T. Role of Tungsten Doping on the Surface States in $BiVO_4$ Photoanodes for Water Oxidation: Tuning the Electron Trapping Process. *ACS Catal.* **2018,** *8* (4), 3331–3342.
69. Sun, Y.; Wang, Z.; Fu, P.; Jiang, Q.; Yang, T.; Li, J.; Ge, X. The Impact of Relative Humidity on Aerosol Composition and Evolution Processes During Wintertime in Beijing, China. *Atmos. Environ.* 2013, *77,* 927–934.
70. Tan, H. L.; Amal, R.; Ng, Y. H. Alternative Strategies in Improving the Photocatalytic and Photoelectrochemical Activities of Visible Light-Driven $BiVO_4$: A Review. *J. Mater. Chem.-A* **2017,** *5* (32), 16498–16521.
71. Tang, J.; Cheng, G.; Zhou, H.; Yang, H.; Lu, Z.; Chen, R. Shape-Dependent Photocatalytic Activities of Bismuth Subcarbonate Nanostructures. *J. Nanosci. Nanotechnol.* **2012,** *12* (5), 4028–4034.
72. Tennyson, E. M.; Gong, C.; Leite, M. S. Imaging Energy Harvesting and Storage Systems at the Nanoscale. *ACS Energy Lett.* **2017,** *2* (12), 2761–2777.
73. Trovarelli, A.; Llorca, J. Catalysts at Nanoscale: How Do Crystal Shapes Shape Catalysis? *ACS Catal.* **2017,** *7* (7), 4716–4735.
74. Vanga, P. R.; Mangalaraja, R.; Ashok, M. Effect of Co-Doping on the Optical, Magnetic and Photocatalytic Properties of the Gd Modified $BiFeO_3$. *J. Mater. Sci.: Mater. Electron.* **2016,** *27* (6), 5699–5706.
75. Vázquez-Núñez, E.; Molina-Guerrero, C. E.; Peña-Castro, J. M.; Fernández-Luqueño, F.; De La Rosa-Álvarez, M. Use of Nanotechnology for the Bioremediation of Contaminants: A review. *Processes* **2020,** *8* (7), 826–830.

76. Wang, D.; Xiao, L.; Luo, Q.; Li, X.; An, J.; Duan, Y. Highly Efficient Visible Light TiO2 Photocatalyst Prepared by Sol–Gel Method at Temperatures <300°C. *J. Hazard. Mater.* **2011,** *192* (1), 150–159.
77. Wu, Y.-C.; Chaing, Y.-C.; Huang, C.-Y.; Wang, S.-F.; Yang, H.-Y. Morphology-Controllable Bi_2O_3 Crystals Through an Aqueous Precipitation Method and Their Photocatalytic Performance. *Dyes Pigments* **2013,** *98* (1), 25–30.
78. Yamasita, D.; Takata, T.; Hara, M.; Kondo, J. N.; Domen, K. Recent Progress of Visible-Light-Driven Heterogeneous Photocatalysts for Overall Water Splitting. *Solid State Ion.* **2004,** *172* (1–4), 591–595.
79. Yan, L.; Zhao, F.; Wang, J.; Zu, Y.; Gu, Z.; Zhao, Y. Safe-by-Design Strategy Towards Safer Nanomaterials in Nanomedicines. *Adv. Mater.* **2019,** *31* (45), 1805391.
80. Yao, X.; Zhao, X.; Hu, J.; Xie, H.; Wang, D.; Cao, X.; Zhang, Z.; Huang, Y.; Chen, Z.; Sritharan, T. The Self-Passivation Mechanism in Degradation of BiVO4 Photoanode. *Science* **2019,** *19,* 976–985.
81. Zhang, L.; Bahnemann, D. Synthesis of Nanovoid Bi_2WO_6 2-D Ordered Arrays as Photoanodes for Photoelectrochemical Water Splitting. *ChemSusChem* **2013,** *6* (2), 283–290.
82. Zhang, Z.; Wang, W.; Shang, M.; Yin, W. Low-Temperature Combustion Synthesis of Bi2WO6 Nanoparticles as a Visible-Light-Driven Photocatalyst. *J. Hazard. Mater.* **2010,** *177* (1–3), 1013–1018.
83. Zhu, H.; Jiang, R.; Xiao, L.; Chang, Y.; Guan, Y.; Li, X.; Zeng, G. Photocatalytic Decolorization and Degradation of Congo Red on Innovative Crosslinked Chitosan/Nano-CdS Composite Catalyst Under Visible Light Irradiation. *J. Hazard. Mater.* **2009,** *169* (1–3), 933–940.
84. Zhu, J.; Wei, S.; Chen, M.; Gu, H.; Rapole, S. B.; Pallavkar, S.; Ho, T. C.; Hopper, J.; Guo, Z. Magnetic Nanocomposites for Environmental Remediation. *Adv. Powder Technol.* **2013,** *24* (2), 459–467.

CHAPTER 4

LOW-COST WASTEWATER TREATMENT TECHNOLOGIES SUITABLE FOR IRRIGATION

YAHAYA HASSAN LABARAN, AUWAL ALHASSAN MUSA, ABUBAKAR USMAN KHALIL, NURA LAWAN, and KHALIL HASSAN IBRAHIM

ABSTRACT

Farmers especially in rural areas are now using wastewater as an alternative source for irrigation due to the unavailability of advanced wastewater (WW) treatment techniques and they also often have shortage and accessibility to the natural water in many countries across the world. However, the use of such water is prohibitively expensive since it must be treated, and as such, many rural farmers are using untreated water directly, particularly for vegetable and fruit productions. Wastewater is posing a threat to the consumer's health and food safety in general. As a result, this chapter focuses on studying the low-cost water treatment technologies for irrigation to assist local farmers in selecting the most appropriate, reliable, and cost-effective water treatment technology. Relevant background information on irrigation systems, low-cost wastewater treatment technologies, and the suitability, and sustainability of using these techniques in irrigation systems and as well to the environment have been discussed in detail. The proposed possible recommendations based on these discussions were developed to serve as benchmark for the employability and reliability of wastewater treatment technologies by the farmers, given more emphasis to the region, availability of construction materials, and cost.

Sustainable Use of Wastewater in Agriculture: Applications, Methods, and Technology. Megh R. Goyal & Gaurav Saini (Eds.)

4.1 INTRODUCTION

Agriculture is one of the essentials for social and economic development in the world. In Africa, farming alone accounts for more than 70% of the livelihoods of poor people.[36] Conventionally, sufficient natural water mainly through rainfall is sufficient for a crop to grow; however, the uneven natural distribution of rainfall necessitates the farmers to devise other artificial means for supplementing water supply to crops during winter season in order to have a sustainable harvest throughout the year. Also, the water competition grows higher among various sectors, such as food industries, urban and environmental usage etc. Then agricultural sector is left with no option rather than diversify and adopt other alternatives through irrigation due to its higher accumulation of water which reaches 80% of the total fresh water available in the world.[33] According to findings from the research work, only 17% of the total farmland is active for irrigation, which contributes to up to 50% of the total food production of the world.[40]

The level of contribution in alleviating food insecurity through implementing irrigation schemes has yielded positive outcomes since the last century. Apart from positive impact of irrigation to the inhabitants in terms of food security, enriching green environment, and conservation of land, there are still several negative effects by such schemes that are posed to the environment. Other challenges faced by the irrigation systems are the threat faced by the water demand by the rapid urbanization in which reasonable quantity of water is diverted to other purposes, and the gap will exponentially rise due to this competition. Irrigation managers should consider and devise alternative means in reducing the compounded surface water diversion, which affects streamflow downstream by establishing small-scale irrigation system that can help in minimizing the challenges, because these will lead to more secured, easily-managed approach toward better irrigation system through minimizing wastes.[9,13,24] Distribution of worldwide irrigated area in 2017 shows that Africa has 39 million acres of irrigated area, which represents only 6% of the cropped lands and 5% of the world's total. Much of the world's irrigated area lies in Asia (71%) followed by Americas (16%).

4.2 CONCEPT OF IRRIGATION

The irrigation system works on process of synthesis of water and transports it using appropriate means in most cases canals to the intended land for this purpose.[1] Irrigation is generally termed as the nonspontaneous or man-made means of supplying sufficient water to the crops for continuous yield, especially in arid and semiarid regions. The process entails supplementing relatively optimal quantity of fresh water through soil layers for the plant using the evapotranspiration phenomenon. However, other regions with uneven distribution of rainfall may be inclusive. The main objectives of irrigation are to have nondrought sessions within the year, help in maintaining food security, reviving and maintaining the nutritional behavior of a particular land.[12]

The irrigation helps in optimizing the compounded natural water for sufficient utilization; this is in line with an unprecedented increasing demand of water for agricultural and other human purposes, and helps in curbing the challenges and difficulties in recharging aquifers, varied water tables regions, land degradation, and leaching through chemical contaminations.[25] However, the most difficult task in irrigation is the management aspects of ensuring the optimal use of water, during a particular specified irrigation period, increasing the output of the yielded crops without altering the natural environmental setups.

To fully understand the concept of irrigation management, it is necessary to have background knowledge on the existing relationships among water, soil, environment/atmosphere, and crops. The principle will give a wider insight based on the three aspects: (a) Weather conditions, which serve as the main driver in determining rate of crop transpiration and soil evaporation, soil physical and chemical properties, and natural response of crop roots to the water. Therefore, design and operating an irrigation system requires professional and experienced personnels, who are able to incorporate factors, such as annual rainfall in the catchment area, percolation rate, evaporation losses and operating conditions, social and cultural setup of the environment and economic gain. These can be achieved by adopting high technologies and innovative approaches of irrigation.[35]

The two conventional sources of water for irrigation are: (a) surface water is mainly obtained from large bodies of water (such as rivers and dams, canal networks, and other reservoirs); and underground water is mainly obtained by using pumps to lift the water from aquifers directly to the wells and tube wells. The process of recharging groundwater aquifers may become difficult due to excessive pumping for irrigation, and deeper percolation is mainly attributed to the irrigation for leaching of unwanted and toxic soluble chemicals that may temper the required properties needed for the irrigation water. Also, irrigation from direct runoff affects the natural quality of soil, which leads to soil adulteration with certain chemicals.[18,20]

4.3 FACTORS FOR SELECTING A SUITABLE IRRIGATION SYSTEM

The factors for selecting most suitable irrigation method for a particular region include quantity and quality of water available, climatic variables (precipitation, temperature, wind, relative humidity, sunshine hours and intensity, geographical location, etc.), land topography and characteristics, nature and characteristics of crops, and the cost of irrigation system.[19] Among all the factors responsible for efficient irrigation farming, the volume of water has major contribution that has tendency of influencing other engineering parameters. Interestingly, an area with uneven precipitation has high yield compared to the areas with abundant precipitation. Also, irrigation helps in contributing to continuous production of certain agricultural produce throughout the year. Although there is an increased cost of labor, energy, and transportation, yet these costs are higher compared with conventional dry farming.

One of the key challenges of irrigation is transition from traditional to modern methods, whereas the former has no control of exact amount of fresh water used, nutrients, energy, leading to land degradation and erosion, while the latter needs additional financial, skills, and technological inputs. It is of outmost importance to elevate and upgrade the conventional irrigation systems by adopting modern technologies, as this is in line with fulfilling global food security while extensive research to correlate the required mandate should be adopted. However, these new technologies should be sustainably bias for the optimal usage by the public. Cognate and underlying principles of irrigation dwelled on proper understanding the mutual behavior of water, soil, and crops should serve as part of background of any irrigation process, infiltration, and redistribution. Therefore, an enhanced modern irrigation system will notably lead to higher crop yield, control the waste of water, save energy, and assist in conserving the land. The farm-based irrigation systems have proved to give high efficiency and sustainability.[34]

4.4 TYPES OF WASTEWATER TREATMENT TECHNOLOGIES

The primary purpose of wastewater treatment is to hasten the natural water purification processes. The selection of wastewater treatment methods is often based on several factors, such as process cost, sustainability, and required water quality. These technologies are divided into primary and secondary treatment processes (Figure 4.1).

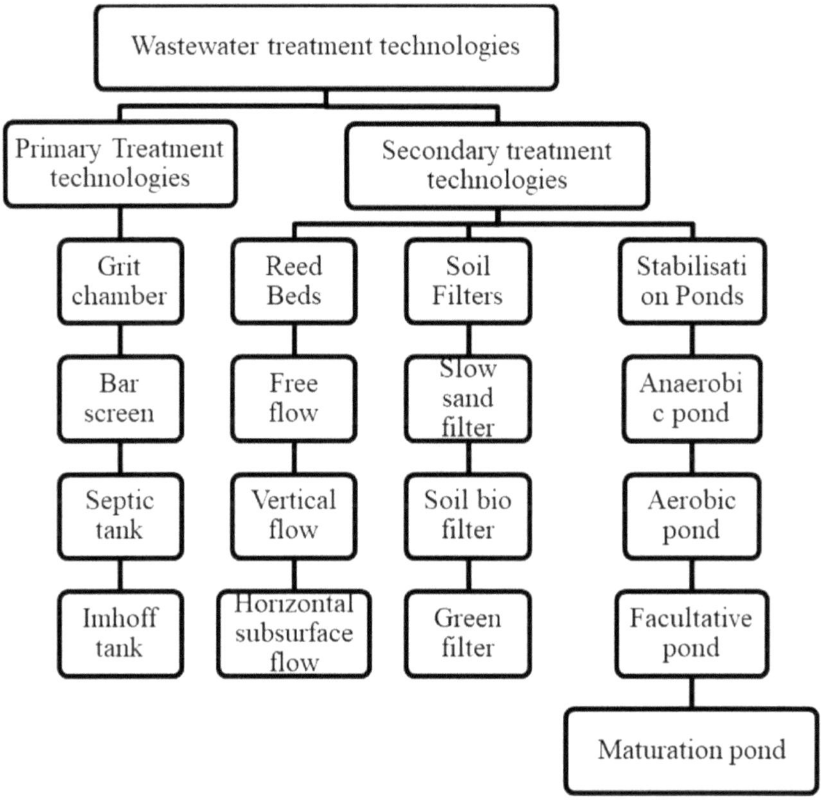

FIGURE 4.1 Wastewater treatment technologies.

4.4.1 PRIMARY TREATMENT TECHNOLOGIES

The primary treatment procedure removes large water-floating debris (such as wood, rags, and scraps) that might damage equipment or block pipes as wastewater passes through a screen while entering a treatment facility. After being screened, the wastewater reaches a "grit chamber," where tiny pebbles, embers, and gravel sink to the bottom of the chamber. After the screening and grit chamber procedures, the wastewater still retains inorganic and organic matter and various other suspended particles that require additional treatments. As a result, a sedimentation tank is typically used to filter these small particulates from the wastewater. While organic wastes (biosolids) are often extracted by pumping them from the tanks, they can then be processed for use as farm fertilizer, burned, or disposed of in landfills.[38,42] Some of the techniques commonly used for wastewater primary treatment are briefly discussed below.

- **Grit chamber** (Sand traps) is a device that extracts tiny particles >0.2 mm in order to safeguard canals, pumps, and pipelines from erosion and silt deposits. Furthermore, it filters gravel, sand, mineral particles, and other organic components that are difficult to break down (such as seeds and grains). It comprises a stationary sedimentation tank with commonly mixed grease and sand traps.[38] The biggest drawback of these devices is that they have variable flow rates, which can alter how much sand could be removed. However, sand traps with regulated speeds were created to fix these issues.[27]
- **Bar Screen** (Coarse screens): The coarser particles, which are typically between 6 and 150 mm in size, are retained by this technique. It comprises a series of parallel curved and straight bars that make a screen filter. It is then angled from 25° to 50° to the direction of water flow. Under certain conditions, this technique is vital to safeguard pipelines, valves, and pumps during subsequent treatment processes.[38] However, only screen filters with manual cleaning features are viable for small-scale treatment systems. While for a bigger setting requiring more than 600 p.e., fine automatically raked screens are usually preferred.[39,42]
- **Septic tank** is the most commonly used wastewater overflow reservoir. It used to be popular due to its affordability and ease of use. However, the septic tank's retention duration and design layout determines its efficiency. Additionally, it does not ensure that the effluent concentration of contaminants is met; therefore, it is typically utilized as a primary treatment step before the subsequent treatment operations.[3]
- **Imhoff tank** is similar to the septic tank. However, Imhoff tanks comprise a single deposit separated into two sections: with sedimentation occurring in the top section while sludge is processed anaerobically in the bottom section. This technique often treats residential black and gray wastewater from low-rise domestic buildings.[38] Although the system usually requires cleaning twice a year, yet it is reliable, uncomplicated, and simple to use.[39]

4.4.2 SECONDARY TREATMENT TECHNOLOGIES

The secondary stage of wastewater treatment, which employs an activated sludge and trickling filter technologies, removes approximately 85% of the total organic materials by utilizing the bacteria in the water. The process starts after the wastewater exits the sedimentation tank during the primary treatment step and is then allowed to flow or be pumped to a plant by one of the following techniques.[43,17]

- **Reed beds (constructed wetlands)** are secondary/tertiary wastewater treatment systems that use a slow filtration technique to treat pretreated wastewater. It is one of the few biological treatment methods with a well-defined flow direction and filter material. The operational concept of this technique is based on the flow of water through a treatment system made up of planted vegetative cover[39] (such as Reed canary grass, cattail, and reed). These constructed wetlands are categorized into three classes based on the flow direction of the wastewater,[27,43] such as:
  - **a) Free flow (FF):** This FF wetland treatment technique resembles the typical natural water treatment process. This technique allows the wastewater to flow freely across the wetland's base basin. The wetland's base foundation comprises gravel, pebbles, and plant roots that cover the basin's bottom. To effectively treat the inflow through biological, chemical, and physical processes, a 10 cm to 50 cm shallow layer of wastewater is allowed to flow slowly across the basin at a time. Thus, harmful chemicals, organic content, suspended particles, viruses, and nutrients are reduced due to the actions of plants, microorganisms, and decantation.[3,27]
  - **b) Vertical flow (VF)** is a wetland treatment process in which the wastewater is allowed to flow vertically through a designated porous medium. In contrast to the free-flow approach, the basin in this method is built entirely of porous materials, such as sand or gravel. The wastewater from the primary treatment step is allowed to flow vertically through the filter material before being discharged to an exit pipe that is located at the basin's bottom.[4,17]
  - **c) Horizontal subsurface flow (HSF):** In this method, the pretreated water is allowed to flow horizontally through a porous medium. The basin is similarly constituted of sand and gravel; however, the diameter of the porous material should be between 3 and 30 mm for this process. Furthermore, the wastewater level should be maintained between 5 and 15 cm underneath the surface of the porous layer to maintain horizontal subsurface flow.[38, 43]

- **Soil filters** are used in the secondary or tertiary stages of the treatment process. This technology, like constructed wetlands, may be separated into filters with radial, horizontal, and vertical flow. However, the soil filters in this scenario are usually built underground.[17] The soil filters are classified into three subclasses as follows:

a) **Slow sand filter (SSF)** is one of the most basic wastewater treatment systems, with sand serving as the filter material. The slow sand filter comprises a chamber, a vessel, and a sand tank. It effectively reduces pathogenic compounds and turbidity, and as a result, it is commonly used for drinking water purification purposes.[38,39]

b) **Soil biofilter** is a soil-based treatment method in which wastewater is allowed to flow directly to the treatment area. The filter medium comprises the soil plant-water ecosystems (comprising of geophagous worms, soil macro, plants, bacteria), and soil minerals. A soil bio-filtration system includes a pump system, raw and processed effluent tanks, and a bioreactor typically made of plants and soil. The primary difference between this method and other sand procedures is that after wastewater treatment, the effluent is collected in a tank where it may be reused.[38,43]

c) **Green filter** is also a soil-based water treatment technique similar to a soil biofilter. The wastewater is allowed to flow directly to the treatment area, consisting of a plant and soil-based filter media. However, unlike soil biofilters, after wastewater is cleaned using this approach, it cannot be re-treated again since the filtrated waters end up in groundwater sources.[16]

- **Stabilization ponds:** Here, treatment is done in stabilization ponds, which may treat up to 15,000 p.e. of wastewater from small-scale companies, communities, hotels, restaurants, and individual residences. In addition, it is used to remediate dirty and pretreated surface water.[39] Some of the wastewater stabilization approaches are:

a) **Anaerobic ponds** are often used to treat large volumes of wastewater. They are conducted at a higher temperature to enhance the efficiency of eliminating organic contaminants. The anaerobic process typically comprises a stabilizing pond and an anaerobic filter.[39] They are typically used to treat agricultural and industrial wastes or as a preliminary stage in municipal treatment systems when industrial effluent substantially contributes to total wastewater generation.[27]

b) **Aerobic ponds** primarily treat wastewater in sunny and hot climates. The main advantage of these ponds is that they generate stable wastewater over short retention intervals while using less energy and land mass. They are usually employed in situations where a high level of BOD treatment is required, but land space is restricted. The disadvantage of this method is that the treated water will retain algae, and unless it is removed, the water will have a high TSS. Furthermore, this approach's operating concept is more sophisticated than facultative ponds.[3]

c) **Facultative pond** treats raw wastewater in small villages, municipalities, and metropolitan areas. They are also employed in industrial operations, mainly in the production chain after anaerobic or aerated ponds, to give extra purification before release.[42]

d) **Maturation pond** is a lagoon technique with a shallow pond at the end of the treatment plant. The ponds are designed to be shallow, allowing sunlight into the pond for a photosynthetic reaction between aerobic microorganisms. As a result, it maintains aerobic conditions while purging the water of pathogens, organic matter, nutrients, and suspended materials.[4]

Cost-effectiveness and water quality are primary objectives of any wastewater treatment initiative. In addition, the selected technology must be suitable for the local setting concerning the availability of local materials, land accessibility, trained labor, and engineering expertise.

4.5 SUSTAINABILITY OF LOW-COST WASTEWATER TREATMENT METHODS

Low-cost wastewater treatment systems to reduce the enormous water scarcity are to establish sustainability by expanding the currently available source of water, especially for irrigation and domestic purposes.[16] The sustainability of such systems can be related to their energy consumption, environmental impact, economic consequences, and life-span. To attain sustainability, more emphasis is given to agricultural and irrigation sectors due to high consumption rate of water in which the sector consumes more than 80% of total freshwater available, which has triggered the sustainability of such sources.[21] Therefore, an alternative source and reinforcement to the current conventional systems should be provided in which treating waste will serve as reliable and sustainable alternatives.[8] The discrepancies associated with the current wastewater treatment are not sustainable due its centralized nature. The composition of such water that contain pathogenic bacteria from domestic, rainwater runoff, and at times industrial waste, causes the system to have limited capabilities of providing high-quality water for human consumption.[22] Most of the wastewater treatment systems in developing countries are not sustainable due to direct transfer and adoption of such technologies from developed countries without adjustment and modification to consider the culture, land, and climate change to suit the actual sewage produced in these regions.[10]

One of the best approaches in coping these challenges and attaining sustainability of wastewater treatment systems is by employing decentralized and low-cost wastewater treatment systems. These will reduce the high concentration of the sewage, reduce the burden on the existing systems, and regulation can be adopted due to the flexibility of the system; also, when one of the systems fails it may not have impact on the entire system.[6] Decentralized nature of these low-cost waste treatment systems will contribute in balancing, reclaiming, and recharging water sources especially in a region, where there is uneven distribution of precipitation.[44]

Unregulated disposal and injection of domestic, industrial, and at times agricultural wastes pose a serious threat to the human life in general and contaminate the freshwater, pollute the environment, and degrade land.[8,23] These are of course associated with the increasing human population and industrial activities. Therefore, as means to sustain the environment especially in developing nations, it is necessary to establish low-cost wastewater treatment systems (i.e., simple to construct, operate and maintain, eco-friendly) for the usage of public especially in rural areas.[14]

4.6 COST EFFECTIVENESS OF LOW-COST WASTEWATER TREATMENT SYSTEMS

The actual cost of wastewater treatment includes the entire cost of constructing, operating (collecting, recycling/treating, reuse, and disposal), and maintaining the facilities. Thus to optimize the cost of constructing such a decentralized and low-cost wastewater treatment, there is a need to understand the nature, source, quantity, content, and level of concentration of the wastewater before embarking on employing appropriate method of treatment. However, as the name implies, low-cost wastewater treatment systems need to have basis of optimizing the general cost from inception to operation and the cost for maintaining the system. In most cases, anaerobic and biofiltration processes are employed especially in areas with consistent aquatic pollution.[30] This is in opposition to most cases across developing countries, where stabilizing ponds are used for treating wastewater rather than using sequences of processes. These conventional means are attributed with longer processing period, larger area requirement, which lead to releasing pollution to the neighboring environments.[11] Almost all low-cost wastewater treatment systems are either decentralized or semi-centralized flexible, with optimized cost of construction and maintenance.[45]

The actual cost of constructing, maintenance, investment cost uses following relationship:[15]

$$TAC = CRF \times IC + OMRC \tag{4.1}$$

where TAC = total annual cost; CRF = capital recovery factor; IC = investment cost; and OMRC = operations, maintenance, and repair cost.

The CRF (capital recovery factor) is given by

$$CRF = \frac{IR(1+IR)^T}{(1+IR)^T} \tag{4.2}$$

where T = Economic life; IR = Interest rate (as a fraction).

Investment cost factor (ICF) is defined as:

$$ICF = CRF \times T \tag{4.3}$$

4.7 ECO-FRIENDLY NATURE OF LOW-COST WASTEWATER TREATMENT TECHNOLOGY

Management of wastewater is a challenging task due to high rise of rural–urban migration, industrialization, climate change, which have posed a major burden to the environment and contributed to point-source pollution. This challenge is concurrent especially in developing nations, and perhaps agriculture is responsible for greater percentage and often the agricultural products are heavily contaminated with the wastewater or untreated water, which has contributed in spreading the diseases.[7,28,41] Protecting the environment from being exposed to the probable pollution and strengthening public health are the fundamental objectives for treating wastewater. However, with the current demand of food security across the world, it is necessary to consider agricultural provision as another reason for treating wastewater.

Characteristics of wastewater are based on physical, chemical, and biological parameters usually considered for determining quality of water; these include biological oxygen demand (BOD), chemical oxygen demand (COD), dissolved oxygen (DO), total suspended solids, nitrate, nitrite and ammonia nitrogen, phosphates, salinity, and a range of other nutrients and trace metals.[5,29] Also, continuous discharging of untreated water to the environment especially the liquid portion of wastewater with highly toxic substances are threat to the environment.[37] Therefore, an effective establishment and implementation of low-cost wastewater treatment will help in the reduction of these challenges by safeguarding the environment and help to yield noncontaminated and agricultural produce with relatively high level of hygiene.[2,26]

Basically, the aim of establishing decentralized and low-cost wastewater treatment systems is to protect humans and improve environmental health. However, another detrimental problem and negative impact on the environment is distortion of watershed boundaries, which need a provision of rigorous infrastructures so as not to expose the environment into adverse effect changes and temperament in the natural ecosystems.[32]

4.8 RECOMMENDATIONS

- The design of such systems should be conformed to the requirements of WHO (1989).
- Critical care should be given priorities in reducing the probable negative impacts from these systems.
- Designing and provision of such infrastructures varies from environment to environment; therefore, on setting such low-cost wastewater treatment systems, factors, i.e., climate conditions, population, intended use of the system, availability of energy source, the quantity and quality of the wastewater should be considered.
- Sessional sensitization, and intensive training should be organized by the government bodies to the localities on technical know-how of the systems.
- To efficiently utilize such treatment systems, it is of paramount importance to have independent systems for domestic, industrial, and agricultural waste due to the nature and concentration of waste generated.
- To sustain most of these systems, the policymakers should include full participation of local communities in regulations for their sustainability.
- On successful establishment and implementation of low-cost wastewater systems, communities should take the responsibility of maintaining the systems.

4.9 SUMMARY

To promptly tap the benefits of employing the low-cost wastewater treatment technologies for agricultural purpose, and to convert the probable menace into actual benefits to the local communities, the following conclusions are drawn:

- To maintain sustainability of the systems, health and hygienic level of environment and inhabitants should be given priorities; conservation of natural and environmental resources should also be considered.
- Choosing a particular low-cost wastewater treatment technology should be a robust, and one should consider all necessary sustainable parameters (technology, social, economic, and environment).
- Comparative cost analysis should be implemented to establish the actual benefits to be tapped by the society, which is mandatory for the policymaking and investment.
- If these decentralized systems were not being designed carefully, then these may create imbalance and create disturbed hydrological and ecological setups; lower constituent of chemicals in treated wastewater may bring about the need to reactivate such and recharge these inputs, which may give low quality of water.
- These might be prone to lower treatment efficiency due to nonskilled or nonqualified personnel in supervising the operation and maintenance of the systems.
- When designing such low-cost wastewater systems, they should be flexible and be associated with better adjustment options and ability of protecting the environment, fitting into different landscapes.
- When need of establishing such a low and decentralized wastewater treatment arises, it is necessary to consider engineering, environmental, social factors; a negative consequential outcome may render the systems inefficient if key considerations are neglected.

KEYWORDS

- **farming**
- **irrigation**
- **low-cost**
- **technology**
- **treatment**
- **wastewater**

REFERENCES

1. Akinbile, C. O.; Ikuomola, B. T.; Ralphael, O. D. Development and Performance Evaluation of low-cost Wastewater Treatment Plant. *Sustain. Water Resour. Manag.* **2008,** *5*, 1217–1226.
2. Akpor, O. B.; Muchie, M. Environmental and Public Health Implications of Wastewater Quality. *Afr. J. Biotechnol.* **2011,** *10* (13), 2379–2387.
3. Alaa Fahad, A. Wastewater and its Treatment Techniques: An Ample Review. *Indian J. Sci. Technol.* **2019,** *12* (25), 1–13.
4. Alice A.; Horton, A. A. Microplastics in Freshwater and Terrestrial Environments: Evaluating the Current Understanding to Identify the Knowledge Gaps and Future Research Priorities. *Sci. Total Environ.* **2017,** *586*, 127–141.
5. Brooks, P. C. Investigation of Temperature Effects on Denitrifying Bacterial Populations in Biological Nutrient Removal (BNR) System (p. 153). MS. Thesis: Virginia Polytechnic Institute and State University, Blacksburg, Virginia – USA, 1996.
6. Chavan, R.; Mutnuri, S. Domestic Wastewater Treatment by Constructed Wetland and Microalgal Treatment System for the Production of Value-added Products. *Environ. Technol.* **2020,** *42*, 3304–3317.
7. Cohen, B. Urban Growth in Developing Countries: A Review of Current Trends and a Caution Regarding Existing Forecasts. *World Dev*. **2004,** *32*, 23–51.
8. Cooper, R. C.; Salveson A. T.; Sakaji R.; Tchobanoglous G.; Requa D. A.; Whitley R. Comparison of The Resistance of MS2 and Poliovirus to UV and Chlorine Disinfection, 2000.
9. Cseko, G.; Hayde, L.; Eds. *Danube Valley: History of Irrigation, Drainage and Flood Control*;.International Commission on Irrigation and Drainage (ICID), New Delhi, India, 2004; p 259.
10. Davies, L. C.; Pedro, I. S.; Ferreira, R. A.; Freire, F. G.; Novais, J. M.; Martins-Dias, S. Constructed Wetland Treatment System in Textile Industry and Sustainable Development. *Water Sci. Technol.* **2008,** *58*, 2017–2023.
11. De Sousa, J. T.; Foresti, E.; Domestic Sewage Treatment in an Up-flow Anaerobic Sludge Blanket Sequencing Batch Reactor System. *Water Sci. Technol.* **1996,** *33* (3), 73–84.
12. Eisenhauer, D. E.; Martin, D. E.; Heeren, D. M.; Hoffman, G. J. (Eds.) Introduction–irrigation. In *Irrigation System Management (Chapter 1); Online; St Joseph–MI: The American Society of Agricultural and Biological Engineers* [Online]. 2021; https://www.asabe.org/Portals/0/aPubs/Books/ISM/ISM1.pdf; (accessed March 30, 2023).
13. FAO. *AQUASTAT database*; Rome, Italy: Food and Agricultural Organization of United Nations (FAO), 2021.
14. Gajewska, M.; Skrzypiec, K.; Józwiakowski, K.; Mucha, Z.; Wójcik, W.; Karczmarczyk, A.; Bugajski, P. Kinetics of Pollutants Removal in Vertical and Horizontal Flow Constructed Wetlands in Temperate Climate. *Sci. Total Environ.* **2020,** *718,* E-article ID 137371.
15. Grau, P. Low-cost Wastewater Treatment. *Wal. Sci. T. Clt.* **1996,** *33* (8), 39–46.
16. GreenArth, Urban Wastewater [Online]. 2012. http://greenarth.com/urbanwastewater.html (accessed March 30, 2023).
17. Hegazy, B. Low-Cost Wastewater Treatment Technology. *J. Appl. Sci.* **2007,** *7* (6), 815–819.
18. Hillel, D.; Vlek, P. The sustainability of irrigation. *Adv. Agron.* **2005,** *87*, 55–84.

19. Holzapfel, Eduardo A.; Alejandro, P.;Ignacio, L.; de Oliveira, A. S. S.; Farka, I. Design and Management of Irrigation Systems. *Chil. J. Agric. Res.* **2009,** *69* (1), 17–25.
20. Hsiao, T. C.; Steduto, P.; Fereres, E. A systematic and Quantitative Approach to Improve Water Use Efficiency in Agriculture. *Irrig. Sci.* **2007,** *25,* 209–231.
21. Hussain, M. I.; Muscolo, A.; Farooq, M.; Ahmad, W. Sustainable use and Management of Non-conventional Water Resources for Rehabilitation of Marginal Lands in Arid and Semiarid Environments. *Agric. Water Manag.* **2019,** *22,* 462–476.
22. Jhansi, S. C.; Mishra, S. K. Wastewater Treatment and Reuse: Sustainability Options. *Consilience: J. Sustain. Dev.* **2013,** *10* (1), 1–15.
23. Kansal, R.; Balakrishnan, L.; Kishore, K. Anaerobic Digestion. *Oral Presented at the California Water Reclamation Meeting*, Santa Rosa – CA, 2003.
24. Kovda, V. A.; VanDenberg, C. R.; Hagan, M.; Gulhati, N. D. *Irrigation/Drainage, and Salinity*.Hutchinson: London, 1973; pp 1–14)
25. Kumar, R.; Singh, J. Regional Water Management Modelling for Decision Support in Irrigated Agriculture. *J. Irrig. Drain. Eng.* **2003,** *129,* 432–439
26. Lübken, K. K. M.; Gehring, I. D.; Wichern, M. fficient Low-cost Anaerobic Treatment of Wastewater Using Biochar and Woodchip Filters. *Water* **2018,** *10,* 818–823.
27. Mahoumed, E. K. In *Low-cost Technology for Wastewater Treatment for Irrigation Reuse*. 6th International Conference on Water Resources and Arid Environments (ICWRAE- 6), 16–17 December, 2014, Riyadh - Saudi Arabia, 2014; pp 101–107.
28. McCasland, M.; Trautmann, N.; Porter, K.; Wagenet, R. Nitrate: Health Effects in Drinking Water; 2008. [Online]. http://pmep.cee.comell.edu/facts.slides–self/facts/nit-heefgrw85 (accessed March 30, 2023).
29. Metcalf, E. Physical Unit Process. In *Wastewater Engineering: Treatment, Disposal, Reuse*, 4th ed.; Tata-McGraw Hill: New York, NY, USA, 2003.
30. Miah, M. S. Cost-effective Treatment Technology on Textile Industrial Wastewater in Bangladesh. *J. Chem. Eng. (IEB)* **2012,** *27* (1), 32–36.
31. Møller, K. A.; Fryd, O.; de Neergaard, A.; Magid, J. Economic, Environmental and Socio-Cultural Sustainability of Three Constructed Wetlands in Thailand. *Environ. Urban* **2012,** *24,* 305–323.
32. Muga, H. E.; Mihelcic, J. R. Sustainability of Wastewater Treatment Technologies. *J. Environ. Manag.* **2008,** *88,* 437–447.
33. Oza, A. Irrigation: Achievements and Challenges. In *India Infrastructure Report – Business for the Future*, online; Oxford University Press, 2008; pp. 1–36; [Online]. http://www.indiaenvironmentportal.org.in/files/INDIAN_INFRASTRUCTURE_REPORT_2008.pdf (accessed March 30, 2023).
34. Pannunzio, A.; Genoud, J. In *Four Trickle Irrigation Treatments in four Varieties of Oranges*, Proceedings of the IV Decennial National Irrigation Symposium, Phoenix, Arizona, USA. 14–16 November. American Society of Agricultural Engineers (ASAE), St. Joseph, Michigan, USA, 2000; pp 629–635.
35. Playán, E.; Mateos, L. Modernization and Optimization of Irrigation Systems to Increase Water Productivity. *Agric. Water Manag.* **2006,** *80,* 100–116.
36. Progress Panel, African. Power, People, Planet. Seizing African's Energy and Climate Opportunities. *Africa Prog. Rep.* **2015,** 182.
37. Qadir, M.; Wichelns, D.; Raschid-Sally, L.; McCormick, P. G.; Drechsel, P.; Bahri, A.; Minhas, P. S. The Challenges of Wastewater Irrigation in Developing Countries. *Agric Water Manag.* **2010,** *97* (4), 561–568.

38. Rodriguez, E. E. *Low-cost Wastewater Treatment Technologies for Agricultural Use*. Wageningen University: Netherlands, 2014; p 142.
39. Rozkošný, M. Natural Technologies of Wastewater Treatment. *Global Water Partnership Central and Eastern Europe (GWPCEE)*, 2014; February; www.gwpcee.org; (accessed March 30, 2023).
40. Ryskulov, D. M. Agricultural Land Improvement: Amelioration and Reclamation: The Economics of Irrigation Systems (p. 448). *Encyclopedia of Life Support Systems (EOLSS: Volume 1)* [Online]. https://www.eolss.net/ebooklib/ebookcontents/E5-09-ThemeContents.pdf (accessed March 30, 2023).
41. Schlosser, C.; Strzepek, K.; Gao, X. The Future of Global Water Stress: An Integrated Assessment. *Earth's Future* **2014,** *2,* 341–361.
42. United States Environmental Protection Agency (EPA). Principles of Design and Operations of Wastewater Treatment Pond Systems for Plant Operators, Engineers, and Managers, EPA/600/R-11/088 (August). 2011; www.epa.gov (accessed March 30, 2023).
43. United States Environmental Protection Agency (EPA). *How Wastewater Treatment Works: The Basics* (May) [Online], EPA 833-F-98-002. 1998.
44. Waly, M. M.; Taha, A.; Abunada, Z.; Slobodan, B.; Thomson, C. Constructed Wetland for Sustainable and Low-cost Wastewater Treatment. *Land* **2022,** *11,* 1388–1395. https://doi.org/10.3390/land11091388; (accessed March 30, 2023).
45. Wendland, C.; Albold, A. Sustainable and Cost-effective Wastewater Systems for Rural and Peri-urban Communities up to 10,000 Population Equivalents. In *Women in Europe for a Common Future (WECF) Guidance Paper*; Rome: FAO, 2010; p 95.

CHAPTER 5

AGRIFUSION OF WATER AND WASTEWATER TECHNOLOGIES: STATUS, MANAGEMENT, AND CHALLENGES

SANUSI MOHAMMED SADIQ, INVINDER PAUL SINGH, MUHAMMAD MAKARFI AHMAD, and SHEHU ABDULMUMIN

ABSTRACT

Water is essential for the ecosystem, economy, and development of humans. Improving rural people' food security, incomes, and standard of living requires access to and security over water resources. For millions of impoverished farmers, particularly those in rainfed regions and also those engaged in irrigated agriculture, reliable access to water continues to be a key obstacle. More farmers are in danger of losing water security and falling back into poverty because of climate change as well as the associated altering rainfall patterns. Therefore, it is imperative to increase the communities' ability to adopt and spread agricultural water management methods. Managing the water used for livestock, crop, and inland fisheries cultivation, all come under agricultural water regulation. The key to improve agricultural water administration in these productive regions is to address both global food security and poverty alleviation. By 2050, up-to-date food production must double to meet the world's food needs. This expansion must come from rainfed agriculture-heavy regions and irrigated agriculture.

Sustainable Use of Wastewater in Agriculture: Applications, Methods, and Technology. Megh R. Goyal & Gaurav Saini (Eds.)

5.1 INTRODUCTION

The world's most significant issue regarding water is its scarcity.[58] Existing water resources in many parts of the world are insufficient to satisfy all environmental, industrial, agricultural, and urban demands.[47] If precipitation surpasses potential evaporation is the main factor in deciding whether a place has a water excess or experiences water scarcity. By 2025, it is predicted that worldwide water consumption will rise by 3800 km^3 annually, with a large portion of this water coming from natural systems.[20] By dramatically limiting the flow of water in many regions, this increase in demand will have an enormous negative effect on the ecology. Water availability may soon be a constraint on food production. In many regions of the world, agricultural water use is unsustainable due to factors such as soil salt accumulation, groundwater overdraft, and the excessive use of surface water resources.[67] Given this condition, it is unclear if there is enough water available to sustain the current population over the long term, let alone the much greater population that will need to be nourished in the latter years of the century.[33]

The rising competition for water resources among domestic, commercial, and agricultural users is putting aquatic and wetland ecosystems under a lot greater stress.[57] Furthermore, the worsening patterns in water quality have greatly aggravated the inadequate environmental water provision in many of the world.[61] Many developed nations have responded to this issue by enacting legislation to ensure environmental supplies, but these assurances are dependent on both the availability of enough food and water for urban requirements. Such assurances cannot be made in underdeveloped nations, which frequently disregard environmental demands and lack the funding necessary to maintain and improve water quality.[1] There does not appear to be any dispute that technology and science must play a crucial part in coming up with the answers needed to deal with the problematic issues brought on by the world's water shortage.[10]

Extrapolating present trends in water use over the world yields a picture that is both alarming and cause for concern.[37] Even though population and economic expansion are driving up the need for new water provision, it is highly improbable that more water would be available to fulfill rising demand due to unsustainable water governance and regional scarcity. In the next two decades, as many nations, including India.[59] The issue will be made worse by the presence of ongoing and widespread groundwater excess, the threat of global warming, and the saltwater intrusion of soils in dry and semiarid regions.[68] In this century, the increased reliance of developing nations on food imports might potentially reach crisis levels.[6] The end effect will unavoidably be an increase in water conflicts between various water-using industries and both upstream and downstream riparians (and nations).[46] The environment will be under increasing stress. It might be very challenging to deal with the concomitant economic turmoil and likely political instability.

80% of all water consumption globally is accounted for by agriculture, making it by far the biggest consumer of water.[52] Irrigated agriculture is significantly more effective than rain-fed agriculture, as has been well-documented. In addition, there is evidence that rural poverty and malnutrition are influenced considerably by the availability of water for agriculture. Other all things being equal, expanding irrigated agriculture would be a sensible reaction to the predicted rise in population and economic expansion.[70] Yet, it appears evident that obtaining more water for agriculture will put great strain on aquatic ecosystems due to demands for more urban supplies and due to current exploitative practices in agriculture, such as groundwater overflow and water supply contamination.[60]

There is a critical amount of water supply below which the ecosystem's health cannot be preserved. As a result, one of the "devil's choices" inherent in the worsening shortage would require accepting the significant expenses associated with declines in environmental sustainability and stability as a portion of the cost paid for increasing the amount of water available for agriculture.[24]

Improving the water efficiency of irrigated agriculture is the obvious answer to this conundrum. There are several ways to increase water's economic productivity to decrease agricultural water use. The matching of the most suitable and productive crops with specific places and climates will be a key component of methods for enhancing productivity.[39] As a result, growing rice in the desert is not usually a good idea. Additionally, there is a chance that production may rise significantly with advances in the on-farm control of farm water, both using cutting-edge irrigation equipment and better irrigation scheduling. Additionally, crops that are moisture-stressed at key times during their life cycle or yearly cycle may produce high-quality harvests with little loss in quantity.[38]

In summary, there is a lot of room for improvement in agricultural water management, and doing so will need both applying already-known scientific knowledge and actively seeking out new knowledge through study.

This chapter focuses on the status, management, and challenges of agrifusion of water and wastewater technologies.

5.2 WATER AND AGRICULTURE

Three of every four of the world's poorest people rely primarily on agriculture as their primary source of income, and it is essential to food safety and economic development in emerging nations.[65] However, a lot of water is needed for food production. Seventy percent of water suctioned from aquifers and rivers worldwide is used for agriculture. Northern Africa, Southern Asia, and the arid areas of sub-Saharan Africa (SSA) are just a few places that are already severely short on water.[25] One of the most urgent problems facing humanity today is a lack of water. River basins with water scarcity are home to over 1.4 billion people, and by 2025, that figure is predicted to rise to around 3.5 billion.[28] It is likely that things will get worse. The world's population is anticipated to reach 9 billion people by 2050, which would result in an up to 70% increase in food consumption,[30] in addition to dietary changes brought on by socioeconomic development.

Living in severe poverty is defined "as earning less than US$1.25 per day for a family of four." Although there is a transition toward urbanization, the majority of the world's poor (about 1 billion people) still live in rural areas, and this is expected to remain the case in the near future.[2] In addition to being heavily regionalized and rural, poverty also disproportionately affects women,[53] particularly as more and more males move to the metropolis in search of alternate sources of income. Women make up about 43% of the labor force in developing nations. In SSA, the agriculture industry employs 62% of the region's economically active women.

The requirement for agricultural products in least developed countries (LDCs) is expected to triple when the overall population of nations that are developing exceeds 75 billion by 2050.[68] The majority of LDC governments rely on their remote communities to increase agricultural production, yet these same communities are resource-poor, low-productive, and have low productivity. The new "rurality" of rural life (which is a change from the previous "rurality") adds to the burden on the poor.[68] As livelihoods change because of globalization, new forms of poverty are arising, and institutional changes brought about by governance and remote service system reforms are also taking place. All these problems raise doubt and risk, and they are likely to disproportionately affect rural poor people's access to and usage of scarce water resources.[56]

5.3 PRESSURE ON WATER RESOURCES

Even though many regions of the world already have limited water supplies, the need for water will rise significantly as more crops for nutrition and energy are grown.[59] Municipalities, industry, agriculture, and the environment are some of the major water-using sectors. As a result, competition for water will eventually increase. Pressure to switch land from food cultivation to energy crops is escalating. With "black clouds" converging toward 2030 and beyond, the variety of difficulties has produced a "perfect storm," resulting in challenges that are far more serious than the aggregate of their individual components.[40] There may be more rivalry for food, water, and energy; higher food costs; and an increase in the number of people heading hungry because most of the population growth will be among those who are already underprivileged in the emerging world.[7]

5.3.1 GLOBAL WARMING

Another "black cloud" on the horizon is the effect of global warming on access to water, agriculture, and, therefore, food supply.[40] Agriculture is responsible for about 18% of global greenhouse gas (GHG) emissions, primarily due to the raising of livestock, modifications in land use, the growing of paddy rice, and the production and use of agricultural chemical products.[56]

In addition to increasing uncertainty and the likelihood of more harsh and recurrent floods and droughts, rising temperatures worldwide will cause harsher dry seasons and higher rainy seasons. Particularly in semi-arid and dry regions, such changes will have an impact on ecosystems and human livelihoods. Groundwater and surface water supplies will be impacted by declining precipitation, especially in regions that are already water-strapped.[51] Initial increases in water supplies from melting glaciers will be followed by sharp decreases. New water infrastructure design will be impacted by this. Usually, design is based on past weather patterns; however, this is no longer useful for forecasting potential future events.

Poor farmers are likely to be harmed by climate change's effects.[56] GHG emissions will rise because of increased food, feed, and bioenergy, which will have a substantial influence on both food availability and nutrition safety.[62] The fact that the most susceptible farmers are in regions with low crop yield and scant access to agricultural expertise and technology means that climate change will also worsen inequality.

5.3.2 THE GREEN REVOLUTION

There are doubters who dispute the projections since they sound pretty dour.[26] Global food production is more significant than multiplied in the second half of the 20th century. Over the past 40 years, agricultural output has increased steadily, and irrigated agriculture is considered one of the accomplishments of the 20th century. Millions of people who would have gone hungry thanks to huge irrigation projects in India, China, Pakistan, and Indonesia now have enough to eat.[54]

The necessity to balance the needs of water for food, humans, industry, and the environment, as well as the significance of water for habitats and their adaptability, were both widely acknowledged in the 1990s.[71] On the basis of environmental sustainability, the notion of a "green revolution" was born. The development of rainfed agriculture was the emphasis of the third "green dimension." In fact, employing conservation agricultural techniques and additional irrigation, many poor nations still have a substantial, untapped resource of rainfall that may be exploited.

5.3.3 EMPHASIS ON WATER TECHNOLOGY

A UK House of Commons research states that by 2030, "*the globe needs to produce 50% more food—"safe food"—on fewer acres of land, with minimal freshwater, using small amounts of energy, fertilizers, and pesticides, while also severely reducing the rate of emissions of greenhouse gases released globally*".[31] Though intimidating, the obstacle is possible and must be overcome.

5.4 SURPLUS/ABUNDANT WATER

A total of 110,000 km^3 of rainfall annually on the surface of the planet, of which 36% ends up in the ocean, and 57% supports forestry, grazing areas, fisheries, and biodiversity.[42] The remaining 0.1% (110 km^3) is used by towns, cities, and industry, and the remaining 7% is used by agriculture (7130 km^3). Water withdrawn for irrigation from rivers, streams, and groundwater accounts for around 22% (1570 km^3) of the water used by agriculture. 5560 km^3 of the water used by agriculture each year is "green water," which refers to water that crops can access from rainfall that has been stored in the soil's root system.[42]

It might be challenging to forecast future water consumption.[23] Since no one could possibly foretell the increase in energy prices, the global economic downturn, and the influence these events would exert on food prices, projections made less than ten years ago have shown to be wrong.[16] The stresses brought on by population expansion and water scarcity, as well as the effects of climate change, are only just now starting to manifest. Only if we take immediate action to improve water consumption, particularly in agriculture, which is the primary consumer, will we have enough water. There is no denying the close relationship between the future of water and food security.

By 2050, the amount of water used worldwide will be nearly doubled if current trends continue. But a more upbeat estimate claims that by 2050, it might increase from 7130 km^3 to 8515 km^3 per year.[35] China has seen the most significant expansion in Asia. The largest producer of meat in 2009 was China, which produced 27% of the world's meat.[50] India contributed 15% of the milk and dairy products consumed worldwide in 2009.

5.5 THE LINKAGE OF IRRIGATION AND RAINFED

Agriculture uses both rainfed and irrigated farmland.[20] Most of the cultivated land worldwide is used for rainfed farming, which accounts for 60% of global food production.[64] Crop yields and output are suitable where there is high and consistent rainfall, such as in northern Europe.[9] The arid regions of Africa, where numerous impoverished individuals live, are an example of a location with little, incompatible, and unpredictable rainfall; yields of grains there standard just a ton/ha, and water consumption is significant because of excessive evapotranspiration levels that range between 2000 as well as 3000 m^3/ton of agricultural product.[31] Only 15–0% of rainfall is used for crop transpiration, and occasionally as little as 5%. The remainder is lost because of ineffective evaporation, drainage, and surface runoff.[8]

Only over 300 million ha, or 20% of the world's arable land, is covered by irrigation practices as of the last decade. Irrigation does, however, play a substantial role in food production, accounting for more than 40% of the global output. According to Singh et al.,[65] South America, Asia, and Africa account for around 84% of all irrigated land. Particularly in areas of SSA, where there is an abundance of water, there is still opportunity for growth. Economic growth and the eradication of poverty are both excellent prospects for irrigated agriculture. When appropriately used, irrigation can "stretch" the wet season and provide efficient methods of water control, lowering potential risks with the unpredictability of rainfed farming in arid areas while increasing cropping intensities in humid and tropical zones. Droughts, which are expected to happen more frequently, can be defended against by it.[22] It can, therefore, ignite the downstream as well as upstream markets, stabilize output, and increase economic activity, all of which can have an enormous positive effect on all economic sectors.

However, there are issues with wastewater, much such as in rainfed agriculture. Less than 20% of the provided water is effectively transpired by crops in several irrigation projects in semi-arid locations, especially among LDCs.[69] Those involved in irrigation are primarily worried about this "inefficiency." Bluewater includes rivers, marshes, lakes, and groundwater.[4]

5.5.1 ASIA

One of the key regions where the relationship between water shortage, agriculture water management (AWM) development, severe poverty, and hunger is Asia. In extreme poverty, there are about 700 million individuals. Asia has approximately 70% of the world's irrigated land, and approximately one-third of its cropland is used for irrigation farming.[11] Supplying timely and dependable water supplies allowed the area to become food self-sufficient, which in turn encouraged higher cropping densities, rice varieties with high yields, and the use of fertilizers to boost production.

Water contamination from fertilizer runoff has been brought on by improper fertilizer and pesticide use, though. Waterlogging and salinization are other effects of poor water management in Asia. In Asia, salinization alone damages more than 40% of irrigated land in arid regions.[66] In countries such as India as well as China, increased use of exterior irrigation has led to water logging, and over the past several decades, increased use of groundwater sources has caused an average annual decline in water tables of 1 meter. In both cases, river flows have dropped. The quality of the nation's water has also become a major issue in China, where 4 million ha, or 7% of the country's irrigated land, is supplied with impure water. In 2008, 46% of China's 26 lakes and reservoirs that were being monitored for their environmental state were undergoing eutrophication, or oxygen depletion, according to the country's Ministry of Environmental Protection.[32]

Poverty, which is a concern notably in South Asia, is made worse by water contamination and excessive use. In lieu of the trend toward urbanization, there will be 1.5 billion more people living in Asia by 2050, 50% of whom will continue to reside in rural areas.[12] As the wealthier population turns to meat and dairy products, which require significantly more water than vegetables, diets are also changing quickly among this group. Over the previous ten years, meat consumption has increased by over 30% throughout Southeast and East Asia.[62] Although there is rainfed farming in the region, there are only a limited amount of land and water resources. Irrigation farming, primarily from existing irrigation systems, is expected to provide the majority of the additional food through higher yields as well as the efficiency of land and water resources. International trade is anticipated to provide some of the world's food supply.[36] The low cost of food and low rates of return prevented further investment in irrigation. Because of this, many of the massive, centrally controlled irrigation systems require upgrading in order to adapt to contemporary farming methods and shifting food demand.[3] Their rehabilitation efforts are uneven.

Low-capacity storage pits, polyvinyl chloride (PVC) tubing, and pumping equipment are just a few examples of locally adapted technologies that multitudes of small-scale farmers in South and Southeast Asia are investing in to access groundwater and have more control over their water supplies.[48] The dependability, promptness, and sufficiency of irrigation are now more in the hands of farmers. The North China Plains and South Asia are currently experiencing a significant rise in this new "water-scavenging" industry. A thriving low-cost Chinese pump sector promotes groundwater abstraction. Small pumps from China now weigh less and cost less, and the country sells about 4 million pumps per year.

More than 60% of India's irrigation requirements are currently met by the method used by smallholder farmers known as "atomistic irrigation"[53] If actions are not taken to promote groundwater recharge, the long-term viability of this nonformal irrigation economy is called into doubt. We can meet some of the needs for food today by using water from the future 18 It is projected that groundwater in northern India loses 54 km^3 in volume each year. Long-term sustainability is a problem since lift-based irrigation uses more energy than gravity-based irrigation systems.

5.5.2 AFRICA

In Africa, another continent, there is a link between water scarcity and poverty. For instance, only in SSA, almost 45% of the population, or about 330 million people, live in utter poverty. Agricultural output has not kept up with population growth, and the area has some of the poorest agricultural yields in the world.[31] The increase in productivity since 1980 has come from increasing the area under crops by more than 80%. On the contrary, in other locations, cropped area has increased by less than 20% yet productivity has increased as a result of technological advancements and innovation. In SSA, this is obviously not the case.[42] Additionally, the agricultural sector in SSA is characterized by rainfed farming, which is mainly oriented toward low-value food crops and has few formal irrigation systems.[31]

Even though rainfed agriculture pervades, rainfall in several of Africa's drier regions is irregular, unreliable, and frequently occurs in extended intervals with little to no warning. African farmers, forced to eke out a life in several of the world's driest locations, frequently lament, "*Just one more decent storm.*" However, both droughts and floods can be dangerous. In the Republic of United of Tanzania, throughout the past several decades, floods have been the cause of more than 40% of all proclaimed disasters compared to droughts (30%), which frequently occur in the same location and during the same period.[42] According to climate change estimates, this could get worse if droughts and floods become more extreme.

The most significant opportunity for increasing output and productivity is in rainfed farming. Even minor technology advancements at cheap cost and small yield increases could have a significant influence on output and the fight against poverty.[24] However, both fossil and renewable groundwater reserves are being severely over-exploited. To achieve self-sufficiency in staple foods, governments provide significant subsidies for energy, pumps, and irrigation machinery. However, this condition simply cannot continue.

The irrigation situation is very different in SSA. Only one-sixth of the worth for Asia and a third of the global average for the cultivated area has irrigation facilities. The history of irrigation investment has not been positive. After underwhelming irrigation investments in the 1960s and 1970s, international funders have not shown much interest in the last 30 years. Even though food production accounts for the most significant share of water consumption in the majority of African nations, national governments have also failed to put water for nutrition on the regional water agenda. There are various and intricate causes behind this. These circumstances severely restrict the economic viability of irrigation development schemes due to the significantly higher development costs than in Asia.[56]

However, Africa has a massive base of renewable water resources per person, implying that there is a sizable undiscovered water gift that might be utilized for irrigated agriculture. In SSA, just 7 million ha of land, or 4% of the entire area under agriculture, have irrigation systems built. If North Africa is taken into consideration, this area nearly doubles, with Egypt accounting for 20% of all artificial soil watering in Africa. According to Hassan and Ersoy,[24] 20% of the irrigated land is thought to be inactive even within this low total. For example, it is estimated that the Fadamas marshland valleys in Nigeria have several hundred thousand hectares of informal irrigation.

However, Africa has a massive base of renewable water resources per person, implying that there is a sizable undiscovered water gift that might be utilized for irrigated agriculture. In SSA, just 7 million ha of land, or 4% of the entire area under agriculture, have irrigation systems built.[56] To address the numerous agroecological areas across the continent, a change in perspective from a "from the top down to a "bottom-up" livelihoods-supported framework that acknowledges the contribution that women make to agriculture is required.[40]

5.5.3 THE CARIBBEAN AND LATIN AMERICA

Though not as severe as in Asia and Africa, the Caribbean and Latin America is the third region where poverty still exists. From 244 million in 1966 to 515 million in 2000, the region's population increased quickly, and by 2030, it is projected to reach 705 million. However, the majority of those in extreme poverty—11 million—live in metropolitan regions as opposed to those in Asia and Africa.[55] Water resources are abundant in most Latin American nations. The area has over 30% of the world's water supply despite having just 9% of the global population. However, there are notable differences both between and within nations. One river, the Amazon, contains the majority of the region's renewable water supply. Particularly the Caribbean islands experience a lack of fresh water. Approximately 60% of the population resides in 20% of the geographical area, which contains just 5% of the recyclable water resources, illustrating the unequal distribution of people in the world.

Even though the amount of land that is irrigated is small compared to Asia and SSA, agriculture is the largest consumer of fresh water. Despite having 13.5 million ha of nonrainfall areas (as of 2009), or about 9% of the world's projected total, Latin America largely relies on rainfed farming. With nearly 6.5 million ha, Mexico is by far the most irrigated land, followed by Brazil with 3.2 million ha, Chile, Argentina, and Bolivia. The quasi-arid Northeast region of Brazil contains around 0.5 million ha and has the worst social and economic statistics overall.[21]

Latin America, as opposed to Asia and Africa, has a long history of private funding in irrigation, with the government serving as both a regulator and an aid to investment. Agriculture-related investments have decreased in recent years. There are worries about the detrimental social and environmental effects of irrigation, rising construction costs, and limited government support for large-scale irrigation initiatives. In general, existing systems are poorly administered.

The amount of irrigation water is under significant pressure due to population increase and fast urbanization.[38] There is a demand to move water from farmland to urban areas where 70–80% of people now resides because of the expanding urban populations. Water quality issues brought on by improperly handled domestic, industrial, and mining wastes exacerbate water scarcity in many locations. While the current infrastructure can potentially handle roughly 35% of the wastewater, only about 20% of the wastewater is treated overall across the Caribbean and Latin America.[21] Additionally, just 40% of the wastewater gathered in Mexico's 1833 plants was processed in 2008. Urbanization and population growth will only make things worse.[38]

In many cases, salinization and water stress exacerbate the effects of ineffective drainage, which impacts broad regions of land. Salinization by irrigation is a growing problem in Brazil, Argentina, Cuba, Mexico, and Peru. Also, Brazil's 1.2 million ha of waterlogged lowlands. Not every part of the region, though, is in bad shape. With 2.8 million ha of irrigated areas and 2.4 million ha of additional irrigation, Mexico, for example, has the most extensive drainage system in the area. Investing in irrigation to deal with salinization and water-logged areas must be a vital component of the region's plan for both water and food security as well as poverty reduction, even though the region's drier regions appear to have ample water supplies generally.

5.6 DRAINAGE

Irrigation and rainfall receive the most attention as they generally are the biggest problems in AWM. However, drainage—the opposite of watering crops—also plays a significant role in many circumstances. To ensure the ideal moisture levels for crops to flourish, extra water is removed from the soil. Although drainage innovations are well recognized and widely used, it is still one of the overlooked aspects of AWM that needs greater attention.

Drainage can stop moisture stress and the accumulation of salts within the soil layer that result from using irrigation water of poor quality in drought and semi-arid regions where irrigation is essential for agriculture. Around 20–30 million ha (7–10%) of the 300 million ha of irrigated land in the world were damaged by salinization in 2002. Approximately 40–60 million ha of land was added in 2008.[17] In contrast to the 40–60% of irrigated land that requires drainage but does not yet have any facilities in place, only 10–20% of irrigated land has drainage facilities. Where irrigation is widely used, and a dry climate predominates, the issue is most severe in Central and South Asia and the Near East.[59]

In the wet and subhumid tropics, where the main goal is to remove extra water from heavy or strong rainfall, drainage is particularly crucial. These regions include East and Southeast Asia and portions of Africa's west and central regions. Significant barriers to agricultural development include poor drainage and inadequate flood protection, which prevent farmers from expanding and varying their cropping. By increasing soil water content and the efficiency of farm mechanization activities, drainage also aids in maximizing production in temperate regions, such as Europe and North America.

5.7 IMPORTANCE OF TECHNOLOGY

Innovative technology use is more than simply a water management component; it is crucial and frequently acts as a catalyst for broader agricultural growth in LDCs.[35] Making the appropriate decisions is crucial for everyone involved in AWM because technological decisions are one of the first to be taken during the development process. The massive public irrigation networks rely on technology for ample storage of water, measurement of flow and management, water lifting, and gathering information on which management decisions are based. The management and distribution of irrigation water would be impossible without these technologies. The absence of advantages for disadvantaged farmers and the hefty costs of major schemes have recently hindered new expansions.[50]

Small-scale farmers who rely on agriculture for their living are now the center of emphasis in many LDCs, instead of designing elaborate irrigation systems. About 80% of people in Africa are smallholder farmers. They serve as the foundation of African agriculture, controlling rainfall and irrigating little farms and backyard gardens that are frequently smaller than 1 ha. The less developed parts of Asia experience a similar problem. Smallholder farmers typically have independent water use decisions and unfettered access to ground or surface water. They combine subsistence and commercial farming, with the family doing most of the labor and the farm serving as their primary source of income.

There are instances of "low-cost" irrigation schemes in SSA where the canal embankments have been improperly engineered, causing leakage and necessitating extensive and expensive maintenance. Ecosystem services must be used by technology in a sustainable and efficient manner. While many of society's needs are met by man-made infrastructure, other services, such as those that keep fisheries healthy and alleviate floods and droughts, are met by "green infrastructure"—healthy rivers and watersheds. Technologies that support and improve such services increase the resilience of our water delivery and water use systems.

Technologies can be broadly categorized into two groups: those that boost water availability, such as water retention to deal with the seasons, gradually variable as well as unanticipated rainfall, flooding, as well as dry-spell, as well as those that enhance the use of readily accessible water. The emphasis is dependent on regional conditions, hence the term "twin-track" method. Traditional blue water resources, for instance, are already overused in many of the drier parts of the world, and the costs of providing more water are rising.

However, merely advancing existing technologies would not stop unsustainable water usage; proper technological solutions must be linked with better water management and effective water use. Additionally, in arid regions, there may be chances to locally increase the amount of green water that is captured.

Although there are new water technologies available, the possibility for rapid use is higher for older technology. Whatever technology is employed, its effectiveness will be more dependent on smallholder farmers' willingness to take chances, experiment, and accept it in environments with inconsistent services, high costs, and unpredictably fluctuating markets. The following subsections list some very futuristic technologies.

5.7.1 STORAGE OF WATER

The most significant opportunity for bettering water management may lie in water storage. History has made use of storage, which is quite old technology.[65]

Dams, environmental issues, and social issues are frequently linked to water storage. To store water, over 45,000 major dams have been constructed worldwide, of which 40% are utilized for irrigation. However, dams are only one type of storage. Storage is characterized by the IWMI as a continuum that includes both surface- and subsurface-based storage. Natural wetlands and reservoirs are examples of surface storage, and groundwater aquifers, soil water storage, tanks, and ponds are examples of subsurface storage.[8,9] In addition to making water that otherwise would have gone to waste available, this also improves irrigation systems by enhancing supply reliability and timeliness, allowing farmers to better plan their irrigation and lower water losses. Similar advantages of groundwater storage have led to the widespread use of groundwater-based "water scavenger" irrigation throughout Asia.[62] The connection between ground and surface water storage is made by water recharge. Today, canals and reservoirs offer chances for groundwater recharge and serve as a buffer between the demand and supply of water for cultivation.

Water can be obtained and used in a variety of ways, and storage solutions have many applications. In some instances, small farmer associations oversee the storage, whereas in others, larger, more formal institutions do so.[51] Regarding technical viability, socioeconomic sustainability, and institutional prerequisites, each has a distinct niche. Storage has a wide range of effects on poverty. Successful water storage systems have been employed in China and India to help farmers get the water they need when they need it, which has improved the regulation of canal irrigation.[56]

The irrigation of night storage canals has a long history in the Sudan. In Nigeria, there are instances of reservoir storage along canal networks. The storage situation in Ghana is complicated. A few reservoirs have resulted in more dependable water supplies, allowed farmers to broaden their crop production, and provided them with more solid financial security. However, similar circumstances at nearby reservoirs with other reservoirs have not resulted in a significant change. The significance of the setting in which technological interventions are undertaken is highlighted by this.

5.7.2 RE-EVALUATION OF CANAL IRRIGATION

Surface flooding is the same as canal irrigation, which creates basins, borders, and furrows. This irrigation technology dominates the entire planet. Surface flooding still accounts for 95% of irrigation; sprinkler irrigation makes up most of the leftover 5%, while trickle watering makes up the remaining 1%. In the next 50–100 years, this balance is not likely to change, especially in the LDCs. Technology that aims to enhance canal irrigation should, therefore, be given great importance. Canal irrigation, especially in Asia, is ineffective. Instead of relying on the unpredictability of canal water, smallholder farmers who formerly relied on the expansive canal systems for their water are seeking solutions by investing in pumps and utilizing the local groundwater, which is frequently refreshed via canal seepage.[27] It is challenging to forsake the massive canal networks and replace them with simple pump systems. Making the current canal system as responsive as irrigation utilizing groundwater presents a barrier to its use.

The typical "top-ender, tail-ender" issue occurs when individuals at the head of a canal frequently take more than their fair share, harming those at the tail end. This makes canals challenging to regulate mechanically. Most large canal systems employ "upstream control" equipment, which exacerbates the "top-ender, tail-ender" issue and is rigid to variations in farmer water demand. When engineers chose the amount of water to supply to farmers, this was tolerable in earlier planned economies. In contrast, farmers seek considerably more management over inputs in today's demand-driven economies.[19] There may be more affordable, locally sourced solutions. In fact, despite initially costing more to construct than canals, pipelines can provide considerably reasonable control over water flows and network systems more adaptable to farmer requirements. The inability to have control over channels would be uncomfortable for most household consumers, which is why pipe systems are used in the majority of domestic water supply systems instead of canals.

It takes more than a technology repair to improve canal irrigation; institutional reforms are also necessary. The government irrigation agencies are offered incentives to match their rewards (for example, performance awards) with those of the farmers, which has resulted in an improvement in China's public canal irrigation schemes. There are additional alternatives for multi-use canal systems, which provide water for environmental, home, industrial, and agricultural uses in addition to agriculture. Significant institutional collaboration between the government's water resources, agriculture, and environment ministries would be necessary for such achievements.

5.7.3 MICROIRRIGATION TECHNOLOGIES

The main rice-growing regions in Southeast and South Asia are not suitable for sprinklers or microirrigation, and neither are they suitable for cultivating staple cereals. The optimum use of the water resources in Africa, where 13 of the world's 18 nations have less than 1000 m^3/capita/day; however, may be made with the help of contemporary technology. When water is expensive, surface irrigation is impractical due to the soil, geography, and water quality, high-value cash crops may be produced and sold, and the farmer wants to maximize his or her revenue, microirrigation can be used to target specific areas.[19,41] They are ineffective for growing common crops, though. Microsystems can save water by reducing wastewater, which frequently happens with other techniques such as surface flooding. However, these advantages are not always achieved in actual use. In fact, the crop uses the same amount of water, whether it comes from a mini, sprinkler, or surface flooding approach. The management of the systems, not the systems themselves, determines a lot.

Since the 1980s, the government has promoted microsystems heavily to commercial and smallholder farmers in India, with varying degrees of success. The farmers' response was muted despite the systems receiving large subsidies—up to 90% of the cost in some cases. Even though the government offered subsidies, other elements such as access to groundwater, crop-specific microirrigation technologies, knowledge, and money were insufficient. The target group of smallholder farmers was also not successfully reached by microsystems. Instead, wealthy commercial "gentlemen farmers" were the ones who embraced the technique the most. Therefore, more needs to be done to encourage smallholder farmers to use the technologies.[41]

However, smallholder farmers have used the technology in several regions of India. Employing multiple crops or switching to higher value crops, smallholder farmers often intensify their production when using microirrigation technologies. For instance, in Maharashtra, those who adopted technology shifted from growing groundnuts and oilseeds to higher value crops such as cotton and bananas that required more water to grow.[65] The economic benefits of drip irrigation must be matched with higher water consumption, which could put further strain on already limited water resources. While drip irrigation can raise yields and accordingly raise revenue, the economic benefits must be matched with this. Microirrigation solutions must take end-user demands into account and operate within societal and environmental limits rather than just applying technologies, which can occasionally make the problem worse.[68]

5.7.3.1 LOW-COST TECHNOLOGIES

Many smallholder farmers may find the up-front expenses and associated risks of current technology to be prohibitive; hence, a number of substitutes—"cheap," technologies—have been created to fill the void.[34] They are reasonably priced, although frequently only on a small scale. It may be affordable to purchase a drip kit that covers 10 m^2 for, for instance, about US$10. On just 1 ha of land, the same amount of funding could cost at least

US$10,000, and that is a very expensive amount of expenditure and would be challenging to justify commercially.

These technologies nevertheless feature low initial investment costs and substantial farm-level returns on capital.[43] Furthermore, the broad application of small-plot irrigation techniques could lead to the creation of rural jobs, both on and off farms. Although they need a fair amount of labor, local business owners can launch operations that create, maintain, and repair irrigation equipment. Farm produce and other nontradable commodities and services are in higher demand as a result of these activities.[5] Fruit trees, miniature cattle, and fishponds are all supported by small-scale rainwater gathering around homes and backyard gardens. One method is to collect rainwater runoff and store it in containers such as off-stream storage reservoirs. This is simple to handle, just needs a small initial investment, and may be constructed beside homes.

5.7.4 WATER LIFTING

In LDCs, gravity or pressurized water supplies are rare for farmers and households.[15,60] A type of water lifting is required for most small-scale irrigation systems and garden irrigation, and these systems are typically distinguished by their energy source, which can include livestock and human energy, dirty energy, electrical, and renewable sources of power, including the sun, wind, and water.

5.7.4.1 LOCAL-POWERED PUMPS

Utilizing pans and other similar containers, many small-scale farmers still move water manually from source to field.[65] Despite being suitable for many, these basic tools are constrained, ineffective, and time-consuming. They keep those in poverty—women, in particular—from pursuing alternate alternatives for work that can provide cash. The majority of manually driven mechanical pumps are not well adapted to irrigation's large water volume requirements because they are constructed for domestic water supply. By shifting the propelling force from the hands to the legs, treadle pumps altered such perspectives through the utilization of human strength. For the purpose of irrigation of rice, they were initially created in the 1980s in Bangladesh to lift relatively large quantities of water through relatively modest lifts of up to 1 m. More than 500,000 pumps are currently utilized every day across the nation, and their adoption by farmers has been praised for being extraordinarily high. The use of treadle pumps is seen as a "steppingstone" between manual pumping and motorized pumping. Investment is modest because the initial capital cost, which ranges from US$50–120, is low.

In the 1990s, treadle pumps were brought from Bangladesh to Africa and are now commonly utilized there. Thousands of pumps are reportedly in operation across sub-Sahara Africa viz. Malawi, Zimbabwe, Zambia, and Niger, while the exact number of installed pumps is unknown.[31] A commercial market with supply chains has been formed in several nations, most notably Kenya, making spare parts and pump repair services accessible. Others view treadle pumps as a further technique of entangling farmers and their households in yet another type of labor. There were issues with the treadle pump technology transfer from Bangladesh to Africa. Now, some treadle pumps can be used with drip and sprinkler irrigation systems.

5.7.4.2 PETROLEUM AND ELECTRIC FUEL PUMPS

The development of low-cost well-drilling technology, the availability of compact, inexpensive gasoline and diesel pumps, economical electric pumps, rural electrification, and subsidized energy have all contributed to a rapid expansion in motorized pumping throughout the world in recent decades. With pumps, smallholder farmers have more independence than they did with the more extensive state-owned programs. When it is suitable to irrigate, which is typically during the day instead of at night, they can water crops as and when they need it.[64]

Electric pumps might be a desirable choice in areas with easy access to electricity close to agricultural areas. Electric pumps, however, are not a practical choice in places with sporadic electrical supplies. Because of economies of scale, motorized pump expenses also typically benefit large-scale farmers, although they are typically unaffordable for some smallholder farmers with scarce resources (land and income).[56] For instance, a treadle pump costs 5.6 times more in Ghana than a motorized pump, requiring owners of small vegetable plots a sizable capital investment. Furthermore, motorized pump operation came at a high expense relative to the benefits. Users frequently must drive great distances to find repair assistance and replacement parts. When choosing to use motorized pumps, maintenance and repair capabilities are crucial factors to consider.

5.7.4.3 PUMPS POWERED WITH RENEWABLE ENERGY

In terms of technical viability and costs, research on renewable sources of energy, such as solar and wind, has produced mixed findings. Because of this, the energy source is frequently unpredictable, and the tools required to gather and utilize enough energy to run an irrigation pump cost a lot of money. However, other studies have shown that certain forms of renewable energy can be more cost-effective than conventional sources of energy in rural areas and for applications on a small scale.[14]

The use of solar electricity in remote locations for applications such as telecommunications and tiny, isolated potable water supplies is a common practice. These applications have relatively low power requirements. They continue to be pricey in relation to their power output despite years of extensive research aimed at creating reliable and inexpensive solar energy gathering systems. Devices that use solar energy, as well as the related machinery used to provide power to a pump, are both very delicate and sensitive. Pump sets are frequently out of service for extended periods of time while waiting for maintenance or spare components.[14]

Photovoltaic (PV) piping system irrigation systems are both economically and technically possible, according to a study on solar-powered agricultural irrigation, but land availability for the solar power plant remains the main barrier.[48] Currently, only low-powered and niche applications make solar-powered equipment economically viable. Even so, they ought to be listed among the prospective technologies, and cost and robustness reductions in the future should increase their competitiveness.

Water has been raised using wind power on a large scale, typically to drain low-lying terrain where there are frequent high winds. They typically are not very dependable or at least require a lot of care and upkeep. Strong winds are another element that can be influenced by location and season. For a large portion of the year, strong winds are not severe over the majority of SSA's arable areas. The use of wind energy resources for crop irrigation and groundwater extraction has, however, been demonstrated in some cases. If certain criteria are met, such as adequate wind resources, farmer revenue, etc., wind power pumps in India have enormous potential for tiny irrigation.[68]

The availability of renewable resources at the site, the number of kWh of electricity needed, and the type of use all have an impact on the decision to adopt renewable energy technologies rather than traditional energy sources. In South Africa and Namibia, wind pumps are notable instances of how renewable energy technologies are being used in rural regions for irrigation. Other uses include miniature biomass facilities for water pumps, tiny hydropower plants, and microirrigation using solar energy.[53]

5.7.5 INFORMATION COMMUNICATION TECHNOLOGY (ICT) FOR WATER MANAGEMENT

Despite not yet having a large impact on agricultural production and AWM, ICTs are becoming more significant in the majority of LDCs.[5,6] Farmers in Ugandan communities, for instance, can call a free phone hotline with queries and have access to a variety of knowledge on the Internet. The local language-speaking operators phone the farmers back after looking up the information they need and giving them details on commodity prices, weather predictions for water management and irrigation, plant diseases, and other topics.

Water management is now being supported in novel ways by geographic information system (GIS) technology.[63] The region's irrigated area might increase by 50% if these indigenous tanks, which were erected approximately 1500 years ago, have been revived to absorb merely 15–20% of local rainfall.[7]

5.7.6 REGULAR AND UNUSUAL WATER SOURCES

5.7.6.1 REUSING WASTEWATER

Most of both residential and commercial water is used and then returned to the watershed through seeping into the ground or directly discharging into rivers.[13] It is beyond financial recovery when released into the ocean or the desert. Water from waste sources can be used again, especially in agriculture. In the Middle East and North African nations where water is in short supply, wastewater reuse is a top priority. Reusing sewage effluent is common in Israel, Egypt, and Syria Arab Republic, with 67% of it being done so for agricultural and environmental reasons in Israel, Egypt, and Syria Arab Republic.

If this water is genuinely "accessible" for extraction, there is an ongoing controversy. For instance, it is unlikely that the annual effluent production of Egypt, which amounts to 0.79 billion m^3, will be sufficiently available for all purposes.[45] According to Egypt's water policy for 2017, more water is consumed than the nation is allotted from the Nile River. This means that the total quantity of water reuse has already been factored into Egypt's hydrological cycle.[44,45] In the Caribbean and Latin America Caribbean, wastewater for agricultural purposes is also evolving into a significant problem. Markets for fresh produce, dairy products, and poultry in big cities use treated wastewater. For example, almost 25% of municipal wastewater in Mexico is utilized to hydrate land of 300,000 ha. The cultural norms and worries about the purity of the treatment, on either hand, can prevent some nations from reusing wastewater. It is not a common practice to grow crops such as fruit and vegetables in wastewater. It might be less controversial to use treated water to cultivate processed goods such as cereals, root crops, and biofuels.

In comparison to treating freshwater, large-scale treatment of wastewater for agricultural purposes might come with a significant cost premium.[49] Even if the crops are not eaten directly, wastewater needs to be treated to prevent health problems. Municipal wastewater is frequently produced in cities and greater cities, where there is an abundance of both industry and people, making it potentially feasible and financially beneficial to invest in the required infrastructure. Waterways and/or pipelines are required to transport the water because communities tend to be far away from the location where the purified water can be used for agriculture. Additionally, the availability of wastewater (which typically flows evenly throughout the year) does not correspond with the need for agricultural water. In order to properly use all the water, some form of water conservation is necessary. The price of reusing water for agriculture may rise significantly as a result of all of this.

The informal and uncontrolled use of wastewater in agriculture is a serious problem in developing nations that requires more attention.[29] Roughly 3–3.5 million ha of agricultural land is being watered with unclean or attenuated wastewater worldwide, which is treble the area of all formally licensed vegetable irrigation networks throughout the Black continent.[42] Fresh water is frequently scarce in low-income nations, and there are few municipal wastewater treatment facilities. As a result, untreated wastewater is frequently the only practical and inexpensive alternative for irrigation. Furthermore, wastewater's nutritional content has resulted in higher yields at reduced prices. As an illustration, using wastewater instead of ordinary water for irrigation increases annual income for farmers in Pakistan by 30–40%. Aside from that, wastewater irrigation works with regional vendors, traders, and other professionals. With over 95% of the region's vegetable vendors being women, SSA also benefits women.[31] On the other side, there may be significant health dangers. This makes a connection between home hygiene and water supply problems, particularly in LDCs, and food security issues.

The question is not whether or whether wastewater should be utilized in agriculture, but rather how wastewater can be used for irrigation in a safe manner using accessible treatment methods.[30] Maximizing benefits while reducing dangers in wastewater utilization is the problem facing policymakers. Adopting safety regulations that are suited for the local environment and spreading basic technology would be required to achieve this. The use of localized drip irrigation, the installation of shallow wells, and water collecting and application techniques that minimize contamination are some of these technologies. There are numerous creative regional options for treating wastewater. Farmers built a wastewater system at the Kikwari hamlet in India, for instance, by utilizing pipelines to link the drainage water to settling tanks. Following filtering, the water was applied to community and school gardens for irrigation. According to observations, farmer field schools can help spread best practices and connect research with extension agencies.[65]

5.7.6.2 DESALINATION

Salinized water must first go through a process called desalination, which takes the salt out.[29] As a result of the significant advancement of the methods for desalination over the past thirty years, the two primary desalination methods, thermal as well as membrane, currently account for almost 98% of the globe's up-to-date desalination capacity for operation, which exceeds presently 35 million m^3 of water per day, most of it in the region known as the Middle East. Drinking water and industry are the two primary uses for desalination. Less than 10% of desalinated water, primarily in Spain, where desalination is strongly subsidized, is thought to be used for agriculture. Both procedures use a lot of energy and result in high-quality water.[56]

Costs for saltwater desalination have dropped by about 14 times since the late 1970s as a result of economies of scale and ongoing membrane technology advancements, but they have stayed relatively unchanged for deployable technologies. Desalination might be a water source for irrigation, but at its present price of between US$0.5 and 1.5 per m^3, the innovation is still viewed as being prohibitively expensive. However, several businesses are working on brand-new technologies that may reduce desalination prices by 50%. Concerns are also raised concerning the water's lack of micronutrients and excessive purity for irrigation.[26] Desalination is still viewed by planners and policymakers as the "magic bullet" answer to water shortages. Desalination will probably make the issue worse by consuming more carbon fuels while increasing the dependence of local water supplies on much expensive fossil fuels unless it is fueled by solar and wind power.[29,40]

A third choice would be to desalinate water using solar energy, although this technology is still extremely new. By reenacting the natural water cycle, solar stills create water vapor. The yields, which rely on solar exposure, are poor, averaging about 2–5 L per day. In these areas, genuine water shortages are at their greatest. Their environmental impact is minimal, they require little upkeep, and they are modest in size.

5.7.7 EFFICIENT AGRICULTURE USING RAINWATER

Rainfed agriculture has the potential for significant advancements, and the technologies already exist. The goal is to get "more produce per drop" by blending water and soil management techniques that are centered on soil fertility, increased rainfall immersion, and water harvesting. The places with the most severe water shortages, as well as most of the hunger and poverty, provide the most promise for development.

To control the abrupt water surpluses and frequent dry spells, creative techniques are needed. By enhancing water penetration and soil water storage, for example, soil and water recycling methods can aid in making better use of rainfall.[16] These methods demand little or no financial outlay. Finding realistic solutions for slow development that can be implemented by inexperienced, part-time farmers who lack access to adequate extension help is a difficulty for the poor.[42,54] Therefore, the fact that these technologies, despite being well recognized, are not being broadly promoted, implemented, and practiced, is both disheartening and extremely concerning.[8]

5.7.8 SUSTAINABLE FARMING

Although conservation agriculture is not specifically a water technology, one of the advantages is better water management.[71] About 95 million ha globally, the rainfed farming technique is used, mainly in Central America; Brazil and Argentina; and North America, though to a considerably lesser amount in Asia and Africa. Instead of exploiting natural resources to increase production, conservation agriculture uses agro-ecosystem and soil resources in a balanced way to maximize crop yields. "Zero-tillage" techniques allow for the permanent maintenance of soil cover with little soil disturbance.

Plant leftovers shield the soil, improving soil organic matter and soil water conservation. As a result, during rainfall events, the soil is better able to infiltrate and store water.

Throughout Kenya, Zambia, and the United Republic of Tanzania, where some farmers have increased grain yields by doubling or even tripling, the technique is just now starting to catch up in Africa. Zambia's vulnerable households have benefited from conservation agriculture's assistance in surviving livestock diseases and drought. These days, over 200,000 farmers employ this method. Conservation agriculture allowed Zambian farmers to harvest one crop during the 2000–2001 droughts, while farmers using traditional farming techniques experienced complete crop failure. Over 350,000 farmers in Ghana practice conservation agriculture.[31]

5.8 RESOLVING WATER ISSUES IN AGRICULTURE

The difficulty of food security can be met with the use of current AWM technologies. But as history has demonstrated, it will be difficult to fully utilize the land and water resources that have been bestowed upon us, and spending money on water alone would not boost food production. Inputs for agriculture must be numerous and diverse. There is a need for complementary investments in a variety of agricultural goods and services, including access to markets, post-harvest infrastructure, good roads, microcredit, seeds, fertilizer, and post-harvest power. When these elements are taken into consideration, the problem of food security becomes incredibly complicated.

Most industrialized nations are equipped with the necessary institutions, infrastructure, and capacity to maintain the existing levels of food and water security. However, the majority of LDCs lack these vital social and physical infrastructures that support sustainable growth. Until recently, international and political agendas did not place a strong priority on agricultural and food security. Water management has not been a priority for agriculture, either. In LDCs, there are minimal institutional frameworks that can support AWM planning, management, and implementation by farmers and professionals.

Essential is financing. In the 1970s, Asian nations started their green revolution by allocating 15% of their annual budgets to agriculture. According to the World Bank, an increase in the agricultural sector's share of the continent's GDP will have a 3–4 times greater impact on poverty reduction than an increase in the nonagricultural sector's share of the economy. However, donor nations only devote 5% of their development contributions to regional agriculture. On the plus side, agriculture is once again at the top of the global agenda, and international investment in the sector is starting to pick up. Resources management, production, marketing, and consumption must all be integrated for agriculture to flourish in a sustainable and economic way. The past mistakes must be avoided because "more of the same" would not be sufficient.

5.8.1 GIVE WOMEN FORCE GREATER ATTENTION

In addition to having little control over the services they get, women are frequently excluded from decision-making processes. They frequently also have restricted access to land, which includes limited access to water. Securing rights to water might result from granting impoverished farmers, especially women, access to land.[65] Due to their restricted access to a variety of physical resources, such as farm inputs, technical skills, and land, women's agricultural production is frequently lower than that of men's. Therefore, removing the challenges that women encounter requires a deeper awareness of their requirements. Women's farm output would rise by 20%–30% if the same resources that are available to men were also made available to them. Agriculture production would increase by 2.5–4% on a global scale. In order to reduce the gender gap in AWM, more resources must be devoted to women because research and extension are primarily conducted by men in LDCs.

Even though women's contributions to agriculture are being acknowledged, many AWM activities still have a male bias. Also, in circumstances when women perform the majority of the labor in the irrigated fields, jobs such as physically applying water to the fields and opening and closing gates are seen as masculine. Additionally, men are typically responsible for caring for income crops and cattle, while women are more likely to tend to staple crops, vegetables, and kitchen gardens. Garden irrigation has consequently taken on significant importance for women who grow vegetables for their own consumption as well as for the local markets. Women dominate the production of vegetables grown under drip irrigation, according to studies conducted in Nepal. The adoption and extension system places a strong emphasis on men and income crops despite the fact that they provide roughly 90% of the overall labor. Information regarding modernized agriculture and technology has either been scarce or nonexistent for women.

Women should participate in managing domestic water resources via Water User Associations (WUA) if irrigation is to meet both men's and women's concerns. In order to ensure that women may profit from water projects, the International Fund for Agricultural Development (IFAD) also suggests innovative business models. This entails adopting a minimum threshold for women's membership in WUAs, establishing minimum allocations for women's land allocation, ensuring equal plot dimensions for men and women, improving women's economic empowerment, incorporating more water facilities such as wells and hand pumps, allowing membership for users of water other than irrigation, and increasing women's financial inclusion.

By offering more than one use for water, multiple-use programs give women the chance to enhance both their own wellness and the wellbeing of society. In reality, communities utilize water for a range of purposes; it was the public sector that artificially created these subsectors and categorized water uses for specific purposes.[8] Water has numerous uses and priorities, including home use, kitchen gardens, animal watering, and fisheries, all of which are historically the women's responsibility. Multiple-use plans acknowledge this.[50]

Aside from adding stairs to irrigation channels to make it easier to reach water for drinking, washing clothes, and other domestic tasks, irrigation systems can also just keep water in periodic irrigation channels all year long for domestic purposes. Likewise, plans initially intended for domestic purposes (sometimes known as "domestic-plus" plans) can evolve into multiple-use plans. For instance, 3 L per person per day are set aside for drinking and food preparation if 50–100 L per person per day are available. A small business, horticulture, or animals may use the water that is left over after meeting home demands. The concurrent roles that women play as entrepreneurs, housewives, livestock keepers, and farmers are also acknowledged through multiple-use plans. Multiple-use programs should, therefore, be community-driven and participative by nature.[23]

Planners consider all users' aspirations for water usage and sources rather than sticking to the government sector's one-time-use mandate. To ensure that the unique issues and requirements of men and women from all social classes are included in the development, usage, and management of water, gender views must be mainstreamed in planning procedures.

5.8.2 EMPHASIS ON CURRENT TECHNOLOGY

Promoting, utilizing, and adapting current technologies to new situations can all be advantageous. This is valid for both drainage and irrigation technologies. Since many technologies are built and constructed without much regard to who will manage them or how they will be handled and maintained, adapting current technologies should embrace "design for management." It was first advocated in the 1980s to enhance irrigation performance, so this is not a brand-new idea and is still very much in demand today.

Still, there are many instances of patterns that are unsuitable for the setting of their consumers. For instance, adjustable gates and flexible canal systems allow for the adjustment of canal flows to accommodate crop water requirements. However, these systems might not always be compatible with the farmers' preference for simpler irrigation technologies. Fixed water management systems may be a better choice in these circumstances because they provide a distribution system that is easier to administer.[51]

To ensure that rural women may fully benefit from programs, design for management must also take gender equality into account. Instances comprise choosing irrigation timings that are more convenient for taking care of family responsibilities and refraining from night irrigation due to gender-based abuse or abuse. Even though women in Zimbabwe were primarily responsible for irrigation, only men were given the task of maintaining and operating the diesel pumps that provide water supply. Women frequently had to bring extra water to make sure their farms were irrigated because they had little control over the supply. Creating plans that take advantage of many technologies already in existence might also lessen tedium and give women extra time for other useful pursuits.[11]

5.8.3 FURTHER STUDY NEEDED FOR ADVANCEMENT AND BETTER COMMUNICATION

AWM studies in LDCs, as well as their application in practice for farmers, frequently do not mesh well together. While specific study outcomes may be helpful but not always reach their intended audience, other areas of study may not be pertinent for farmers. There is a need for new communication strategies that take end-user needs into account. In addition to being intended for farmers, the material should be presented in a way that is appropriate for policymakers, agro-entrepreneurs, extension specialists, and the public. One such example is the promotion of research uptake by researchers in East Africa. Social scientists must join the team of natural resource investigators for their work to be more gender and end-user relevant. The integration of research and practice requires the support of NGOs. They benefit from having direct contact with the neighborhood. To better comprehend local needs, get input, and exchange knowledge, researchers should work more closely with NGOs.

5.8.4 MORE INTELLIGENT USE OF WATER

Utilizing information technologies such as sensors, control devices, meters, operating systems, and mobile phones in more inventive ways represents one of the most significant unrealized prospects for more competent water governance in all sorts of businesses. These may appear to be high-tech possibilities, but given the fast-growing usage of mobile phones in LDCs, there is an opportunity to give farmers in remote locations that currently have a link to extension workers helpful information and guidance. Given the crucial role played by women in agriculture, further efforts should be taken to close the gender imbalance in ICT access and use.

5.8.5 CREATE NEW INSTITUTIONS

It is necessary to create new institutional arrangements that improve user ownership and engagement while centralizing water regulation and decentralizing management duties. Monitoring, gathering, and synthesizing information on water assets is essential for regulating and controlling water resources at the national level. Communication between government agencies charged with water management is also essential. The different ministries that handle water, food, agriculture, the environment, and money need to be connected via bridges.

Many LDC governments promote decentralization, but effective local administration depends on strong local institutions that can engage local communities, articulate their demands, and analyze, create, and put into practice new policies and innovations. Social capital, which is the foundation of these organizations, must be strengthened if decentralization is to be successful. Although these concepts are well accepted by international organizations, there is still a long way to go before national planning adopts them and turns them into practical and context-specific activities.

5.8.6 CAPACITY BUILDING OF AWM

Farmers in general, governments' agencies, banks, donation institutions, nonprofit organizations, neighborhood-based teams, private companies, and academic institutions must all work together in a protracted, systematic process of interaction, collaboration, engagement, and collaborative learning to build capacity. In most LDCs, a severe capacity shortage at all levels is a significant barrier to the development of water for agriculture. Developing organizational structures, notably as hydro abstractor groups and extension aid services, and creating a climate that is conducive to improved food supply and agriculture water investment are all part of capacity development. This goes beyond simply providing training for farmers, localized professionals, and government-based scientific and extension service staff who assist farmers. For instance, lowering tariffs on imported irrigation and soil enhancement technology and pump sets will assist in reducing expenses and increase the profitability of agriculture.

5.8.7 ENCOURAGE PUBLIC-PRIVATE COLLABORATION

It's not necessary for these to be two distinct economic sectors. Instead, there are prospects for collaboration between the two, allowing smallholder farmers to work alongside commercial farmers to create a system where everyone wins.

Cooperation between beginning commercial farmers and smallholder farmers is promoted in Zambia. Smallholder farmers can gain by participating in the value chain and learning more about cutting-edge farming practices and managerial abilities. Commercial farmers gain from scale economies, the ability to buy crops from nearby small-scale farmers, and value additions such as maize milling and biofuel processing. Costs associated with each unit of water can be decreased by incorporating smallholder farmers into commercial irrigation plans.

The inadequate commercial and technical capacity of government organizations to collaborate in PPP with private entities and financial partners is just one of the difficulties with the method. The need to tackle the shortage of experienced private companies in the irrigation domain and the necessity of making sure the initiatives are large enough (250–1000 ha) to be professionally managed and commercially viable are crucial lessons for success that have already been taught, though, have already been learnt.

5.8.8 ENCOURAGE BUSINESS DEVELOPMENT

The foundation of many AWM initiatives addressing poverty is aid. However, there are many chances for the private industry to supply water management solutions on a commercial basis. Farmers and African AWM decision-makers might take note of this because there are ample opportunities to introduce suitable, inexpensive pumps that are made or constructed domestically. Similar opportunities include the distribution of drip packages, treadle pumps to be used, and the development of distribution networks that offer support and spare parts. Products need to fulfill specific client criteria, such as matching the tiny plots generally tended by smallholder farmers, to be economically successful. They also need to be affordable, have a very quick payback period, and significantly increase the revenue and yield of the people who buy them.

5.8.9 INCREASE FOCUS ON YOUNG PEOPLE

More than 25% of the globe's population is under the age of 24 years, and 35% of that populace is in several African nations. Many of these people are born into impoverished rural homes. Despite having a substantial capacity to support growth in the economy and food security, youth are largely "invisible" in the development of natural resources. It is claimed that now is the time to include young people in natural resource policy reforms and to reject the "accepted wisdom" that claims they are uninterested in using water and land resources for a living.

5.8.10 EXPAND TRADE IN WATER AND FOOD

Half a billion farms around the world consume about 85% of the water to generate food products that stay inside the producer economies. International trading in embedded water accounts for only 15% of agricultural output. This "virtual water" trade between countries with abundant water supplies and others with scarce water resources may become more important in facilitating improved food distribution to nations that struggle to produce enough staple food crops.

5.9 SUMMARY

Because agriculture has improved to a degree where there is now enough food supply to fulfill the requirements of the world's rising population, it is expected that there is unlikely to be a worldwide food crisis in the ensuing decades. The success of contemporary agriculture has largely been attributed to investments in water resources. With the advancement of water technology, farmers were able to plant more crops each year and increase crop yields. Increasing production per acre, decreasing poverty, and meeting needs for food security are the three major difficulties facing agriculture today. These challenges are all tied to water management. One of the most important concerns of the twenty-first century is ensuring the best management of the world's water resources. Over the next 50–75 years, there will be an increase of 3 billion or more in the world's population, and urban populations will almost triple. The majority of global population growth will take place in underdeveloped nations where there is already a severe lack of water and many people live in poverty. Therefore, increasing the consistency of the water provision for agriculture is a necessary but insufficient prerequisite for lowering malnutrition and hunger and fostering quicker agricultural expansion.

KEYWORDS

- **climate**
- **conservation**
- **innovation**
- **irrigation**
- **SDG**
- **smart**
- **water**

REFERENCES

1. Aivazidou, E. Water Management in Agriculture and Industry: Challenges, Trends, and Opportunities. *Sustainability* **2021,** *14* (1), 66–70.
2. Alobid, M.; Szűcs, I. Appraising the Water Status in Egypt Through the Application of the Virtual Water Principle in the Agricultural Sector. In *Environmental Footprints of Crops*; Springer: Singapore, 2022; pp 27–50.
3. Araújo, S. O.; Peres, R. S.; Barata, J.; Lidon, F.; Ramalho, J. C. Characterising the Agriculture 4.0 Landscape-Emerging Trends, Challenges and Opportunities. *Agronomy* **2021,** *11* (4), 667–675.
4. Casadei, S.; Peppoloni, F.; Ventura, F.; Teodorescu, R.; Dunea, D.; Petrescu, N. Application of Smart Irrigation Systems for Water Conservation in Italian Farms. *Environ. Sci. Pollut. Res.* **2021,** *28* (21), 26488–26499.

5. Cavazza, F.; Galioto, F.; Raggi, M.; Viaggi, D. The Role of ICT in Improving Sequential Decisions for Water Management in Agriculture. *Water* **2018,** *10* (9), 1141–1148.
6. Cavazza, F.; Galioto, F.; Raggi, M.; Viaggi, D. Digital Irrigated Agriculture: Towards a Framework for Comprehensive Analysis of Decision Processes Under Uncertainty. *Future Internet* **2020,** *12* (11), 181–189.
7. Chidambaram, S.; Prasanna, M. V.; Venkatramanan, S.; Nepolian, M.; Pradeep, K.; Panda, B.; Thilagavathi, R. Groundwater Quality Assessment for Irrigation by Adopting New Suitability Plot and Spatial Analysis Based on Fuzzy Logic Technique. *Environ. Res.* **2022,** *204,* 111729.
8. Chilinda, Z. B.; Wahlqvist, M. L.; Lee, M. S.; Huang, Y. C. Optimal Household Water Access Fosters the Attainment of Minimum Dietary Diversity Among Children Aged 6–23 Months in Malawi. *Nutrients* **2021,** *13* (1), 178–185.
9. Chungloo, D.; Tisarum, R.; Samphumphuang, T.; Sotesaritkul, T.; Singh, H. P.; Takabe, T.; Chaum, S. Mitigation of Water-Deficit Stress, Physio-Morphological Adaptation, and Elevation of Andrographolide in *Andrographis paniculata* Using Foliar Glycine Betaine. *J. Plant Growth Regul.* **2022,** *2022,* 1–13.
10. Correa-Porcel, V.; Piedra-Muñoz, L.; Galdeano-Gómez, E. Water–Energy–Food Nexus in the Agri-Food Sector: Research Trends and Innovating Practices. *Int. J. Environ. Res. Public Health* **2021,** *18* (24), 12966.
11. Dagher, L. A.; Hassan, J.; Kharroubi, S.; Jaafar, H.; Kassem, I. I. Nationwide Assessment of Water Quality in Rivers Across Lebanon by Quantifying Fecal Indicators Densities and Profiling Antibiotic Resistance of *Escherichia coli. Antibiotics* **2021,** *10* (7), 883–892.
12. Dash, S.; Kalamdhad, A. S. Hydro-Chemical Dynamics of Water Quality for Irrigation Use and Introducing a New Water Quality Index Incorporating Multivariate Statistics. *Environ. Earth Sci.* **2021,** *80* (3), 1–21.
13. De Clercq, M.; Vats, A.; Biel, A. Agriculture 4.0: The Future of Farming Technology. *Proc. World Gov. Summ., Dubai - UAE* **2018,** *2018,* 11–13.
14. Dharminder, R. K. S.; Kumar, V.; Devedee, A. K.; Mruthyunjaya, M.; Bhardwaj, R. The Clean Water: The Basic Need of Human and Agriculture. *IJCS* **2019,** *7* (2), 1994–1998.
15. Dhawan, V. Water and Agriculture in India. In *Background Paper for the South Asia Expert Panel During the Global Forum for Food and Agriculture*; Rome: FAO, 2017; Vol. 28.
16. Es'Haghi, S. R.; Karimi, H.; Rezaei, A.; Ataei, P. Content Analysis of the Problems and Challenges of Agricultural Water Use: A Case Study of Lake Urmia Basin at Miandoab, Iran. *SAGE Open* **2022,** *12* (2), 21582440221091247.
17. Evans, A. E.; Mateo-Sagasta, J.; Qadir, M.; Boelee, E.; Ippolito, A. Agricultural Water Pollution: key knowledge Gaps and Research Needs. *Curr. Opin. Environ. Sustainability* **2019,** *36,* 20–27.
18. Fayaz, M.; Rabani, M. S.; Wani, S. A.; Thoker, S. A. Nano-Agriculture: A Novel Approach in Agriculture. In *Microbiota and Biofertilizers*; Springer, Cham, 2021; pp 99–122.
19. Friha, O.; Ferrag, M. A.; Shu, L.; Maglaras, L.; Wang, X. Internet of Things for the Future of Smart Agriculture: A Comprehensive Survey of Emerging Technologies. *IEEE/CAA J. Autom. Sin.* **2021,** *8* (4), 718–752.
20. Galioto, F.; Chatzinikolaou, P.; Raggi, M.; Viaggi, D. The Value of Information for the Management of Water Resources in Agriculture: Assessing the Economic Viability of New Methods to Schedule Irrigation. *Agric. Water Manag.* **2020,** *227,* 105848.

21. Giordano, M.; Barron, J.; Ünver, O. Water Scarcity and Challenges for Smallholder Agriculture. In *Sustainable Food and Agriculture*;Academic Press: San Diego–CA, 2019; pp 75–94.
22. Gunawan, M. N. G.; Ismail, Y. I. Study of Rainwater Harvesting Implementation. *JIE Sci. J. Res. Appl. Ind. Syst.* **2020,** *5* (1), 37–47.
23. Halder, S.; Kumar, P.; Das, K.; Dasgupta, R.; Mukherjee, A. Socio-Hydrological Approach to Explore Groundwater–Human Wellbeing Nexus: Case Study from Sundarbans, India. *Water* **2021,** *13* (12), 1635=1643.
24. Hassan, A. M.; Firat-Ersoy, A. Statistical Assessment of Seasonal Variation of Groundwater Quality in Çarşamba Coastal Plain, Samsun (Turkey). *Environ. Monitor. Assess.* **2022,** *194* (2), 1–29.
25. Heggy, E.; Sharkawy, Z.; Abotalib, A. Z. Reply to Comment on 'Egypt's Water Budget Deficit and Suggested Mitigation Policies for the Grand Ethiopian Renaissance Dam filling Scenarios. *Environ. Res. Let.* **2022,** *17* (8), 088002.
26. Joseph, N.; Ryu, D.; Malano, H. M.; George, B.; Sudheer, K. P. Investigation into Sustainable Water Use in India Using Combined Large-Scale Earth System-Based Modelling and Census-Based Statistical Data. *J. Hydrol.* **2020,** *587,* 124930.
27. Kamienski, C.; Soininen, J. P.; Taumberger, M.; Dantas, R.; Toscano, A.; Salmon Cinotti, T.; Torre Neto, A. Smart Water Management Platform: IoT-Based Precision Irrigation for Agriculture. *Sensors* **2019,** *19* (2), 276–282.
28. Khairy, S.; Shaban, M.; Negm, A. M.; Eldeen, O. W.; Ramadan, E. M. Drainage Water Reuse Strategies: Case of El-Bats Drain, Fayoum Governorate, Egypt. *Ain Shams Eng. J.* **2022,** *13* (5), 101681.
29. Kumar, R.; Ahmed, M.; Bhadrachari, G.; Thomas, J. P. Desalination for Agriculture: Water Quality and Plant Chemistry, Technologies and Challenges. *Water Sci. Technol.: Water Supply* **2018,** *18* (5), 1505–1517.
30. Laassilia, O.; Ouazar, D.; Bouziane, A.; Hasnaoui, M. D. Simulation-Optimization Model for Dynamic and Sustainable Management of the Water Transfer Between Two Interconnected Reservoirs: Case Study in Morocco. *J. Environ. Account. Manag.* **2022,** *10* (02), 177–190.
31. Libala, N.; Griffin, N.; Nyingwa, A.; Dini, J. Freshwater Ecosystems and Interactions with the SDG 2030 Agenda: Implications for SDG Implementation in South Africa. *Afr. J. Aquat. Sci.* **2022,** *47* (3), 353–368.
32. Liu, Y.; Ma, X.; Shu, L.; Hancke, G. P.; Abu-Mahfouz, A. M. From Industry 4.0 to Agriculture 4.0: Current Status, Enabling Technologies, and Research Challenges. *IEEE Trans. Ind. Inform.* **2020,** *17* (6), 4322–4334.
33. Mainardis, M.; Cecconet, D.; Moretti, A.; Callegari, A.; Goi, D.; Freguia, S.; Capodaglio, A. G. Wastewater Fertigation in Agriculture: Issues and Opportunities for Improved Water Management and Circular Economy. *Environ. Pollut.* **2021,** 118755.
34. Malakar, A.; Snow, D. D.; Ray, C. Irrigation Water Quality—A Contemporary Perspective. *Water* **2019,** *11* (7), 1482–1488.
35. Maldet, M.; Schwabeneder, D.; Lettner, G.; Loschan, C.; Corinaldesi, C.; Auer, H. Beyond Traditional Energy Sector Coupling: Conserving and Efficient Use of Local Resources. *Sustainability* **2022,** *14* (12), 7445–7451.
36. Melero-Meraz, V.; Zegbe, J. A.; Herrera, M. D.; Guzmán-Maldonado, S. H.; Medina-García, G.; Sánchez-Toledano, B. I.; Servín-Palestina, M. On-Farm Supplemental Irrigation of 'Roja-Lisa' Cactus Pear: Pre-and Postharvest Effects. *Horticulturae* **2022,** *8* (6), 483–488.

37. Miao, R.; Khanna, M. Harnessing Advances in Agricultural Technologies to Optimize Resource Utilization in the Food-Energy-Water Nexus. *Annu. Rev. Resour. Econ.* **2020,** *12,* 65–85.
38. Minh, H. V. T.; Van Ty, T.; Avtar, R.; Kumar, P.; Le, K. N.; Ngan, N. V. C.; Downes, N. K. Implications of Climate Change and Drought on Water Requirements in a Semi-Mountainous Region of the Vietnamese Mekong Delta. *Environ. Monitor. Assess.* **2022,** *194* (2), 1–20.
39. Mohamed, M. A. H.; El-Mahdy, M. E. S. Evaluation of Climate Change Impact on Extreme Temperature Variability in the Blue Nile Basin, Ethiopia. *Geosci. Instrument., Methods Data Syst.* **2021,** *10* (1), 45–54.
40. Mondal, B.; Bauddh, K.; Kumar, A.; Bordoloi, N. India's Contribution to Greenhouse Gas Emission from Freshwater Ecosystems: A Comprehensive Review. *Water* **2022,** *14* (19), 2965.
41. Monteleone, S.; De Moraes, E. A.; Maia, R. F. Analysis of the Variables That Affect the Intention to Adopt Precision Agriculture for Smart Water Management in Agriculture 4.0 Context. In *2019 Global IoT Summit (GIoTS)*; IEEE, 2019; pp 1–6.
42. Mugejo, K.; Ncube, B.; Mutsvangwa, C. Infrastructure Performance and Irrigation Water Governance in Genadendal, Western Cape, South Africa. *Sustainability* **2022,** *14* (19), 12174.
43. Nazari, B.; Liaghat, A.; Akbari, M. R.; Keshavarz, M. Irrigation Water Management in Iran: Implications for Water Use Efficiency Improvement. *Agric. Water Manag.* **2018,** *208,* 7–18.
44. Negm, A. M. Ed. *Conventional Water Resources and Agriculture in Egypt*; Springer: Heidelberg - Germany, 2019.
45. Negm, A. M.; Omran, E. S. E.; Abdel-Fattah, S. Update, Conclusions, and Recommendations for the Unconventional Water Resources and Agriculture in Egypt. *Rep. Unconvent. Water Resour. Agric. Egypt* **2018,** *2018,* 509–532.
46. Pal, S.; Talukdar, S.; Ghosh, R. Damming Effect on Habitat Quality of Riparian Corridor. *Ecol. Indicat.* **2020,** *114,* 106300.
47. Patle, G. T.; Kumar, M.; Khanna, M. Climate-Smart Water Technologies for Sustainable Agriculture: A Review. *J. Water Climate Change* **2020,** *11* (4), 1455–1466.
48. Pearce, J. M. Agrivoltaics in Ontario Canada: Promise and Policy. *Sustainability* **2022,** *14* (5), 3037.
49. Pereira, L. S. Water, Agriculture and Food: Challenges and Issues. *Water Resour. Manag.* **2017,** *31* (10), 2985–2999.
50. Prakash, P.; Elanthendral, G.; Vedanarayanan, V.; Kumar, J. A.; Samrot, A. V.; Jaiganesh, R. A Sustainable Approach of Artificial Neural Network for Prediction of Irrigation, Pesticides, Fertilizers, and Crop Yield. In *Artificial Intelligence Applications in Agriculture and Food Quality Improvement*;IGI Global: New Delhi, 2022; pp 169–189.
51. Pramanik, M.; Khanna, M.; Singh, M.; Singh, D. K.; Sudhishri, S.; Bhatia, A.; Ranjan, R. Automation of Soil Moisture Sensor-Based Basin Irrigation System. *Smart Agric. Technol.* **2022,** *2,* 100032.
52. Pueppke, S. G. Ancient WEF: Water–Energy–Food Nexus in the Distant Past. *Water* **2021,** *13* (7), 925–935.
53. Rajput, J.; Kothari, M.; Bhakar, S. R.; Dimple, Kushwaha, N. L.; Singh, P. K.; Rana, L. Evaluation of Water Delivery Performance of Right Main Canal of Bhimsagar Medium Irrigation Scheme, Rajasthan. *ISH J. Hydraul. Eng.* **2022,** *2022,* 1–11.

54. Ram, S. A.; Irfan, Z. B. Application of System Thinking Causal Loop Modelling in Understanding Water Crisis in India: A Case for Sustainable Integrated Water Resources Management Across Sectors. *Hydro Res.* **2021,** *4,* 1–10.
55. Ricart, S.; Rico, A. M. Assessing Technical and Social Driving Factors of Water Reuse in Agriculture: A Review on Risks, Regulation and the Yuck Factor. *Agric. Water Manag.* **2019,** *217,* 426–439.
56. Roy, P.; Pal, S. C.; Chakrabortty, R.; Chowdhuri, I.; Saha, A.; Shit, M. Climate Change and Groundwater Overdraft Impacts on Agricultural Drought in India: Vulnerability Assessment, Food Security Measures and Policy Recommendation. *Sci. Total Environ.* **2022,** *849,* 157850.
57. Sadiku, M. N.; Ashaolu, T. J.; Musa, S. M. Emerging Technologies in Agriculture. *Int. J. Sci. Adv.* **2020,** *1* (1), 31–34.
58. Schwindenhammer, S.; Gonglach, D. SDG Implementation Through Technology? Governing Food-Water-Technology Nexus Challenges in Urban Agriculture. *Polit. Governance* **2021,** *9* (1), 176–186.
59. Shahzaman, M.; Zhu, W.; Bilal, M.; Habtemicheal, B. A.; Mustafa, F.; Arshad, M.; Iqbal, R. Remote Sensing Indices for Spatial Monitoring of Agricultural Drought in South Asian Countries. *Remote Sens.* **2021,** *13* (11), 2059.
60. Shin, S.; Aziz, D.; El-sayed, M. E.; Hazman, M.; Almas, L.; McFarland, M.; Burian, S. J. Systems Thinking for Planning Sustainable Desert Agriculture Systems with Saline Groundwater Irrigation: A Review. *Water* **2022,** *14* (20), 3343.
61. Shortle, J.; Ollikainen, M.; Iho, A. *Water Quality and Agriculture*; Springer International Publishing: Cham, 2021.
62. Singh, D. K.; Sobti, R. Long-Range Real-Time Monitoring Strategy for Precision Irrigation in Urban and Rural Farming in Society 5.0. *Comput. Ind. Eng.* **2022,** *167,* 107997.
63. Singh, R. K.; Berkvens, R.; Weyn, M. AgriFusion: An Architecture for IoT and Emerging Technologies Based on a Precision Agriculture Survey. *IEEE Access* **2021b,** *9,* 136253–136283.
64. Singh, S.; Singh, N.; Pandey, Y. Water Productivity of Vegetables Crop Under Temperate Condition of kashmir Valley. *Int. J. Curr. Microbiol. App. Sci* **2020,** *9* (7), 3152–3168.
65. Singh, Y.; Pandey, S.; Goswami, A. K. Ensuring Water Availability in Future Through Revival of Indian Traditional Water Culture. In *Environmental Management-Pollution, Habitat, Ecology, and Sustainability* [Online]; IntechOpen, 2021a.
66. Sonone, S. S.; Jadhav, S.; Sankhla, M. S.; Kumar, R. Water Contamination by Heavy Metals and Their Toxic Effect on Aquaculture and Human Health Through Food Chain. *Lett. Appl. Nano Bio Sci.* **2020,** *10* (2), 2148–2166.
67. Srivastav, A. L.; Dhyani, R.; Ranjan, M.; Madhav, S.; Sillanpää, M. Climate-Resilient Strategies for Sustainable Management of Water Resources and Agriculture. *Environ. Sci. Pollut. Res.* **2021,** *28* (31), 41576–41595.
68. Varma, D. S.; Nandanan, K.; PC, V. R.; Soundharajan, B.; Pérez, M. L.; Sidharth, K. A.; Ramesh, M. V. Participatory Design Approach to Address Water Crisis in the Village of Karkatta, Jharkhand - India. *Technol. Forecast. Soc. Change* **2021,** *172,* 121002.
69. Villa-Cox, G.; Cavazza, F.; Jordan, C.; Arias-Hidalgo, M.; Herrera, P.; Espinel, R.; Speelman, S. Understanding Constraints on Private Irrigation Adoption Decisions Under Uncertainty in Data Constrained Settings: A Novel Empirical Approach Tested on Ecuadorian Cocoa Cultivations. *Agric. Econ.* **2021,** *52* (6), 985–999.

70. Zhuang, J.; Gill, T.; Löffler, F. E.; Jin, M.; Sayler, G. S. Can Food–Energy–Water Nexus Research Keep Pace with Agricultural Innovation? Engineering [Online], 2022. https://doi.org/10.1016/j.eng.2022.08.014; (accessed 30 April 2023)
71. Zobeidi, T.; Yaghoubi, J.; Yazdanpanah, M. Developing a Paradigm Model for the Analysis of Farmers' Adaptation to Water Scarcity. *Environ. Dev. Sustainability* **2022,** *24* (4), 5400–5425.

PART II

UPTAKE BY PLANTS: HEAVY METALS AND EMERGING CONTAMINANTS

CHAPTER 6

UPTAKE OF HEAVY METALS BY PLANTS IRRIGATED WITH TREATED WASTEWATER: A REVIEW

MANJEETA PRIYADARSHI and ATHAR HUSSAIN

ABSTRACT

In the past few years, the tremendous upsurge in population and rapid industrialization resulted in water pollution, thus resulting in high stress on natural water bodies. This ultimately resulted in increased demand for water at an alarming rate. Wastewater reuse for irrigation across the world is a very common practice for enormous amounts of food production thus eradicating water scarcity from the system. Therefore, using such type of water for irrigation pollutes the soil and contaminants are transferred to the edible parts of the crops. The crops grown on such soil uptake these contaminants that ultimately get incorporated into the food chain. Such toxic contaminants in irrigation water are heavy metals that accumulate in soil, and, thereafter, in flora. Thereby, long-use of wastewater for irrigation proves to have a potentially harmful impact on human beings through the transfer of contaminants in the food chain. The essential element required in human intake is vegetables due to the enrichment of vitamins, fibers, and minerals, which are essential for a healthy life. Heavy metal contamination is gaining attention around the globe and is regarded as a significant ecological issue as they are toxicants that affect human organs through accumulation in their bodies via the food chain. As per the plant cell formula, some of the heavy metals are also vital to all plants as stimulants for their reproduction and growth. However, at high concentrations, they possess typically damaging effects. Therefore, the intake of such vegetables that are grown

Sustainable Use of Wastewater in Agriculture: Applications, Methods, and Technology. Megh R. Goyal & Gaurav Saini (Eds.)

in contaminated water may contain heavy metal concentrations that result in human health problems. The heavy metals causing contamination include Cd, Cr, Hg, Zn, Cu, Fe, Mn, As, Ni, and Co, which interrupt biochemical cycles that can threaten the persistence of human, plant, and animal lives.

6.1 INTRODUCTION

The tremendous increase in population and industrial sectors in the last few decades has resulted in the exploitation of natural resources, thus aggravating the environment in terms of pollution and human health aspects. Water is another such a vital resource that has been exploited majorly by human beings as well as by industries. Lock manufacturing facilities and steel, paper, textile, chemical, and petroleum refining companies are the biggest water users in these sectors, using about 80% of the water available. This ultimately results in a large quantum of untreated or partially treated wastewater consisting of many toxic pollutants that are ultimately released in surface water bodies. This ultimately either joins the groundwater table or is consumed by people directly or indirectly.

In India, industrial units are supposed to spill tremendous amounts of heavy metals into the sewage as effluents. Agriculture nowadays is facing a new threat from industrial effluents containing harmful heavy metals contaminating the soil in farmed fields. Most wastes and effluents include heavy metals in concentrations that are harmful to crop plants. When such crops are ingested, the accumulated heavy metals enter the food chain and ultimately result in various severe health issues for human beings and animals.

It is pertinent to mention that due to the swift population increase in urban areas, a high demand for public water supply on a daily basis resulted in the generation of high volume of wastewater. Due to the increase in demand for potable water supply, most of the cities are facing challenges in freshwater supply. As a result, farmers in several developing nations, including India, started using industrial effluents for crop irrigation. However, municipal sewage treatment plant effluents can contain significant concentrations of macro and micronutrients in addition to heavy metals. Heavy metal soil contamination in agronomic fields irrigated by effluents from industries become a new danger. The degradation of the soil is brought on by the permeation of the contaminants through the soil. The spreading of pollutants on the soil's surface is the leading cause of soil contamination. Surface pollution originates from a variety of causes, including spills, current farming methods, and the percolation of surface contaminants via unsaturated soil.[13]

Any living thing needs food and water to survive, both of which are sources of numerous nutrients and metals. They are also the significant ways that we get exposed to different types of hazardous metals. Since they contain proteins, carbs, vitamins, minerals, and trace elements, vegetables play a substantial part in the human diet. Vegetable crops watered with harmful heavy metals contaminated wastewater may provide a significant dietary intake pathway for these contaminants. The lives of inhabitants who eat the fruits and/or vegetables cultivated in contaminated areas are in danger due to the continuous heavy metals accumulation in agronomic soils when irrigated with wastewater.

This chapter reviews the uptake of heavy metals by plants that are irrigated with wastewater.

6.1.1 WATER AND WASTEWATER POLLUTION

The amount of freshwater used by the commercial, home, and industrial sectors has increased due to the continuing growth of the urbanized population and the rising water consumption, which also causes an increase in the amount of wastewater produced. Due to the lack of an alternate source, farmers in several developing nations, such as India, irrigate the crops utilizing industrial effluents. In addition to providing water to plants, municipal sewage treatment plant effluent frequently contains very high concentrations of macro- and micronutrients as well as heavy metals. Copper, zinc, manganese, iron, and boron are examples of trace elements that are required for plant reproduction and growth. These substances may be hazardous to plants in high quantities; hence they are also regarded as heavy metals.[14]

Any water for domestic usage in any business or for any industrial purpose—a large part of it—ultimately will result in wastewater. Water has been used once for any purpose, or water consumed for household purposes such as cooking, drinking, cleaning, or any usage results in sewage. Also, the water that has been used for commercial, industrial, or irrigation, stormwater with surface runoff, or any other means, such as any inflow into the sewerage system or infiltration all contribute to wastewater. Urbanization and industry are the two leading causes of water contamination. Any runoff from urban areas, such as small towns or big cities, and from industries carries the heavy metals that accumulate in the sediments and deposits of such water bodies. Despite traces of such pollutants being carried to bodies of water, they may still be extremely harmful to people and other organisms. The effect of these heavy metals generally depends upon various factors that include the type of metal; chemical properties and their role in biodegradability of that metal;

and the nature of the metal to react with its surrounding environment. It also depends on the time of exposure to the organism, along with the immunity of the organism to resist the toxicity of the metal it is exposed to.[6]

6.1.2 SOIL POLLUTION

Any biological or chemical alteration of soil that is damaging to living things is referred to as soil pollution. Most of the pollution in the soil is a straight outcome of contamination sources on the surface of the ground. Soil degradation results from pollutants on the surface seeping through the soil. Most soil pollution is brought on by surface pollutants. There are lots of sources of surface pollution, including agricultural activities, spills, the disposal of trash (solid or liquid), and the filtration contaminants on the surface through unsaturated soil strata.[26]

6.1.3 SOURCES OF CONTAMINATION

The creation of sludge, energy consumption, chemical use, and atmospheric emissions at wastewater treatment plants, which are a significant source of concern for ecological treatment systems, all have an impact on the environment (Figure 6.1). The most significant by-product of wastewater treatment facilities is sludge, and the disposal of this substance is one of the utmost complex environmental issues in the industry. Water and wastewater treatment processes produce sludge that can be utilized as a by-product. Wastewater biosolids are another name for the sludge produced by biological treatment processes.

When agricultural land is directly irrigated with wastewater emerging from municipal areas or from sources that are polluted, as well as by the continuous use of fertilizers and agrochemicals that contain heavy metals, it may become heavy metals polluted even if it is not originally harmed. It has been demonstrated that vegetables and other crops grown in such soils are polluted with heavy metals and are accumulated with more heavy metals than those grown in unpolluted soil.[3]

6.1.4 SOIL IRRIGATION

Heavy metals can build up in soil for years without portraying any overt symptoms of their perilous harmful effect. In contrast, soils have a limited ability to filter and cleanse substances. Above a certain point, soils lose their capacity to do so and start to emit pollution. Toxic substances are released into the environment, ingested by cultivated plants and crops, integrated into vegetative and reproductive organs, and then enter the food chain, where they have long-term negative effects. Targeted field experiments can precisely and reliably study the relationship between heavy metals, soils, and plants.[10]

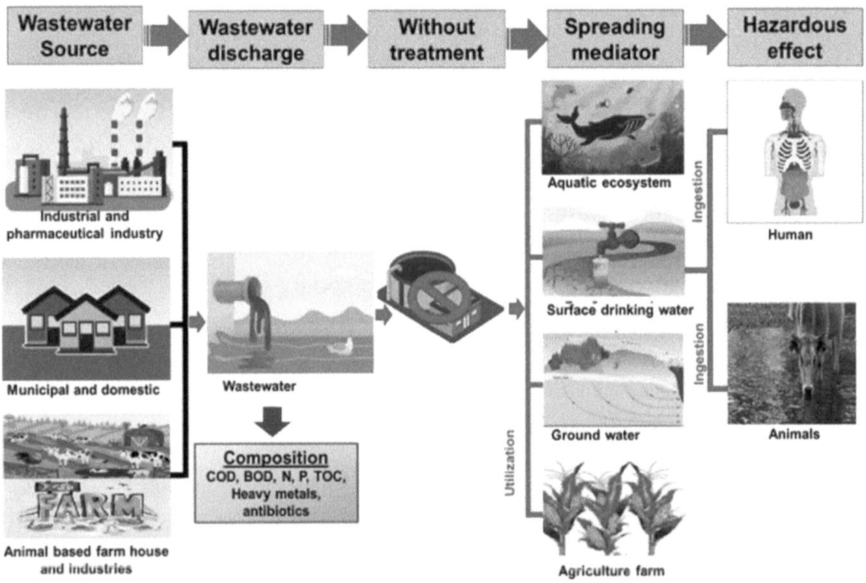

FIGURE 6.1 Sources of contamination of the environment.

Pollution due to heavy metal affects large tracts of agricultural land and is primarily caused by industrial emissions, sewage sludge application, or past or present mining activity. The soil mainly consists of metals in four different forms, namely, sorbed form (bonded to soil), organic form, mineral, and dissolved form. The third-largest pool of metals is sorbed metals, which are typically very closely linked to soil surfaces. Plants can gradually release metals into solution even though they do not immediately absorb mineral, organic, and sorbed metals. Metals are described and categorized by different terminologies, such as trace metals, hazardous metals, transition metals, heavy metals, and micronutrients. Many of these descriptions are subjective, and these terms are often employed in literature in a broader sense to refer to things that do not quite fit the definitions given.

6.1.5 HEAVY METALS

Due to rapidly expanding industrialization and agricultural operations, heavy metal contamination of soil has resulted in a universal environmental problem. Although heavy metals occur naturally in the soil, many man-made activities have raised their content, thus threatening the environment. Unnecessary use of agrochemicals, sludge produced from sewage, industrial wastewater, biological solids, and compost solutions are some heavy metal sources in soil..[5]

Any element with a silvery shine and good conductivity for heat and electricity is considered a metal. Metals can be classified by their known essentiality to living organisms as essential nonliving elements. The Co, Cr, Mo, Cu, Fe, Se, Mn, Zn, and Ni are needed by organisms on some level, but at more significant exposure levels, they become poisonous. Pb, Hg, Sb, Cd, Sn, Tl, and Ag are poisonous nonessential metals that are never needed by an organism at any level. A typical dose–response curve for trace minerals, both essential and not.

6.1.5.1 HEAVY METALS POLLUTION

Heavy metals are harmful or lethal at lower concentrations and have a comparatively high density. The persistence, ubiquity, and toxicity of heavy metals have stoked widespread concern throughout the world. Through percolation, heavy metal pollution in soil reduces soil quality overall, yield, and water quality. An important problem that might cause long-term environmental issues is pollution due to heavy metals in agro-based soils. When these metals are transferred from soil to crops, it creates an atmosphere of the chief source of such metals' acquaintance to humans.[11] Heavy metals that enter rivers from a wide range of sources are adsorbed onto suspended particles, where they combine to form soluble chemical complexes and free metal ions that aquatic species can take up. Rivers carry varying levels of several hazardous metals, cadmium, arsenic, mercury, copper, lead, nickel, chromium, and zinc that are present in untreated or purportedly treated industrial effluents.

6.1.5.2 HEAVY METAL SOURCES

The earth's crust contains heavy metals naturally, and they are released into the soil through numerous anthropological activities, raising the number of metalloids in the soil to dangerous levels. Pb, Cr, As, Zn, Cd, Cu, and Hg are the most excellent frequent heavy metals to be found in contaminated soils, in that order. However, many natural processes in the environment also contribute to heavy metals pollution in soils.[2]

- **Natural sources:** Also, weathering of the underlying bedrock, heavy metals are naturally present in soils. These are usually derived from ore via various mineral processing techniques. Heavy metals are found in rocks in a variety of chemical forms called ores, and these metals are then extracted as minerals. Heavy metals generally occur in ore forms as oxides of Mn, Se, Al, and Sb, as well as sulfides of all heavy metals. Therefore, these metals are commonly recovered or mined as oxide along with ores of sulfide from soils. Typically, metal sulfides (such as As, Hg, Pb, and Cd) coexist in natural form in soils with copper sulfides.[17]
- **Anthropogenic sources:** The reprocessing and mining of ores and the industries manufacturing pesticides, batteries, paper and tannery, and fertilizer manufacturing industries. This also includes municipal solid waste disposal and the sludge produced during the sewage treatment process, including wastewater used for irrigation and vehicle emissions, which are all examples of man-made sources of pollution contaminating the soil by heavy metals. Generally, the metalloids come into the environment through various industrial processing and activities of mining. Inorganic and organic compounds with elemental species of heavy metals are released. In other cases, long after these activities have stopped, metals produced during these processes keep building up in the soil and other parts of the ecosystem. According to Peplow (1999), it typically takes 5–15 years to completely deplete minerals from hard rocks after mining, while metal contamination of the soil lasts for several years after the mining operation is through.[17]

6.1.5.3 METAL SOURCES IN WATER AND SOIL

The toxicity of heavy metal poisoning to living things through bioaccumulation has drawn attention from all around the world.[9] Heavy metals are mainly classified as metalloids or metals based on their density. The various examples pertaining to natural activities of heavy metals contamination are volcanoes, erosion of soil, and disintegration of natural rocks (Figure 6.2;

Table 6.1). However, the various examples related to anthropogenic sources resulting from fossil fuels are incomplete burning, extraction of minerals, decomposition in landfills, refining of metals, the manufacture of electronic goods, manufacturing and usage of dyes, chemicals manufactured for agricultural usage, military operations, and vehicle releases. HMs that are toxic to humans, such as Cr, Cu, Hg, Pb, As, and Cd, frequently exist in municipal sewage, alloys, dyes, and wastewater from mines.[28]

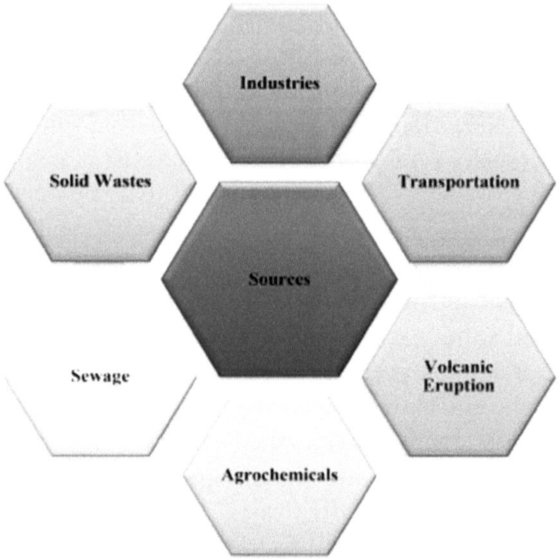

FIGURE 6.2 Different sources of accumulation in water and soil.

TABLE 6.1 Standard Limits of Heavy Metals in Soil, Food, Water, and Standard Soil.[29]

Heavy metal	Soil (µg/Kg)	Food (mg/Kg)	Water (mg/L)	EU Standard Soil (µg/Kg)
Cd	3–6	1.5	0.01	3
Cr	-	20	0.05	150
Cu	135–270	30	0.05	140
Fe	-	-	0.03	-
Ni	75–150	1.5	-	75
Pb	250–500	2.5	0.1	300
Zn	300–600	50	5	300
As	-	1.1	0.05	-
Mn	-	-	0.1	-

6.1.5.4 VEGETABLES AND HEAVY METALS

Vegetables absorb heavy metals and build them up in both edible and nonedible sections in higher concentrations are enough to harm both animals and people clinically. Vegetables operate as a buffering agent for the acid produced during digestion, and some of the metals they contain are even crucial from a biochemical and psychological standpoint for wellness. Some elements, such as Cd, Pb, and As, are highly poisonous to humans. These include metals such as iron (Fe), cobalt (Co), copper (Cu), selenium (Se), and manganese (Mn). Due to their ability to impair biochemical systems such as metabolism, photosynthesis, and stomata opening, heavy metals may also negatively impact the growth of plants.

Being a rich source of vitamins, minerals, and fiber, vegetables are regarded as a crucial part of the human diet. The consumption of plants tainted with too many heavy metals, however, could be harmful to human health. The bioavailability of trace elements from vegetables is essential for the human body as the continuous supply of the elements and metals is obligatory for a vigorous and day-to-day life.[25]

Due to the hazardous effects and build-up in the food supply chain and eventually in the humanoid body, heavy metal contamination provides a severe environmental issue. Numerous soil parameters, including soil clay content, pH, EC, and organic matter content, as well as their physical and mechanical qualities, all affect how well ions may be absorbed and accumulated by plants. Through a variety of chemical processes, including absorption, ionic exchange, redox reactions, and others, plant extracts heavy metals from soil.

6.1.6 HEAVY METALS IN FOOD CHAIN

The development of efficient remediation systems requires a deeper understanding of the mechanisms governing the transport of nutrients from soil to food crops. To achieve this goal, this paper first discusses various heavy metal contamination sources in soil systems, their significance to dangers to human health, the fundamental mechanisms driving such pollution, and the ensuing dietary consumption of contaminated crops. To clarify the physiological/molecular mechanisms involved in the uptake of metallic pollutants inside food crops (Figure 6.3), the ramifications of those subsystems for the environment and human health are examined in detail.

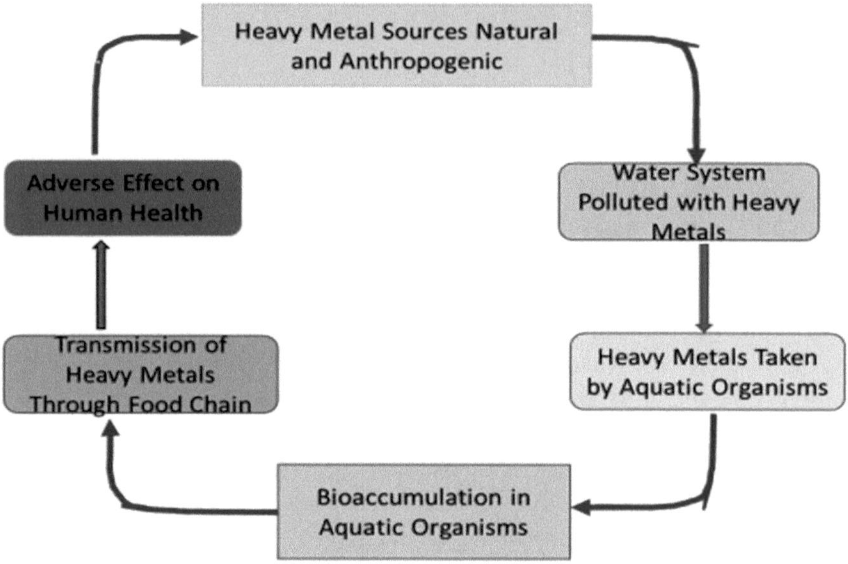

FIGURE 6.3 Heavy metals transmission through the food chain.

Source: Adapted from Ref. [31]

6.1.7 HEAVY METALS SOURCES IN CROP-SOIL SYSTEMS

The primary source of many heavy metals is agriculture and the soil in the environment through air deposition, manure, and livestock irrigation with sewage or contaminated water. Also, herbicides or metallo-pesticides, sewage sludge-based contaminants, and phosphate-based fertilizers contribute to heavy metal pollution.

6.1.8 HEAVY METALS CONTAMINATION IN SOIL–FOOD CROP SYSTEMS

The global picture for heavy metal pollution of crops in relation to their numerous anthropogenic sources, including the impacts of ecotoxicology on human health (Figure 6.4). Major global health problems relate to the bioaccumulation of these metals in food crops along with the impact on the health of individuals. Although the source of metallic pollutants has received limited attention, knowledge of geographic trends may help us comprehend how their impact on issues with human health can differ from nation to country.

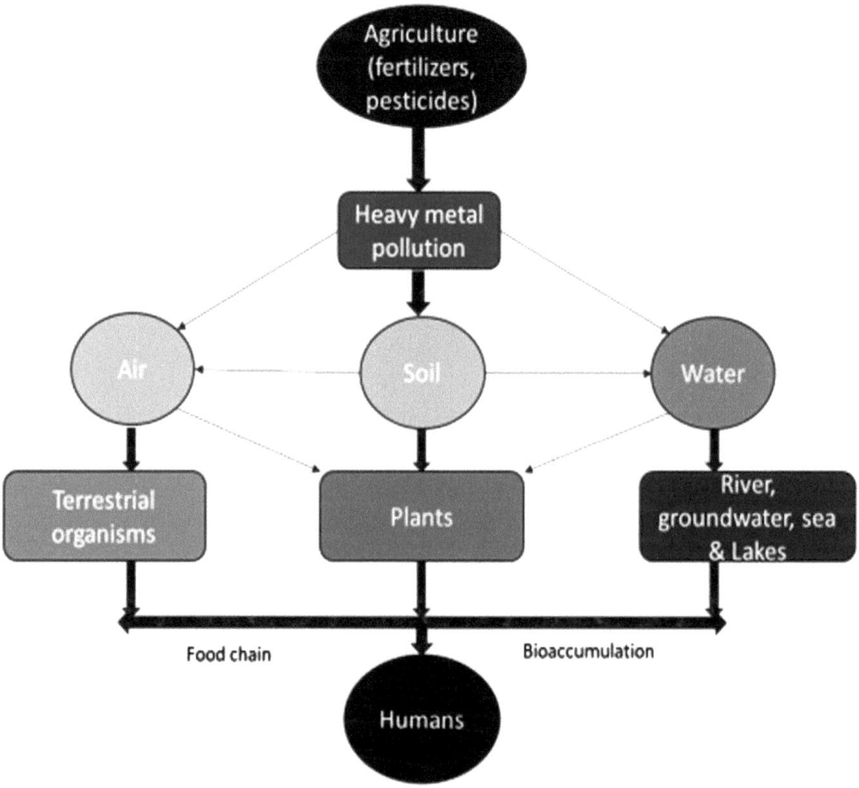

FIGURE 6.4 Schematics representation depicting the source of heavy metal pollution (soil–crop).

6.2 TYPES OF HEAVY METALS & THEIR SOURCES

According to a review of the literature, all heavy metals, including cadmium, iron, arsenic, chromium, lead, cobalt, mercury, copper, and zinc, are those that are regarded as heavy metals. Studies by Kim have divided heavy metals into two categories: essential and nonessential (Table 6.2). Essential At low quantities, heavy metals are little damaging and function as coenzymes in biodegradation processes. For instance, iron is present in myoglobin and hemoglobin, and cobalt is present in vitamin B12. Heavy metals that are not necessary for life are highly hazardous, although at extremely small concentrations. They are also not biodegradable and have harmful impacts on all living things.

TABLE 6.2 Heavy Metal Sources and Health Effects.[24]

S. No.	Heavy metals type	Source	Health Effects
1.	Zinc (Zn)	• Oil Refining • Plumbing • Brass manufacturing	Gastrointestinal disorders, kidney & Liver abnormal functioning
2.	Copper (Cu)	• Copper polishing • Plating • Printing	Abdominal disorders, Metabolic activity abnormalities
3.	Iron (Fe)	• High intake of iron supplements & oral consumption	Vomiting Diarrhea Abdominal pain Dehydration & lethargy
4.	Cobalt (Co)	• Hip alloy replacement case	Haemotological Cardiovascular Hepatic Endocrine
Non-Essential Heavy Metals			
1.	Chromium (Cr)	• Steel fabrication Electroplating • Textile	Lung disorders (bronchitis, cancer), Renal and reproductive system
2.	Lead (Pb)	• Batteries • Coal combustion • Paint industry	Serious effect on mental health (Alzheimers disesase), Nervous system
3	Arsenic (As)	• Atmospheric deposition Mining pesticides	Highly effects dermal region (Cancer), Brain & Cardiac problems.
4.	Mercury (Hg)	• Coal combustion Fish Mining Paint Industry Paper Industry Volcanic eruption	Sclerosis Blindness Minamata disease Deafness Gastric problems Renal disorders
5.	Cadmium (Cd)	• Plastic Fertilizers pesticides	• Osteo related problems Prostate cancer • Lung diseases • Renal issues

6.3 HEAVY METALS TOXICITY

Although several metals are necessary for certain biological processes in humans, the dosage at which they are consumed can have unexpectedly dangerous health effects, including their physiological system. Investigations indicate that heavy metals are acting as carcinogenic agents despite having positive health impacts.

Dissolved forms of these metals are causing severe harm to the cellular structure and are increasing the danger of cancer by joining the food chain as air, water, and soil contaminants, then ending up in humans. As per the investigations by the International Agency for Cancer Research, nonessential heavy metals (As, Cd, and Cr) are major cancer-causing agents.[18] Various heavy metals have different levels of toxicity. Insoluble metals travel through the body with little to no harm. The most harmful metals are those that immediately make you sick and do not give you time to seek medical attention. The worst metals, such as mercury, are those that bioagglomerate and bioamplify in the food chain. Bioaccumulation, the retention of some poisons in the tissues of creatures that can neutralize others, is a property of these organisms. As the food chain progresses through the trophic levels, the biomagnified concentration of hazardous metals gradually rises. By making food unfit for eating, such biomagnification may result in disease.

6.4 BIOACCUMULATION OF HEAVY METALS THROUGH THE FOOD CHAIN

The main route through which living things are exposed to heavy metals is through the food chain. Most often, irrigation with contaminated water from industries and sewage treatment plants causes metal pollution of agricultural soil. These metals accumulate in agricultural soils and the parts of crops after wastewater is applied over an extended period of time, which can restrict the soil functionality, decrease the nutritive value of vegetables, be harmful to all types of crops, and contaminate the food supply.[7]

The abundance of helpful and necessary minerals and nutrients found in vegetables makes them a crucial component of the human diet. However, both edible and inedible components of vegetables can absorb and accumulate heavy metals above permissible limits.[27] Once these dangerous metals enter the food chain, it is difficult to get rid of them at the top, which causes them to spread across the entire food web. Both humans and animals eat a lot of hyper-accumulated plants. Therefore, the cycle of contamination from soil to human beings through a plant and back to soil there afterward to the top consumers pass away provides a way for heavy metals to remain supported in the atmosphere for long durations, causing various negative impacts.

This cycle of the food chain pertaining to heavy metals starts from various types of sources. For example, the heavy metal Cd enters the plant body from the soil itself through roots (Table 6.3). However, Pb enters the plants' body as it is absorbed from the air dust into them. The presence of arsenic (As) and mercury (Hg) in the food chain is through sewage water. Also, heavy metals have a tendency to accumulate in the body tissues of top consumers, including muscles, brain, kidney, liver, intestines, and other parts during their transfer from one section to another.[22]

TABLE 6.3 Toxic Heavy Metals Negative Effects on Both Plants and People.

Heavy metals	Effects on plants	Effects on humans	References
As	Seed germination and seedling height reduction in rice crop	Cardiovascular, neurologic, and developmental anomalies	[1]
Cd	Seed germination reduction, decrease in nutrient use efficiency, and reduced shoot and root length in wheat crop	Bone demineralization, renal dysfunction	[20]
Co	Antioxidant enzyme activities reduction; decrease in plant sugar, starch, amino acids, and protein content in mung bean	Severe health effects on the lungs, including asthma and pneumonia	[15]
Cr	Inhibition of germination process; plant biomass reduction in garlic	Multiorgan toxicity, nose ulcers, and cancer of the respiratory tract	[23]
Pb	Reduction in germination percentage; suppressed growth; reduced plant biomass; decrease in plant protein content in maize crop	Cause kidney failure, affect CNS and PNS, and damage the reproductive systems	[12]

6.5 TOXICITY OF HEAVY METALS IN PLANTS AND HUMANS

- **Mercury (Hg):** All trace elements can be harmful if consumed for a large enough duration or in elevated concentrations. When absorbed by roots, Hg travels from the soil to plants. Through the stomata (leaf pores), plants can also absorb mercury vapor from the surrounding air. Mercury enters the tissues of animals because it is soluble in fatty acids, which are the building blocks of all cell membranes inside a tissue. Mercury biomagnifies and bioaccumulates as it enters people via the food chain, primarily through fish intake. It can be found in fish in the form of methylmercury.
- **Cadmium (Cd):** Plants that require zinc intake cadmium together with zinc because cadmium occurs with Cadmium and therefore enters the food chain. Rice, potatoes, wheat, and other grains all contain cadmium.

Also high in cadmium are seafood products. Both chewing tobacco and smoking tobacco expose humans to cadmium. Cadmium is absorbed by tobacco leaves through irrigation water. People who live close to or work for mines and smelters that produce zinc and electroplate have a high risk of developing cadmium poisoning. As cadmium is water soluble, it can get to the miners of zinc.

- **Lead (Pb):** The impacts of lead on the human body are harmful. The body can be exposed to it in the following ways. Cans of food and beverages can introduce lead into the human body. When exposed to acidic medium or high temperatures, lead leaches from glazed ceramics. Lead is also present in *surma* used in the eyes. Lead-containing compounds from automotive exhaust enter people through skin contact and breathing. Children in crowded areas have elevated blood levels of lead. Even the breast milk of women who reside on sidewalks contains significant levels of lead. Lead oxides collect in the soil, water, and green crops, where they are easily ingested by humans and enter the food chain.
- **Copper (Cu):** Copper, despite being potentially harmful, is a vital metal for the growth and reproduction of plants. It is also considered an important micronutrient for the development of plants. It is also an important nutrient that plays a crucial role in ATP production and uptake of CO_2. However excessive enrichment of copper in the soil thus results in cytotoxic effects and stress in plants, and their growth is hampered. The presence of leaf chlorosis indicates its effect on plant growth. Oxidative stress and ROS are produced when plants are exposed to too much copper. The disruption of metabolic pathways and damage to macromolecules are results of oxidative stress. Rhodes grass's root growth is slowed by copper (*Chloris gayana*). The black bindweed (*Polygonum convolvulus*) is another indicator of copper toxicity that reduces plant death, amount of biomass, and reduction in seed output. The accumulation of Cu in *Phaseolus vulgaris* (bean) plants, particularly in the roots of the plant, thus results in malformation of roots and its retardation.[4]
- **Zinc (Zn)** contributes to chlorophyll production in plants. When soil is deficient in zinc, the growth of the plants is inhibited, and leaves become discolored. Zinc deficiency results in chlorosis, a disease that results in leaf discoloration in the vein tissue turning yellow. The leaf base near the stem is frequently affected by chlorosis brought on by a zinc shortage. Chlorosis first manifests on the lower leaves before progressively spreading up the plant. In extreme cases, the lower

leaves get darkened or become purple in color and die off, while the higher leaves turn chlorotic.

- **Chromium (Cr):** The hazardous element chromium is well known to harm both plants and animals. Lipid peroxidation, which severely harms cell membranes in plants, is generated by chromium-induced oxidative stress. Chromium-induced oxidative stress starts to degrade photosynthetic pigments, which slows growth. High chromium concentrations can disrupt the ultrastructure of the chloroplast, which in turn disrupts the photosynthetic activity. The germination of seed to in a medium containing Cr will indicate its tolerance level to this metal. The germination of seed is an initial physiological function that is generally influenced by Cr.
- **Cobalt (Co):** A transition metal called cobalt is a necessary part of several enzymes and coenzymes. The presence of cobalt and its concentration in the soil and rhizosphere has varying degrees of effects on the development of plants and their metabolism. It also combines with other substances, which ultimately results in various complexes. However, the plants generally absorb small amounts of Co from the soil. Also, the data pertaining to the phytotoxic effects of Co is very scarcely available. Some investigations pertaining to the phytotoxicity of cobalt on oilseed rape tomato and barley showed inverse aftereffects on the biomass and shoot growth of such plants.
- **Nickel (Ni):** For plants, nickel is a vital nutrient. The quantity of Ni needed, however, is incredibly small for normal plant growth. It is crucial to comprehend the efficient functions and hazardous consequences of Ni in plants because of the rising amount of Ni pollution in the environment. Numerous physiological changes, including chlorosis and necrosis, are brought on by excessive Ni^{2+} levels in the soil, which affects rice, among other plant species. Growing plants on soil with elevated Ni^{2+} levels resulted in disturbed cell membrane functions and nutrient imbalance.
- **Arsenic (As):** Arsenic affects the weight of the fresh leaves and the fruit output of the tomato (*Lycopersicon esculentum*). Arsenic, on the other hand, causes stunted development, chlorosis, and wilting in *Brassica napus* (canola). Additionally, arsenic in Oryza sativa (rice) decreases the germination of seeds, the height of the seedlings, the area of leaves, and the disposition of dry matter.
- **Iron (Fe):** The primary function of iron in plants is during photosynthesis. Iron is easily available in soil under low pH, and the accessibility of the micronutrient to the roots of plants relies on the soil pH.

Both iron and manganese are vital for the development and reproduction of plants, but they frequently compete for nutrient absorption because an excess of one completes the other little accessible to the roots of the plant. It is also a very important element of the cellular redox systems, which include iron, sulfur proteins such as ferredoxin, aconitase, and superoxide dismutase (SOD), as well as proteins that include cytochromes and leghemoglobin. Iron plays several significant organic roles in activities as miscellaneous as chloroplast growth, photosynthesis, and chlorophyll formation. Iron is a vital element for all types of plants.

6.5.1 HAZARD QUOTIENT (HQ)

Metal contamination of vegetables poses a high hazard to the health of human beings. Hazard quotient (HQ) is the measurement of the probable harm to human health from eating vegetables contaminated with metals (U.S. Environmental Protection Agency US EPA 1989). HQ represents the contact to the reference oral dosage ratio (RFD). No visible risk will exist if the ratio is less than one (1). An estimation of the probable risk of such heavy metals to human health (*HQ*) via ingesting vegetables can be assessed using the following equation:[16]

$$HQ = (D_{iv}) \times C_{metal} / R_f D \times B_o \qquad (6.1)$$

where (D_{iv}) = daily vegetable intake (kg/day);

C_{metal} = heavy metal concentration in vegetable (mg/kg);

$R_f D$ = reference dose of metal (mg/kg) of a body weight per day; and

B_o = weight of the human body (kg).

6.5.2 TRANSFER FACTOR

The heavy metals concentration in agricultural soil extract, as well as other plants is assessed on a dry weight basis. The transfer factor (TF) of a plant can be computed as

$$\text{TF} = C_{plant} \times C_{soil} \qquad (6.2)$$

where $C_{plant} \times C_{soil}$ represents the toxic amount of metal in plants being extracted and in soils on a dry weight basis, respectively.

6.5.3 DAILY INTAKE OF METAL (DIM)

The existence of some pathogenic microorganisms in the wastewater was discovered by a microbiological analysis; as a result, the farmer may be advised to exercise caution when doing irrigation operations. One option is to use the root as an accumulator of heavy metal. However, under such conditions it is suggested that the roots of such plants should be discarded while the plant leaves are important as it has a high nutrition content. Also, it contains lower concentrations of heavy metals and therefore is suitable for human consumption. The daily intake of metal (DIM) can be assessed by using the following equation.[25]

$$DIM = C_{\text{metal}} \times C_{\text{factor}} \times D_{\text{food intake}} / B_{\text{average weight}} \tag{6.3}$$

where C_{metal}, C_{factor}, $D_{\text{food intake}}$, and $B_{\text{average weight}}$ represent the concentrations of heavy metals in plants (mg/kg), its factor of conversion, vegetables daily intake, weight of mass averaged out, respectively.

6.6 ENVIRONMENTALLY FRIENDLY TECHNIQUES FOR HEAVY METAL TOXICITY PREVENTION

Preventive measures must be put into operation to protect human life and other creatures from disasters brought on by the poisonous effects of heavy metals. A distinct Ministry of Environment and Forests, which deals with environmental issues, was established by the Indian government in January 1985. Numerous nongovernmental organizations are also hard at work informing people about the negative consequences of metal and other pollutants and how to stop them. Businesses have been urged to take action to dispose of industrial waste properly. For the removal of effluents, special devices have been created.

The government has made the decision to penalize businesses that fail to install the necessary equipment and discharge waste in rivers or on the ground. Leaded gasoline usage is being phased out over time through ongoing efforts. Now, motor vehicles are being run using lead-free petrol; such vehicles are designed with catalytic converters that use lead-free petrol.

6.6.1 HEAVY METAL REMOVAL TECHNIQUES

Removal of heavy metals is difficult as they are present in a very low concentration, therefore two types of designing systems are available for the removal of such metal pollution:

- Single metal removal system
- Various metals removal in a single system

However, when rivers and their sediments are polluted with heavy metals, the following processes can be applied:

- The contaminated sediment is placed under the uncontaminated clean soil layers it may since the metal that contains sediment may not be washed away by river water bodies.
- The pH of the sediment can be increased using $CaCO_3$ thus immobilizing the heavy metals.
- Detoxifiers in the form of limestone, iron sulfate, activated charcoal, and gypsum can be used.
- The *pistia* and *hydrilla* can be used as such plants can uptake mercury from water bodies thus reducing mercury concentration from river water.

6.6.2 REMEDIATION TECHNIQUES FOR HEAVY METALS POLLUTED SOIL

The thermal desorption process is one of the physical techniques for reclaiming soil. Therefore, the concentration of such pollutants can be lowered, replacement of soil can be used to replace polluted soil with soil that is free of contaminants.

6.6.2.1 SOLIDIFICATION AND STABILIZATION

Cleanup techniques such as stabilization and solidification prevent or delay the discharge of pollutants such as heavy metals. By binding, these techniques stop heavy metals from leaching. By combining polluted material with an agent that binds it, solidification results in the polluted material fixing to one another.[8]

6.6.2.2 VITRIFICATION

Using vitrification to clean up heavy metal and solid industrial waste-contaminated soil. Heavy metals' mobility is decreased by heating the soil contaminated with metal at an elevated temperature during vitrification. High temperatures in the process result in the volatilization of such organic molecules and some metals, including mercury, which should be recovered, while the other materials are transformed into an inert product that is a glass-like product.[8]

6.6.2.3 SOIL WASHING

Solvents and mechanical techniques are used in the remediation process known as "soil washing" to remove heavy metal contamination from soil. With this method, the amount of contaminated soil is reduced since the fine and coarse portions of the soil are separated.[19]

6.6.3 SUSTAINABLE BIOLOGICAL TECHNIQUE FOR POLLUTED SOIL REMEDIATION

The various sustainable biological remediation technique to recover the heavy metals from such contaminated soils includes phytoremediation and bioremediation. However, both above techniques can be combined.

6.6.3.1 BIOREMEDIATION

The elimination of toxic pollutants, specifically heavy metals, from less dangerous chemicals and organic materials such as water, nitrogen gas, and carbon dioxide, is known as bioremediation. Metals will be immobilized, and their bioavailability will be reduced through microbial reclamation of heavy metal-polluted soil. Microorganisms cannot break down heavy metals, but they can change them into another form by changing their physical and chemical characteristics. The heavy metal types and the microorganisms have an impact on the efficiency of microbial remediation. Another tool for treating soil with heavy metal contamination is microbial leaching.[19]

6.6.3.2 PHYTOREMEDIATION

It is an environmentally friendly technique that can be utilized for readily growing plants and is generally utilized to eliminate and uptake the contaminants from soil or water. Phytoremediation is a technique that helps in the eradication of heavy metal ions through plants' metabolic pathways. The alternate terms used for phytoremediation are botano-remediation, agro-remediation, or green remediation. It is considered an eco-friendly approach and a sustainable process for the reclamation of soil compared to conservative soil remediation approaches.[21]

6.6.3.3 PHYTOSTABILIZATION

A plant-based inactivation technique copes with metal-contaminated soil, and it is known as phytostabilization, phytoimmobilization, or phytorestoration. This method is used to decrease the movement and biological availability of heavy metals. The technique also restricts the heavy metals leaching consequently and restricts access to groundwater results in its elimination from the food chain, respectively.[17]

6.6.3.4 PHYTOTRANSFORMATION

Phytotransformation is just the same as phytodegradation, which is the organic compound breakdown either by biological developments of plants or by plants' enzymes and it does not rely on the microorganism. Plants are sometimes referred to as the biosphere's "green liver." The only benefit of phytodegradation is the hydrolysis of biological molecules. Phytotransformation, also known as phytodegradation, is where biological pollutants are break down and scavenged by plants via a metabolic process internally or the action of substances such as enzymes produced by the plant.[19]

6.6.3.5 PHYTOFILTRATION

Utilizing plant roots in the phytofiltration process, low-contaminant wastewater can be recovered from surface and groundwater sources. Plants are first given contaminated water to get used to it, and then they are moved to a contaminated site to be cleaned up. When roots become soaked, they are harvested.[30]

6.6.3.6 PHYTOEXTRACTION

Phytoextraction is a process employed in rapidly developing plants to eliminate the heavy metals from the affected water and such soils of the environment. Two methods used in phytoextraction are (i) Continuous or natural phytoextraction and (ii) Chemically induced phytoextraction.[30]

6.7 SUMMARY

The use of water is significant, and both humans and businesses have heavily relied on it. Any living thing needs food and water to survive, both of which are sources of numerous nutrients and metals. They are also the significant ways that we get exposed to different types of hazardous metals. The lives of locals who eat the fruits and/or vegetables grown in polluted areas are at threat because of the build-up of high concentrations of such metals in soils of agricultural usage from irrigation wastewater. As treated wastewater has more nutrients than tap water does, plants that are irrigated with it produce more fruit and vegetables per unit of water. The two main routes through which humans are exposed to food chain routes include soil pollution and soil via plant to humans, and soil–human; metal contamination in soil continues to exist (incidental soil ingestion).

Due to the higher nutritional content of reclaimed wastewater in comparison with tap water, plants, when irrigated using such water, produce a high yield. Among the most severe dangers to soil and human health are heavy metals. Different causes cause the discharge of these metals into the environment. Traditional clean-up methods are expensive and environmentally harmful. The best technology-based method for eliminating contaminants from polluted sites without harming the soil's structure is phytoremediation of metals. A phytoremediation technique's effectiveness is largely reliant on the plants' ongoing access to metals.

KEYWORDS

- **crops**
- **food chain**
- **heavy metals**
- **irrigation**
- **toxicity**
- **wastewater**

REFERENCES

1. Abbas, G.; Murtaza, B.; Bibi, I.; Shahid, M.; Niazi, N. K.; Khan, M. I.; Amjad, M.; Hussain, M.; Natasha. Arsenic Uptake, Toxicity, Detoxification, and Speciation in Plants: Physiological, Biochemical, and Molecular Aspects. *Int. J. Environ. Res. Public Health* **2018,** *15* (1). https://doi.org/10.3390/ijerph15010059.
2. Abou-Shleel, S. M.; Abdel-Kareem, M. S. Effect of Marine Algae on Bio-Accumulation of Heavy Metals from Polluted Soil by Some Leafy Vegetables. *Nat. Sci.* **2015,** *13* (3). http://www.sciencepub.net/nature
3. Alrawiq, N.; Khairiah, J.; Talib, M. L.; Ismail, B. S.; Anizan, I. Accumulation and Translocation of Heavy Metals in Paddy Plant Selected from Recycled and Non-Recycle Water Area of MADA Kedah, Malaysia. *Int. J. ChemTech Res.* **2014,** *6* (4), 2347–2356.
4. Asati, A.; Pichhode, M.; Nikhil, K. Effect of Heavy Metals on Plants. *Int. J. Appl. Innov. Eng. Manag.* **2016,** *5* (03), 56–66.
5. Ashraf, S.; Ali, Q.; Zahir, Z. A.; Ashraf, S.; Asghar, H. N. Phytoremediation: Environmentally Sustainable Way for Reclamation of Heavy Metal Polluted Soils. *Ecotoxicol. Environ. Saf.* **2019,** *174* (February), 714–727. https://doi.org/10.1016/j.ecoenv.2019.02.068
6. Briffa, J.; Sinagra, E.; Blundell, R. Heavy Metal Pollution in the Environment and Their Toxicological Effects on Humans. *Heliyon* **2020,** *6* (9), e04691. https://doi.org/10.1016/j.heliyon.2020.e04691
7. Chen, Z. F.; Zhao, Y.; Zhu, Y.; Yang, X.; Qiao, J.; Tian, Q.; Zhang, Q. Health Risks of Heavy Metals in Sewage-irrigated Soils and Edible Seeds in Langfang of Hebei Province, China. *J. Sci. Food Agric.* **2010,** *90* (2), 314–320. https://doi.org/10.1002/jsfa.3817
8. Dada, E.; Njoku, K.; Osuntoki, A.; Akinola, M. A Review of Current Techniques of *in situ* Physico-chemical and Biological Remediation of Heavy Metals Polluted Soil. *Ethiop. J. Environ. Stud. Manag.* **2015,** *8* (5), 606. https://doi.org/10.4314/ejesm.v8i5.13
9. Ekmekyapar, F.; Şabudak, T.; Şeren, G. Assessment of Heavy Metal Contamination in Soil and Wheat (Triticum Aestivum L.) Plant Around the Çorlu-Çerkezkoy Highway in Thrace Region. *Glob. Nest J.* **2012,** *14* (4), 496–504. https://doi.org/10.30955/gnj.000838
10. Fodor, L.; Szabó, L. Study of Heavy Metal Leaching in the Soil. *Soil Conserv.* **2004,** *3,* 4–7.
11. Gaurav, V. K.; Kumar, D.; Sharma, C. Assessment of Metal Accumulation in the Vegetables and Associated Health Risk in the Upper-Most Ganga-Yamuna Doab Region, India. *Am. J. Plant Sci.* **2018,** *09* (12), 2347–2358. https://doi.org/10.4236/ajps.2018.912170
12. Hussain, A.; Abbas, N.; Arshad, F.; Akram, M.; Khan, Z. I.; Ahmad, K.; Mansha, M.; Mirzaei, F. Effects of Diverse Doses of Lead (Pb) on Different Growth Attributes of *Zea-Mays* L. *Agric. Sci.* **2013,** *04* (05), 262–265. https://doi.org/10.4236/as.2013.45037
13. Hussain, A.; Priyadarshi, M.; Dubey, S. Experimental Study on Accumulation of Heavy Metals in Vegetables Irrigated with Treated Wastewater. *Appl. Water Sci.* **2019,** *9* (5), 1–11. https://doi.org/10.1007/s13201-019-0999-4
14. Hussain, A.; Priyadarshi, M.; Said, S.; Negi, S. Effect of Wastewater on the Soil and Irrigation Process: A Laboratory Study. *J. Geograph. Stud.* **2017,** *1* (1), 46–55. https://doi.org/10.21523/gcj5.17010105
15. Jaleel, C. A.; Jayakumar, K.; Jaleel, C. A.; Azooz, M. M. Phytochemical Changes in Green Gram (*vigna radiata*) Under Cobalt Stress. *Glob. J. Mol. Sci.* **2008,** *3* (2), 46–49.
16. Jolly, Y. N.; Islam, A.; Akbar, S. Transfer of Metals From Soil to Vegetables and Possible Health Risk Assessment. *SpringerPlus* **2013,** *2* (1), 1–8. https://doi.org/10.1186/2193-1801-2-385

17. Khalid, S.; Shahid, M.; Niazi, N. K.; Murtaza, B.; Bibi, I.; Dumat, C. A Comparison of Technologies for Remediation of Heavy Metal Contaminated Soils. *J. Geochem. Explor.* **2017,** *182,* 247–268. https://doi.org/10.1016/j.gexplo.2016.11.021
18. Kim, J. J.; Kim, Y. S.; Kumar, V. Heavy Metal Toxicity: An Update of Chelating Therapeutic Strategies. *J. Trace Elem. Med. Biol.* **2019,** *54,* 226–231. https://doi.org/10.1016/j.jtemb.2019.05.003
19. Liu, L.; Li, W.; Song, W.; Guo, M. Remediation Techniques for Heavy Metal-Contaminated Soils: Principles and Applicability. *Sci. Total Environ.* **2018,** *633,* 206–219. https://doi.org/10.1016/j.scitotenv.2018.03.161
20. Loi, N. N.; Sanzharova, N. I.; Shchagina, N. I.; Mironova, M. P. The Effect of Cadmium Toxicity on the Development of Lettuce Plants on Contaminated Sod-Podzolic Soil. *Russ. Agric. Sci.* **2018,** *44* (1), 49–52. https://doi.org/10.3103/s1068367418010111
21. Mahar, A.; Wang, P.; Ali, A.; Kumar, M.; Hussain, A.; Wang, Q.; Li, R.; Zhang, Z. Ecotoxicology and Environmental Safety Challenges and Opportunities in the Phytoremediation of Heavy Metals Contaminated Soils: A Review. *Ecotoxicol. Environ. Saf.* **2016,** *126,* 111–121. https://doi.org/10.1016/j.ecoenv.2015.12.023
22. Malav, L. C.; Kumar, S.; Daripa, A.; Yadav, S. Heavy metal Contamination: A Serious Hazard to Food-chain. *Food Sci. Rep.* **2020,** *1* (4), 1–12.
23. Medda, S.; Mondal, N. K. Chromium Toxicity and Ultrastructural Deformation of Cicer Arietinum with Special Reference of Root Elongation and Coleoptile Growth. *Ann. Agrarian Sci.* **2017,** *15* (3), 396–401.
24. Mokhena, T.; Mochane, M.; Tshwafo, M.; Linganiso, L.; Thekisoe, O.; Songca, S. Liveness-detection in Biometric. In *Advanced Biometric Technologies*, Intech Open, 2016; pp 225–240. https://www.intechopen.com/books/advanced-biometric-technologies/liveness-detection-in-biometrics; (accessed July 31, 2023).
25. Parveen, T.; Hussain, A.; Rao, M. S. Growth and Accumulation of Heavy Metals in Turnip (*Brassica rapa*) Irrigated with Different Concentrations of Treated Municipal Wastewater. *Hydrol. Res.* **2015,** *46* (1), 60–71.
26. Patil, S. S.; Gandhe, H. D.; Ghorade, I. B. Physico-chemical Prosperities: Soil Pollution, Ecosystem and Irrigation. *Environment* **2014,** *3* (4), 81–82.
27. RC, T. Heavy Metals Contamination in Vegetables and Its Growing Soil. *J. Environ. Anal. Chem.* **2015,** *02* (03). https://doi.org/10.4172/2380-2391.1000142; (accessed August 13, 2023).
28. Sabreena; Hassan, S.; Bhat, S. A.; Kumar, V.; Ganai, B. A.; Ameen, F. Phytoremediation of Heavy Metals: An Indispensable Contrivance in Green Remediation Technology. *Plants* **2022,** *11* (9), 1–28. https://doi.org/10.3390/plants11091255
29. Selvi, A.; Rajasekar, A.; Theerthagiri, J.; Ananthaselvam, A.; Sathishkumar, K.; Madhavan, J.; Rahman, P. K. S. M. Integrated Remediation Processes Toward Heavy Metal Removal/recovery from Various Environments-A Review. *Front. Environ. Sci.* **2019,** *7* (May). https://doi.org/10.3389/fenvs.2019.00066; (accessed August 13, 2023).
30. Yan, A.; Wang, Y.; Tan, S. N.; Mohd Yusof, M. L.; Ghosh, S.; Chen, Z. Phytoremediation: A Promising Approach for Revegetation of Heavy Metal-Polluted Land. *Front. Plant Sci.* **2020,** *11* (April), 1–15.
31. Sonone, S. S.; Jadhav, S.; Sankhla, M. S.; Kumar, R. Water Contamination by Heavy Metals and Their Toxic Effect on Aquaculture and Human Health Through Food Chain. *Lett. Appl. Nano Bio Sci.* **2020,** *10* (2), 2148–2166.

CHAPTER 7

UPTAKE OF EMERGING CONTAMINANTS AND PATHOGENS BY PLANTS: USE AND IMPACT OF WASTEWATER

ANUPRIYA VERMA and GAURAV SAINI

ABSTRACT

Water is a scarce resource for several geographical regions around the world and this causes a shortage of water supply for irrigation purposes. A suitable alternative for achieving water management and security, without compromising agricultural productivity, can be the use of treated wastewater (TWW) with some degree of treatment. However, this practice may contribute to the dissemination of emerging contaminants (ECs) into the environment, a matter of global concern. Uptake of ECs (such as microplastics, antibiotic resistance genes (ARGs), pharmaceuticals, personal care products, etc.) by the edible plant parts may result in their entry in the food chain, with associated human exposure. The use of poorly treated or untreated wastewater for the irrigation of agricultural crops results in the uptake and deposition of ECs and other toxic elements in plants and soils. This practice also exposes the soil, plants (and farmers) to pathogens, resulting in health hazards to human beings.

7.1 INTRODUCTION

It has been common practice for generations to use wastewater and water of poor quality for irrigation.[54] In the early prehistoric period, brick conduits

Sustainable Use of Wastewater in Agriculture: Applications, Methods, and Technology. Megh R. Goyal & Gaurav Saini (Eds.)

were utilized in Crete, an island in Greece, to transport wastewater to fields where it was used to irrigate crops. Around 600 BC, urine from public restrooms was collected by the Romans and traded to dyers and tanners.[95] There are various reports associated with the use of wastewater for irrigation in various countries such as India, Brazil, China, Greece, the Czech Republic, Mexico, Vietnam, France, Morocco, Spain, Egypt, Scotland, and Turkey.[57]

During the past few decades, rapid population growth, industrialization, changes in consumption habits, and other factors have significantly increased municipal wastewater generation.[79] For instance, over the past 20 years, total wastewater production by municipalities in Australia has increased by up to 36%.[28] For Belarus, Columbia, China, Cuba, India, and Egypt, the equivalent gains were 18, 15, 158, 68, 215, and 97%, respectively. The total amount of municipal wastewater produced over that time increased by 72, 285, 17, 11, 353, and 50% in Israel, Indonesia, the Netherlands, Switzerland, Nicaragua, and Tunisia, respectively.

The human population is expected to increase steadily from 2017 to 2050.[59] Urbanization has increased quickly over the past few decades due to the tremendous growth in the human population.[76] As a result, there is a significant increase in the generation of wastewater.[122] Discharging wastewater spreads water-borne diseases and pollutes the receiving water bodies. Managing enormous amounts of wastewater properly is a challenge on a global scale.[78,91]

More than 60% of the world's population lives in coastal regions with a significant wastewater disposal issue.[44] It is widespread practice around the world to use treated wastewater (TWW) for agricultural operations.[88,107] TWW irrigation reduces environmental degradation and increases farm profitability.[17] Untreated wastewater is used to irrigate over 20 Mha of land globally.[100] Future growth in this sector is anticipated to be greater.[77]

The pressure on water supplies has increased due to the world's continuously expanding population, especially in arid and semiarid countries, where water scarcity is already a significant problem.[35] Currently, agriculture uses 70% or more of all freshwater, making it the industry with the highest water consumption.[31] The Food and Agricultural Organization (FAO) advises expanding agriculture irrigation along with enhancing water management.[31] As a result, improving TWW reuse in agricultural irrigation should be the primary objective of any water reuse program.

This chapter discusses the impacts on agricultural products by using wastewater/poorly TWW.

7.2 TREATMENT OF WASTEWATER FOR REUSE

Wastewater consists of 99% of water and 1% of dissolved, colloidal, and suspended particles.[47] Inorganic matter, harmful compounds, pathogens, organic matter, and nutrients (N, P, and K) are all present in municipal wastewater.[4] Untreated municipal wastewater comprises the following characteristics: total solids, total dissolved solids, volatile solids, total suspended solids, biochemical oxygen demand (BOD), chemical oxygen demand (COD), total organic carbon, nitrogen (total N), phosphorous (total P), organic, and inorganic between 350–1200, 250–850, 105–325, 100–350, 110–400 (5 days), 250–1000, 80–290, 20–85, 4–15, 1–5, and 3–10 mg/L, respectively.[34]

Without adequate treatment, wastewater is typically released into rivers and lakes in developing nations' municipalities, which greatly contaminates water supplies.[120] Three steps are commonly included in traditional wastewater treatment: (1) temporary sewage holding to remove contaminants in suspensions; (2) eliminating dissolved elements; and (3) ultimate treatment. Table 7.1 illustrates the conventional treatment process for wastewater, including the different unit operations utilized at different levels of treatment.

TABLE 7.1 Wastewater Treatment Processes.

Type of treatment	Typical unit operations	Potential applications
Primary treatment	Screening, sedimentation, and filtration	Unfit for irrigation
Secondary treatment	Trickling filters, activated sludge, and membrane bioreactors	Groundwater recharge, garden irrigation, and wetlands
Tertiary treatment	Adsorption, media filtration, and membrane filtration	Food crop irrigation and landscape irrigation
Advanced treatment	Activated carbon and reverse osmosis	Drinking purposes, recharge of surface, and groundwater bodies

It demonstrates that only first-stage wastewater treatment is not suitable for agricultural irrigation. Watering gardens, groundwater replenishment, irrigating landscapes, irrigation of food crops, etc., all can utilize secondary and tertiary TWW. In contrast, wastewater undergoing advanced treatment is utilized to recharge surface and groundwater bodies and for indirect drinking. Hydrogen peroxide is used in the advanced oxidation process to break down organic pollutants into carbon dioxide. Ozone, electrolysis, solar, ultrasound, wet air oxidation, and other processes are also used in this treatment. Similarly, pathogens and harmful soluble contaminants are removed at the macro or micro levels under membrane filtration. Wastewater collection and effective treatment are required for the application of TWW, and both are cost-sensitive. More expenditure is required in more advanced treatments, something that is not a viable option for most of the developing and underdeveloped countries.

Industrial and urban water treatment procedures generate a sizable amount of effluent or water of poor quality.[26] These waters are often discharged into sewage and groundwater systems in metropolitan areas, where they represent potential dangers to the environment and human health.[50] In areas with limited access to high-quality water, these waters can be used to irrigate plants.[29] By doing this, the associated issues of their emissions into sewage and groundwater systems in cities can also be somewhat avoided. Given the scarcity of freshwater in metropolitan areas, halophytes in these regions can be watered with low-quality saline wastewater. In one of the studies, saline-treated water was applied on deeply rooted halophytes available locally in a region named Twentynine Palms located in California, United States.[36]

7.3 CROP IRRIGATION WITH WASTEWATER AND POORLY-TREATED WASTEWATER

Wastewater is frequently used to irrigate crops in water-scarce places. One such study assessed the effects of wastewater irrigation on the yield of tomatoes and broccoli as well as the characteristics of the soil in Italy.[58] To compare the outcomes, several regimes for irrigation, including TWW and freshwater were investigated. When compared to freshwater irrigation, the results showed greater values of chemicals present in soil by using wastewater irrigation. Additionally, nitrate and sodium ion concentrations were marginally over the threshold limits provided for vegetables.[107]

Crops growing in Mediterranean locations with limited water supplies can be safely irrigated with wastewater that has been appropriately cleaned. The study did find marginally greater electrical conductivity and sodium adsorption ratio in wastewater-irrigated locations.[106]

7.4 EMERGING CONTAMINANTS

Emerging contaminants (ECs) are described as substances with a broad human application (e.g., microplastics, genes for antibiotic resistance, pharmaceuticals, personal care items, etc.) that have the potential to negatively impact the environment's living creatures.[64,92] Although often present in the environment at negligibly low concentrations (ng/L to g/L), it is unclear exactly what their immediate and long-term impacts are.[74] After being used by humans, ECs can enter TWW, where they are usually ineffectively removed using conventional treatment procedures.[45]

The two main reasons responsible for spreading ECs in agroecosystems are sludge land application and irrigation with recovered water. Additionally, surface water has been found to contain ECs. ECs are not regularly monitored in the environment. These pollutants are also referred to as contaminants of emerging concern because of their adverse effect on environmental and/or human health effects.

7.4.1 PHARMACEUTICALS AND PERSONAL CARE PRODUCTS (PPCPS)

PPCPs are mostly released into the environment as a result of human activity and discharges by wastewater treatment plants (WWTPs). Additionally, animal farming and aquaculture are significant sources, particularly for antibiotics that are administered both therapeutically and subtherapeutically, as well as for hormones that are either naturally secreted or synthesized and utilized to control animal development and reproduction. These substances have been found in drinking water groundwater, surface water, and WWTP effluents across the world.[5,6,65,75,86,99,109,121]

Pharmaceuticals make up a substantial portion of PPCPs. Antibiotics, lipid regulators, analgesics, psychotropic medications, steroid hormones, and β-blockers are among the drugs that pose a threat to the environment.[20] Atenolol, estrone, carbamazepine, gemfibrozil, naproxen, meprobamate, phenytoin, trimethoprim, sulfamethoxazole, and ibuprofen are medications that are often discovered in drinking water.[63,73] The presence of pharmaceuticals in drinking water is of great worry to scientists since it is unclear what consequences long-term exposure to these substances will have on human beings and the environment.

Compounds related to personal care are generally transmitted to the aqueous environment through wastewater treatment works. These include alkyl esters of *p*-hydroxybenzoic acid known as parabens, which have been utilized as bacteriostatic and fungistatic agents in medications, cosmetics, and food products since the 1930s; tonalide and galaxolide (known as polycyclic musks) are employed as perfumes in a variety of cleaning and personal care products; *N,N*-diethyl-meta-toluamide, also known as DEET, is the most widely used ingredient in insect repellents; triclosan is a common ingredient in home items including toothpaste, soap; and anti-microbial sprays as a bactericide and antifungal agent.[85]

In surface water, triclosan and its metabolite methyl triclosan were detected in Switzerland, and the metabolite was considered to be persistent.[71] Effluents from wastewater treatment works contain tonalide [acetyl hexamethyl tetralin (AHTN)], HHCB-lactone, and galaxolide (HHCB).[51] These compounds were utilized as indicators for wastewater in surface water.[12] In another study, researchers analyzed the findings of research on synthetic musk concentrations detected in sewage sludge, sewage, surface water, biota samples, and aquatic sediment in terms of metabolism in fish, bioaccumulation, and human and environmental risk assessment.[49] The majority of substances utilized as sunscreens are conjugated aromatic, lipophilic compounds and are detected in aqueous environments.[56]

7.4.2 ANTIBIOTIC RESISTANCE GENES

One of the most effective treatments for both humans and animals is antibiotics. Antibiotic resistance genes (ARGs) in bacteria and in soil environments may be promoted as a result of the introduction of antibiotics into the environment. As a result, additional research has been done in this domain. The 10 most extensive surveys of antibiotics were recently reported in U.S. seafood.[24] 4-Epioxytetracycline, oxytetracycline, virginiamycin, and sulfadimethoxine ormetoprim were the five antibiotics that were detected at very low concentrations (0.3–8.6 ng/g fresh weight).

All samples met government regulatory limits for the presence of antibiotics in seafood, although extended exposure to these low amounts of antibiotics may be favorable to encourage the formation of ARGs. This study emphasized the need for more research clarifying the relationship between antibiotic usage in aquaculture, human exposure to ARGs, and the emergence of ARGs. In another study, tetracycline and sulfonamide, ARGs were found in animal waste and agricultural soils that had been altered with manure. Vegetables cultivated in soil modified with manure have also been shown to have ARGs.[29]

7.4.3 ENGINEERED NANOMATERIALS (ENMS)

Engineered nanomaterials (ENMs) have sparked worries about the effects of nanotechnology-based goods on the environment.[9] A total of 1814 nanotechnology-based products were recorded in the consumer product inventory, according to an inventory conducted by the project on Emerging Nanotechnologies in March 2015[115] On the basis of in vitro and in vivo investigations, extensive research has been conducted globally to establish the toxic effects of these compounds along with their potential for plant uptake.[84,111]

However, studies on the prevalence of ENMs in natural habitats are still limited, as analytical detection of ENMs is challenging. As a substitute, predictive exposure methods, such as life cycle modeling, can be used to estimate the ENM concentrations occurring in the environment.[68] Various treatment technologies have been studied to remove the projected levels of ENMs from WWTPs, namely, polymeric membranes and biological treatment.[33,62,66,120] Recent publications explored the existence of ENMs in various environmental compartments in more detailed review works.[11,87]

7.4.4 MICROPLASTICS (MPS)

Pharmaceutical chemicals, which are chemical products, designed to have biological effects and may thus interact with the broadest variety of living forms, can be transported by MPs.[15,45] It is essential to understand the changes and the methods of interaction caused by factors such as natural degradation processes, which add another stage to the research process.

MPs can enter the aquatic environment from two different sources: primary sources, where they are produced in microscopic sizes, or secondary sources, where plastic waste has been broken down into minute pieces in the environment owing to various physical, chemical, or biological processes.[19] Due to their potential to harm biota and marine species, these issues have drawn a lot of attention as a focus of recent studies. Polyvinyl chloride (PVC), polyvinyl chloride terephthalate (PET), polystyrene (PS), polypropylene (PP), polyamide (PA), and polyethylene (PE) are the manmade polymers that are most frequently discovered in the water.[2,38,41,72] Types of ECs, along with their chemical name, are listed in Table 7.2.

7.5 PATHOGENS

The term "pathogen" refers to any organism that can cause a medical condition, including diseases. These are generally microscopic in dimensions, although more prominent pathogens are also known. These include parasitic organisms as well. The pathogens, once inside a healthy body, can thrive and develop while avoiding the host's immune system and can depart and infect a new individual. The rate of spread of pathogens and mode of spread can vary from pathogen to pathogen. The spread can take place through aerial route, dermal contact, exposure to bodily fluids, etc. Table 7.3 presents a brief list of the various pathogens that are present in poorly treated or untreated wastewater and diseases associated with their exposure.

TABLE 7.2 Different Types of Emerging Contaminants (ECs).

Emerging contaminants	Chemical name	References
Engineered nanomaterials (ENMs)	C_{60}; silver; gold; titanium dioxide; carbon nanotubes (CNTs); zinc oxide and, cerium dioxide	[21]
Life-style compounds	Nicotine, caffeine	[38, 101, 110, 114]
Microplastics	Polyvinyl chloride (PVC), polyvinyl chloride terephthalate (PET), polystyrene (PS), polypropylene (PP), polyamide (PA), and polyethylene (PE)	[70]
Personal care	Bacteriocide and antifungal agents, polycyclic musk, parabens, DEET, UV filters/sunscreen,	[12, 32, 49]
Pharmaceuticals	Chemotherapy drugs: Cyclophosphamide, 5-fluorourcil; Illicit drugs: Amphetamines and cocaine	[12, 60, 61, 80, 104, 124]
Water treatment by-products	Haloacetic acids, trihalomethanes, iodo-THMs, epichlorohydrin, and polyacrylamide	[8, 96]

TABLE 7.3 Diseases Associated with Pathogen/Microorganisms.[107]

Pathogen/organisms	Concentration in wastewater (n/L)	Associated diseases
Bacteria	$1–10^{10}$	Diarrhea, gastroenteritis, cholera, typhoid fever, and arthritis
Helminths	$1–10^{3}$	Clonorchiasis, hookworm infection, fascioliasis ascariasis, and taeniasis
Protozoa	$1–10^{4}$	Dysentery, fever, diarrhea, amoebiasis, cryptosporidiosis, and giardiasis
Viruses	$1–10^{6}$	Fever, gastroenteritis, paralysis, rash, heart diseases, and infectious hepatitis

7.6 MODELING APPROACH

In addition to field monitoring, modeling is required to handle the issue of wastewater irrigation more adequately and monitor the contaminants entering the soil and crops. Along with that benefits of wastewater irrigation can be achieved by controlling specific parameters. Different management methods were developed by researchers and utilized to address the problems associated with wastewater irrigation. By using simulation and optimization models, these problems can be resolved.

7.6.1 SIMULATION MODELING

For crop production to be sustainable, an appropriate model for the optimal use of water and nutrients must be developed. However, it is difficult under a variety of climatic and agricultural circumstances.[50] Different management practices may be assessed using mathematical models.[81,97] Various models, such as SUNDIAL, RZWQM, SOILN, and NLEAP, have been employed recently to evaluate irrigation and nutrient management applications.[7,10,14,102]

7.6.1.1 SUNDIAL MODEL

Recent studies have shown that SUNDIAL's forecasting capability makes it effective in the simulation of nitrogen absorption in cropping systems.[1,30] SUNDIAL predictions, on the other hand, rely on weather estimations, and the real weather typically differs from the estimates.

7.6.1.2 RZWQM MODEL

The RZWQM model has been widely used to evaluate optional management options and their impact on agriculture. The RZWQM has shown the ability to logically simulate water dynamics and fertilizer as well as the development of plants in a variety of global environments.[13,52,48,67] However, according to recent research, the original RZWQM model overestimates the amount of nitrogen in the crop.[27]

7.6.1.3 SOILN MODEL

The SOILN model is effective in simulating carbon and nitrogen fluxes as well as the subsequent processes that take place in soils and plants.[116] The model includes a regular time step that necessitates extensive data and is impractical in areas with limited infrastructure since it might be difficult to regularly get the necessary input data in these areas, such as radiation, temperature, moisture, and evaporation conditions. A novel plant uptake model was developed for predicting lettuce contaminations in a recent study.[42] The results from the model and the field data were quite similar, indicating that the model was effective. According to the survey, leaf contamination was over 80% greater than root contamination.

7.6.1.4 NLEAP MODEL

The model NLEAP was created initially to mimic the carbon and nitrogen cycle, just like SOILN.[94] The model has, however, been improved and is frequently used to predict nitrogen management situations for various cropping systems.[23] The user-supplied data on soil types, soil management techniques, climatic factors, and soils are interpreted by the model NLEAP into predicted nitrogen budgets and potential leaching via the root zone.

The research demonstrates a strong correlation between groundwater concentrations and the leaching predicted by NLEAP. However, the locations with a high potential for nitrate leaching had higher nitrate levels.[25] To provide coordinated sewage system monitoring, the model incorporates distributed smart sensors. Results showed a method for data inquiry to support synchronized flood forecasting. The model was used to identify the key water system variables that have a substantial impact on wastewater agriculture and to examine the consequences of altering agricultural circumstances by adjusting the key variables.[106]

7.6.2 OPTIMIZATION MODELS

In recent years, researchers have employed a variety of optimization strategies to address waste-related issues.[93] For instance, few models concentrated on energy generation while others concentrated on transportation infrastructure. Because of the combination of social and environmental factors, the cultivation of food crops increased dramatically with WWI.[113] Despite the advantages of this approach, it poses risks to human health, primarily through pathogen contamination.[55]

The mathematical model was proposed by RIRA to assist planners in determining the risk that WWI posed to edible crops.[46] The model was created to present a wide variety of microbiological hazard assessment situations. The model can also undertake risk analyses in accordance with the approaches recommended in the primary wastewater reuse standards because of its adaptable structure.

The nitrogen and water management strategies were evaluated and employed in the treated WWI in Beijing, China, using the model RZWQM.[119] On a planned experimental plot in the research region, four irrigation treatments were tested in the field between 2001 and 2003. The collected data was used to calibrate the model's hydrologic and agricultural growth parameters. The investigation showed how the model could simulate various cropping system characteristics under various irrigation and fertilizer conditions. Before making a reliable decision about wastewater management, several criteria must be taken into account. Before making a choice, factors including the rate of wastewater generation, human well-being, and environmental harm might be taken into account. Such decisions can be made using a multicriteria decision analysis technique.[119] This method begins by categorizing various wastewater management scenarios. Before determining the weighting, different tactics are categorized, assessment standards are established, and qualitative and quantitative measurements are assessed. After that, the situations are analyzed and ranked.[119]

7.7 ENVIRONMENTAL IMPACTS

The deleterious impact of wastewater or poorly treated water used in irrigation on the health of soil and plants as a result of the build-up of emerging pollutants and other dangerous compounds in soil and plants is highlighted. Infections and harmful chemical compounds have a negative impact on both human and animal health. This section also covers vector-borne diseases associated with water.

7.7.1 SOIL

Due to the water limitations, the farm-production system now routinely uses TWW for irrigation.[58] However, its usage might also result in environmental issues with agroecosystems.[91] For instance, over time, it may result in an increase in N and P concentrations as well as other hazardous components in the soil profile.[22,89,103] It can also change the soil's physical characteristics, which makes it harder for hazardous salts to diffuse through the roots. Crop yields within the TWW irrigated plots varied insignificantly. Monitoring of the soil after WWI treatment is required to maintain agricultural quality considerations.

According to a recent analysis, nations such as Tunisia that lack access to clean water are increasingly concerned about the availability of water resources of low-quality such as wastewater, drainage water, and low-quality groundwater.[43] More recently, heavy metal build-up in plants and soil profiles because of ongoing WWI was observed.[118] The study also detailed changes in soil characteristics, including salinity, pH, etc. Along with having a detrimental effect on the ecosystem, these changes collectively have a negative impact on soil health and agricultural productivity.[43] Because of the increasing use of water in agriculture and depleting water supplies, irrigation is seen to be a promising solution for utilizing wastewater.[18]

Improperly treated sewage increases the harmful components in soils and plants, which might have a severe effect on the ecosystem. Wastewater also includes vital plant nutrients.[3] When managing such soils, it is best to use nontoxic substances such as lime. Additionally, organic materials can be employed to clean up these contaminated soils.[83] The immediate water source contamination brought on by the release of untreated wastewater into water bodies is remedied by using it directly for agricultural productivity.[40] However, it could cause further problems by altering the soil characteristics of agricultural regions, such as their hydraulic conductivity.[108]

7.7.2 LIVING ORGANISMS AND PATHOGENS

Untreated or improperly treated sewage includes a variety of expelled creatures, including pathogens, protozoa, bacteria, helminths, viruses, and others that can lead to a number of illnesses.[90,98] Helminth eggs may cause illnesses such as clonorchiasis and hookworm infection. Similarly, protozoa are responsible for amoebiasis, giardiasis, cryptosporidiosis, dysentery, diarrhea, and fever. Additionally, "water-related vector-borne illnesses," including dengue fever, malaria, yellow fever, etc. can spread because of improperly or insufficiently treated sewage.[105] Protozoa and helminth eggs are the most dangerous of these diseases, and they are also the most difficult to cure.[47]

Depending on the source and concentration of the wastewater, there can be 1–10^{10} bacteria in 1 L. Pathogens in wastewater are seldom directly studied because of their fluctuating concentrations and challenging analytical procedures. As a result, the pathogens are identified using proxies that have comparable characteristics. A listing of the various pathogens and diseases caused by them is presented in Table 7.3.

The inadequately treated sewage also carries toxic organic compounds (TOCs) and ECs, which pose significant environmental and health dangers.[90] These toxins contaminate water sources by penetrating through the root zone and into the groundwater system.[90] It was recently revealed that WWI accelerated the mobility of TOCs in the soils.[40] Pesticides and other chemicals used in fields can move to different plant parts, influencing plant growth and exposing people and animals who eat those plants.[112] The use of such water sources causes a variety of health problems in both people and animals. Climate, plant characteristics, and the characteristics of the TOCs all affect how well plants absorb TOCs.[82]

When consumed in large doses, heavy metals with high absorption rates are also fatal and pose a health risk.[47] The effects of exposure to certain metals through the food chain could be quite harmful. The negative impacts of wastewater use do not just affect humans; they also have an influence on other animals, such as fish and frogs.[69] In a recent study, one investigated how TWW affected fish health in a dry area of Israel's Negev Desert.[123]

Histopathology was utilized to assess the overall state of fish health while the study tracked the levels of organic micropollutants in fish tissues. The investigation concluded that fish exposed to TWW had varying degrees of infection. Similar impacts of sewage on fish health were reported.[16]

7.8 SUMMARY

The growing worldwide population has significantly increased the production of wastewater, and the proper management of this wastewater has become a global concern. Reusing TWW is encouraged by the United Nations' Sustainable Development Goal (SDG) 6 to "secure the availability of water for everyone." TWW includes vital plant nutrients and can potentially increase agricultural production and decrease the exploitation of freshwater sources.

Unfortunately, human development has inadvertently also caused the presence of a significant number of ECs and pathogens in the wastewater. The application of poorly treated or untreated wastewater for irrigation exposes the soil, crops, groundwater, human beings, animals, and the environment to such contaminants. Since the number of ECs is increasing at an alarming pace, detailed and rapid studies into their spread, potential risks, and appropriate treatment/containment/disposal are needed.

Adequate funding and research infrastructure, joint interdisciplinary efforts, coupled with public awareness campaigns, are needed to properly use our water resources without compromising on agricultural irrigation requirements. Similar sustained efforts are required to improve our water management infrastructure, including treatment plants, conveyance networks, and disposal standards. Efforts are also needed to develop and incorporate treatment techniques to tackle ECs and incorporate them into existing infrastructure.

Farmers in economically underdeveloped countries need to be educated and made aware of the serious environmental and health dangers associated with using untreated or inadequately treated effluent for irrigation. Future policy directives for wastewater reuse should incorporate it.

KEYWORDS

- **bioaccumulation**
- **food web**
- **hazard**
- **health impacts**
- **wastewater treatment**
- **water management**

REFERENCES

1. Abukhdeir, N. M.; Vlachos, D. G.; Katsoulakis, M.; Plexousakis, M. Long-Time Integration Methods for Mesoscopic Models of Pattern-Forming Systems. *J. Comput. Phys.* **2011,** *230* (14), 5704–5715.
2. Akdogan, Z.; Guven, B. Microplastics in the Environment: A Critical Review of Current Understanding and Identification of Future Research Needs. *Environ. Pollut.* **2019,** *254.* https://doi.org/10.1016/j.envpol.2019.113011
3. Aquilino, M.; Martínez-Guitarte, J. L.; García, P.; Beltrán, E. M.; Fernández, C.; Sánchez-Argüello, P. Combining the Assessment of Apical Endpoints and Gene Expression in the Freshwater Snail Physa Acuta After Exposure to Reclaimed Water. *Sci. Total Environ.* **2018,** *642,* 180–189.
4. Asano, T.; Smith, R. G.; Tchobanoglous, G. Municipal Wastewater: Treatment and Reclaimed Water Characteristics. In *Irrigation With Reclaimed Municipal Wastewater–A Guidance Manual*; 2019, 2-1-2–26.

5. Batt, A. L.; Kincaid, T. M.; Kostich, M. S.; Lazorchak, J. M.; Olsen, A. R. Evaluating the Extent of Pharmaceuticals in Surface Waters of the United States Using a National-Scale Rivers and Streams Assessment Survey. *Environ. Toxicol. Chem.* **2016,** *35* (4), 874–881.
6. Bayen, S.; Zhang, H.; Desai, M. M.; Ooi, S. K.; Kelly, B. C. Occurrence and Distribution of Pharmaceutically Active and Endocrine Disrupting Compounds in Singapore's Marine Environment: Influence of Hydrodynamics and Physical-Chemical Properties. *Environ. Pollut.* **2013,** *182,* 1–8.
7. Bergström, L.; Johnsson, H.; Torstensson, G. Simulation of Soil Nitrogen Dynamics Using the SOILN Model. *Fertilizer Res.* **1991,** *27* (2–3), 181–188.
8. Boorman, G. A.; Dellarco, V.; Dunnick, J. K.; Chapin, R. E.; Hunter, S.; Hauchman, F.; Gardner, H.; Mike, C.; Sills, R. C. Drinking Water Disinfection Byproducts: Review and Approach to Toxicity Evaluation. *Environ. Health Perspect.* **1999,** *107* (Suppl. 1), 207–217.
9. Bour, A.; Mouchet, F.; Silvestre, J.; Gauthier, L.; Pinelli, E. Environmentally Relevant Approaches to Assess Nanoparticles Ecotoxicity: A Review. *J. Hazard. Mater.* **2015,** *283,* 764–777.
10. Bradbury, N. J.; Jenkinson, D. S. J.; Whitmore, A. P.; Hart, P. B. S. Modelling the Fate of Nitrogen in Crop and Soil in the Years Following Application of15N-Labelled Fertilizer to Winter Wheat. *J. Agric. Sci.* **1993,** *121* (3), 363–379.
11. Bu, Q.; Wang, B.; Huang, J.; Deng, S.; Yu, G. Pharmaceuticals and Personal Care Products in the Aquatic Environment in China: A Review. *J. Hazard. Mater.* **2013,** *262,* 189–211.
12. Buerge, I. J.; Buser, H. R.; Müller, M. D.; Poiger, T. Behavior of the Polycyclic Musks HHCB and AHTN in Lakes, Two Potential Anthropogenic Markers for Domestic Wastewater in Surface Waters. *Environ. Sci. Technol.* **2003,** *37* (24), 5636–5644.
13. Cameira, M. R.; Fernando, R. M.; Ahuja, L.; Pereira, L. Simulating the Fate of Water in Field Soil-Crop Environment. *J. Hydrol.* **2005,** *315* (1–4), 1–24.
14. Cannavo, P.; Recous, S.; Parnaudeau, V.; Reau, R. Modeling N Dynamics to Assess Environmental Impacts of Cropped Soils. *Adv. Agron.* **2008,** *97,* 131–174.
15. Caruso, G. Microplastics as Vectors of Contaminants. *Mar. Pollut. Bull.* **2019,** *146,* 921–924.
16. Cazenave, J.; Bacchetta, C.; Rossi, A.; Ale, A.; Campana, M.; Parma, M. J. Deleterious Effects of Wastewater on the Health Status of Fish: A Field Caging Study. *Ecol. Indicat.* **2014,** *38,* 104–112.
17. Chowdhury, S.; Al-Zahrani, M. Fuzzy Synthetic Evaluation of Treated Wastewater Reuse for Agriculture. *Environ. Dev. Sustainability* **2014,** *16* (3), 521–538.
18. Christou, A.; Agüera, A.; Bayona, J. M.; Cytryn, E.; Fotopoulos, V.; Lambropoulou, D.; Manaia, C. M.; Michael, C.; Revitt, M.; Schröder, P.; Fatta-Kassinos, D. The Potential Implications of Reclaimed Wastewater Reuse for Irrigation on the Agricultural Environment: The Knowns and Unknowns of the Fate of Antibiotics and Antibiotic Resistant Bacteria and Resistance Genes—A Review. *Water Res.* **2017,** *123,* 448–467.
19. Cole, M.; Lindeque, P.; Halsband, C.; Galloway, T. S. Microplastics as Contaminants in the Marine Environment: A Review. *Mar. Pollut. Bull.* **2011,** *62* (12), 2588–2597.
20. Daughton, C. G.; Ternes, T. A. Pharmaceuticals and Personal Care Products in the Environment: Agents of Subtle Change? *Environ. Health Perspect.* **1999,** *107* (Suppl. 6), 907–938.
21. De Marchi, L.; Coppola, F.; Soares, A. M. V. M.; Pretti, C.; Monserrat, J. M.; Torre, C. Della, & Freitas, R. Engineered Nanomaterials: From Their Properties and Applications, to Their Toxicity Towards Marine Bivalves in a Changing Environment. *Environ. Res.* **2019,** *178.* https://doi.org/10.1016/j.envres.2019.108683

22. De Oliveira, J. F.; Rodrigues, F. N.; Fia, R.; Mafra, D. C. B.; Landim, D. V. Percolate Quality in Soil Cultivated with Application of Wastewater from Swine Slaughterhouse and Dairy Products. *Engenharia Agricola* **2017,** *37* (6), 1222–1235.
23. Delgado, J. A.; Ascough, J. C.; Lighthart, N.; Neer, D. Potential Use of a New Nitrogen Trading Tool to Assess Nitrogen Management Practices to Protect Groundwater Quality. *Comput. Electron. Agric.* **2020,** *169*. https://doi.org/10.1016/j.compag.2019.105195
24. Done, H. Y.; Halden, R. U. Reconnaissance of 47 Antibiotics and Associated Microbial Risks in Seafood Sold in the United States. *J. Hazard. Mater.* **2015,** *282,* 10–17.
25. Edmondson, V.; Cerny, M.; Lim, M.; Gledson, B.; Lockley, S.; Woodward, J. A Smart Sewer Asset Information Model to Enable an 'Internet of Things' for Operational Wastewater Management. *Autom. Construct.* **2018,** *91,* 193–205.
26. Egbuikwem, P. N.; Naz, I.; Saroj, D. P. Appraisal of Suspended Growth Process for Treatment of Mixture of Simulated Petroleum, Textile, Domestic, Agriculture and Pharmaceutical Wastewater. *Environ. Technol. (U. K.)* **2020,** *41* (25), 3338–3353.
27. Fang, Q. X.; Ma, L.; Trout, T. J.; Comas, L. H.; DeJonge, K. C.; Ahuja, L. R.; Sherrod, L. A.; Malone, R. W. Modeling N Concentration and Uptake for Maize Hybrids Under Growth Stage-Based Deficit Irrigations. *Trans. ASABE* **2017,** *60* (6), 2067–2081.
28. FAO. Global Information System on Water and Agriculture [Online]. Food and Agriculture Organization of the United Nations (FAO), 2019. http://www.fao.org/nr/water/aquastat/data/query/index.html?lang=en
29. FengHua, W.; Min, Q.; Zheng, C.; JianQiang, S.; YongGuan, Z. Antibiotic Resistance Genes in Manure-Amended Soil and Vegetables at Harvest. *J. Hazard. Mater.* **2015,** *299,* 215–221.
30. Flattery, P.; Fealy, R.; Fealy, R. M.; Lanigan, G.; Green, S. Simulation of Soil Carbon Efflux from an Arable Soil Using the ECOSSE Model: Need for an Improved Model Evaluation Framework? *Sci. Total Environ.* **2018,** *622–623,* 1241–1249.
31. Frenken, K.; Gillet, V. Irrigation Water Requirement and Water Withdrawal by Country[Online]. Food and Agriculture Organization of the United Nations # 264; 2012. http://www.fao.org/nr/water/aquastat/water_use_agr/IrrigationWaterUse.pdf%0A http://www.fao.org/nr/water/aquastat/water_use_agr/index.htm
32. Fromme, H.; Otto, T.; Pilz, K. Polycyclic Musk Fragrances in Different Environmental Compartments in Berlin (Germany). *Water Res.* **2001,** *35* (1), 121–128.
33. Ganesh, R.; Smeraldi, J.; Hosseini, T.; Khatib, L.; Olson, B. H.; Rosso, D. Evaluation of Nanocopper Removal and Toxicity in Municipal Wastewaters. *Environ. Sci. Technol.* **2010,** *44* (20), 7808–7813.
34. Gerba, C. P.; Pepper, I. L. *Municipal Wastewater Treatment*; Environment and Pollution Science, 2019.
35. Gerland, P.; Raftery, A. E.; Ševčíková, H.; Li, N.; Gu, D.; Spoorenberg, T.; Alkema, L.; Fosdick, B. K.; Chunn, J.; Lalic, N.; Bay, G.; Buettner, T.; Heilig, G. K.; Wilmoth, J. World Population Stabilization Unlikely This Century. *Science*, **2014,** *346* (6206), 234–237.
36. Glenn, E. P.; Mckeon, C.; Gerhart, V.; Nagler, P. L.; Jordan, F.; Artiola, J. Deficit Irrigation of a Landscape Halophyte for Reuse of Saline Waste Water in a Desert City. *Landscape Urban Plann.* **2009,** *89* (3–4), 57–64.
37. Gloaguen, T. V.; Forti, M. C.; Lucas, Y.; Montes, C. R.; Gonçalves, R. A. B.; Herpin, U.; Melfi, A. J. Soil Solution Chemistry of a Brazilian Oxisol Irrigated with Treated Sewage Effluent. *Agric. Water Manag.* **2007,** *88* (1–3), 119–131.

38. Godfrey, E.; Woessner, W. W.; Benotti, M. J. Pharmaceuticals in on-Site Sewage Effluent and Ground Water, Western Montana. *Ground Water* **2007,** *45* (3), 263–271.
39. Godoy, V.; Martín-Lara, M. A.; Calero, M.; Blázquez, G. The Relevance of Interaction of Chemicals/Pollutants and Microplastic Samples as Route for Transporting Contaminants. *Process Saf. Environ. Protect.* **2020,** *138,* 312–323.
40. Gonçalves, R. A. B.; Folegatti, M. V.; Gloaguen, T. V.; Libardi, P. L.; Montes, C. R.; Lucas, Y.; Dias, C. T. S.; Melfi, A. J. Hydraulic Conductivity of a Soil Irrigated with Treated Sewage Effluent. *Geoderma* **2007,** *139* (1–2), 241–248.
41. Gong, W.; Jiang, M.; Han, P.; Liang, G.; Zhang, T.; Liu, G. Comparative Analysis on the Sorption Kinetics and Isotherms of Fipronil on Nondegradable and Biodegradable Microplastics. *Environ. Pollut.* **2019,** *254.* https://doi.org/10.1016/j.envpol.2019.07.095
42. González García, M.; Fernández-López, C.; Polesel, F.; Trapp, S. Predicting the Uptake of Emerging Organic Contaminants in Vegetables Irrigated with Treated Wastewater–Implications for Food Safety Assessment. *Environ. Res.* **2019,** *172,* 175–181.
43. Guo, H.; Nasir, M.; Lv, J.; Dai, Y.; Gao, J. Understanding the Variation of Microbial Community in Heavy Metals Contaminated Soil Using High Throughput Sequencing. *Ecotoxicol. Environ. Saf.* **2017,** *144,* 300–306.
44. Hafsa, H.; Eman, H.; Particular, I. Assessment of Water Quality Along the Coast of Alexandria – Egypt. *J. Mar. Sci. Eng.* **2008,** *9* (1328), Online; DOI:10.3390/jmse9121328
45. Halling-Sørensen, B.; Nors Nielsen, S.; Lanzky, P. F.; Ingerslev, F.; Holten Lützhøft, H. C.; Jørgensen, S. E. Occurrence, Fate and Effects of Pharmaceutical Substances in the Environment—A Review. *Chemosphere* **1998,** *36* (2), 357–393.
46. Hamilton, A. J.; Stagnitti, F.; Kumarage, S. C.; Premier, R. R. RIRA: A Tool for Conducting Health Risk Assessments for Irrigation of Edible Crops with Recycled Water. *Comput. Electron. Agric.* **2007,** *57* (1), 80–87.
47. Hanjra, M. A.; Blackwell, J.; Carr, G.; Zhang, F.; Jackson, T. M. Wastewater Irrigation and Environmental Health: Implications for Water Governance and Public Policy. *Int. J. Hyg. Environ. Health* **2012,** *215* (3), 255–269.
48. Hanson, J. D.; Rojas, K. W.; Shaffer, M. J. Calibrating the Root Zone Water Quality Model. *Agron. J.* **1999,** *91* (2), 171–177.
49. Heberer, T. Occurrence, Fate, and Assessment of Polycyclic Musk Residues in the Aquatic Environment of Urban Areas-A Review. *Acta Hydrochim. Hydrobiol.* **2003,** *30*(5–6), 227–243.
50. Hong, P. Y.; Julian, T. R.; Pype, M. L.; Jiang, S. C.; Nelson, K. L.; Graham, D.; Pruden, A.; Manaia, C. M. Reusing Treated Wastewater: Consideration of the Safety Aspects Associated with Antibiotic-Resistant Bacteria and Antibiotic Resistance Genes. *Water (Switzerland)* **2018,** *10* (3). https://doi.org/10.3390/w10030244
51. Horii, Y.; Reiner, J. L.; Loganathan, B. G.; Senthil Kumar, K.; Sajwan, K.; Kannan, K. Occurrence and Fate of Polycyclic Musks in Wastewater Treatment Plants in Kentucky and Georgia, USA. *Chemosphere*, **2007,** *68* (11), 2011–2020.
52. Hu, C.; Saseendran, S. A.; Green, T. R.; Ma, L.; Li, X.; Ahuja, L. R. Evaluating Nitrogen and Water Management in a Double-Cropping System Using RZWQM. *Vadose Zone J.* **2006,** *5* (1), 493–505.
53. Hu, W.; Guo, Y.; Tian, J.; Chen, L. Eco-Efficiency of Centralized Wastewater Treatment Plants in Industrial Parks: A Slack-Based Data Envelopment Analysis. *Resourc. Conserv. Recycl.* **2019,** *141,* 176–186.

54. Hussain, M. I.; Muscolo, A.; Farooq, M.; Ahmad, W. Sustainable Use and Management of Non-Conventional Water Resources for Rehabilitation of Marginal Lands in Arid and Semiarid Environments. *Agric. Water Manag.* **2019,** *221,* 462–476.
55. Islam, M. A.; Romić, D.; Akber, M. A.; Romić, M. Trace Metals Accumulation in Soil Irrigated with Polluted Water and Assessment of Human Health Risk from Vegetable Consumption in Bangladesh. *Environ. Geochem. Health* **2018,** *40* (1), 59–85.
56. Jeon, H. K.; Chung, Y.; Ryu, J. C. Simultaneous Determination of Benzophenone-Type UV Filters in Water and Soil by Gas Chromatography-Mass Spectrometry. *J. Chromatogr. A* **2006,** *1131* (1–2), 192–202.
57. Jeong, H.; Bhattarai, R.; Adamowski, J.; Yu, D. J. Insights from Socio-Hydrological Modeling to Design Sustainable Wastewater Reuse Strategies for Agriculture at the Watershed Scale. *Agric. Water Manag.* **2020,** *231.* https://doi.org/10.1016/j.agwat.2019.105983
58. Jesse, S. D.; Zhang, Y.; Margenot, A. J.; Davidson, P. C. Hydroponic Lettuce Production Using Treated Post-Hydrothermal Liquefaction Wastewater (PHW). *Sustainability (Switzerland)* **2019,** *11* (13). https://doi.org/10.3390/su11133605
59. Jing, X.; Yao, G.; Liu, D.; Liang, Y.; Luo, M.; Zhou, Z.; Wang, P. Effects of Wastewater Irrigation and Sewage Sludge Application on Soil Residues of Chiral Fungicide Benalaxyl. *Environ. Pollut.* **2017,** *224,* 1–6.
60. Johnson, A. C.; Belfroid, A.; Di Corcia, A. Estimating Steroid Estrogen Inputs into Activated Sludge Treatment Works and Observations on Their Removal from the Effluent. *Sci. Total Environ.* **2000,** *256* (2–3), 163–173.
61. Kasprzyk-Hordern, B.; Dinsdale, R. M.; Guwy, A. J. The Occurrence of Pharmaceuticals, Personal Care Products, Endocrine Disruptors and Illicit Drugs in Surface Water in South Wales, UK. *Water Res.* **2008,** *42* (13), 3498–3518.
62. Kiser, M. A.; Ladner, D. A.; Hristovski, K. D.; Westerhoff, P. K. Nanomaterial Transformation and Association with Fresh and Freeze-Dried Wastewater Activated Sludge: Implications for Testing Protocol and Environmental Fate. *Environ. Sci. Technol.* **2012,** *46* (13), 7046–7053.
63. Kleywegt, S.; Pileggi, V.; Yang, P.; Hao, C.; Zhao, X.; Rocks, C.; Thach, S.; Cheung, P.; Whitehead, B. Pharmaceuticals, Hormones and Bisphenol A in Untreated Source and Finished Drinking Water in Ontario, Canada—Occurrence and Treatment Efficiency. *Sci. Total Environ.* **2011,** *409* (8), 1481–1488.
64. Kümmerer, K. Emerging Contaminants. *Treatise Water Sci.*, **2011,** *3,* 69–87.
65. Kong, L.; Kadokami, K.; Wang, S.; Duong, H. T.; Chau. H. T. C.. Monitoring of 1300 Organic Micro-Pollutants in Surface Waters from Tianjin, North China. *Chemosphere* **2015,** *122,* 125–130.
66. Ladner, D. A.; Steele, M.; Weir, A.; Hristovski, K.; Westerhoff, P. Functionalized Nanoparticle Interactions with Polymeric Membranes. *J. Hazardous Mater.* **2012,** *211–212,* 288–295.
67. Landa, F. M.; Fausey, N. R.; Nokes, S. E.; Hanson, J. D. Plant Production Model Evaluation for the Root Zone Water Quality Model (RZWQM 3.2) in Ohio. *Agron. J.* **1999,** *91* (2), 220–227.
68. Lazareva, A.; Keller, A. A. Estimating Potential Life Cycle Releases of Engineered Nanomaterials from Wastewater Treatment Plants. *ACS Sustainable Chem. Eng.* **2014,** *2* (7), 1656–1665.
69. Li, C.; Chen, Q.; Zhang, X.; Snyder, S. A.; Gong, Z.; Lam, S. H. An Integrated Approach with the Zebrafish Model for Biomonitoring of Municipal Wastewater Effluent and Receiving Waters. *Water Res.* **2018,** *131,* 33–44.

70. Li, Z.; Wen, X.; Hu, C.; Li, X.; Li, S.; Zhang, X.; Hu, B. Regional Simulation of Nitrate Leaching Potential from Winter Wheat-Summer Maize Rotation Croplands on the North China Plain Using the NLEAP-GIS Model. *Agric. Ecosyst. Environ.* **2020,** *294*. https://doi.org/10.1016/j.agee.2020.106861
71. Lindström, A.; Buerge, I. J.; Poiger, T.; Bergqvist, P. A.; Müller, M. D.; Buser, H. R. Occurrence and Environmental Behavior of the Bactericide Triclosan and Its Methyl Derivative in Surface Waters and in Wastewater. *Environ. Sci. Technol.* **2002,** *36* (11), 2322–2329.
72. Liu, G.; Zhu, Z.; Yang, Y.; Sun, Y.; Yu, F.; Ma, J. Sorption Behavior and Mechanism of Hydrophilic Organic Chemicals to Virgin and Aged Microplastics in Freshwater and Seawater. *Environ. Pollut.* **2019,** *246,* 26–33.
73. Benotti, M. J.; Trenholm, R. A.; Vanderford, B. J.; Holady, J. C.; Stanford, B. D.; Snyder, S. A. Pharmaceuticals and Endocrine Disrupting Compounds in US Drinking Water. *Environ. Sci. Technology*, **2009,** *43,* 597–603.
74. Mandaric, L.; Celic, M.; Marcé, R.; Petrovic, M. Introduction on Emerging Contaminants in Rivers and Their Environmental Risk. In *Emerging Contaminants in River Ecosystems: Occurrence and Effects Under Multiple Stress Conditions*; Springer International Publishing, 2015.
75. Martín, J.; Camacho-Muñoz, D.; Santos, J. L.; Aparicio, I.; Alonso, E. Occurrence of Pharmaceutical Compounds in Wastewater and Sludge from Wastewater Treatment Plants: Removal and Ecotoxicological Impact of Wastewater Discharges and Sludge Disposal. *J. Hazard. Mater.* **2012,** *239–240,* 40–47.
76. Maryam, B.; Buyukgungor, H. Wastewater Reclamation and Reuse Trends in Turkey: Opportunities and Challenges. *J. Water Process Eng.* **2019,** *30,* 230–235.
77. Mendoza-Espinosa, L. G.; Burgess, J. E.; Daesslé, L.; Villada-Canela, M. Reclaimed Water for the Irrigation of Vineyards: Mexico and South Africa as Case Studies. *Sustainable Cities Soc.* **2019,** *51*. https://doi.org/10.1016/j.scs.2019.101769
78. Menegassi, L. C.; Rossi, F.; Dominical, L. D.; Tommaso, G.; Montes, C. R.; Gomide, C. A.; Gomes, T. M. Reuse in the Agro-Industrial: Irrigation with Treated Slaughterhouse Effluent in Grass. *J. Cleaner Prod.* **2020,** *251*. https://doi.org/10.1016/j.jclepro.2019.119698
79. Mishra, B. K.; Regmi, R. K.; Masago, Y.; Fukushi, K.; Kumar, P.; Saraswat, C. Assessment of Bagmati River Pollution in Kathmandu Valley: Scenario-Based Modeling and Analysis for Sustainable Urban Development. *Sustainability Water Q. Ecol.* **2017,** *9–10,* 67–77.
80. Moldovan, Z. Occurrences of Pharmaceutical and Personal Care Products as Micropollutants in Rivers from Romania. *Chemosphere* **2006,** *64* (11), 1808–1817.
81. Moretti, M.; Van Passel, S.; Camposeo, S.; Pedrero, F.; Dogot, T.; Lebailly, P.; Vivaldi, G. A. Modelling Environmental Impacts of Treated Municipal Wastewater Reuse for Tree Crops Irrigation in the Mediterranean Coastal Region. *Sci. Total Environ.* **2019,** *660,* 1513–1521.
82. Motoki, Y.; Iwafune, T.; Seike, N.; Otani, T.; Akiyama, Y. Relationship Between Plant Uptake of Pesticides and Water-Extractable Residue in Japanese Soils. *J. Pestic. Sci.* **2015,** *40* (4), 175–183.
83. Murtaza, G.; Ghafoor, A.; Qadir, M.; Owens, G.; Aziz, M. A.; Zia, M. H.; Saifullah. Disposal and Use of Sewage on Agricultural Lands in Pakistan: A Review. *Pedosphere* **2010,** *20* (1), 23–34.

84. Navarro, D. A.; Bisson, M. A.; Aga, D. S. Investigating Uptake of Water-Dispersible CdSe/ZnS Quantum Dot Nanoparticles by *Arabidopsis Thaliana* Plants. *J. Hazard. Mater.* **2012,** *211–212,* 427–435.
85. Noguera-Oviedo, K.; Aga, D. S. Lessons Learned from More Than Two Decades of Research on Emerging Contaminants in the environment. *J. Hazard. Mater.* **2016,** *316,* 242–251.
86. Padhye, L. P.; Yao, H.; Kung'u, F. T.; Huang, C. H. Year-Long Evaluation on the Occurrence and Fate of Pharmaceuticals, Personal Care Products, Andendocrine Disrupting Chemicals in an Urban Drinking Water Treatment Plant. *Water Res.* **2014,** *51,* 266–276.
87. Pal, R.; Megharaj, M.; Kirkbride, K. P.; Naidu, R. Illicit Drugs and the Environment - A Review. *Sci. Total Environ.* **2013,** *463–464,* 1079–1092.
88. Paranychianakis, N. V.; Salgot, M.; Snyder, S. A.; Angelakis, A. N. Water Reuse in EU States: Necessity for Uniform Criteria to Mitigate Human and Environmental Risks. *Crit. Rev. Environ. Sci. Technol.* **2015,** *45* (13), 1409–1468.
89. Paudel, I.; Cohen, S.; Shaviv, A.; Bar-Tal, A.; Bernstein, N.; Heuer, B.; Ephrath, J. Impact of Treated Wastewater on Growth, Respiration and Hydraulic Conductivity of Citrus Root Systems in Light and Heavy Soils. *Tree Physiol.* **2016,** *36* (6), 770–785.
90. Peña, A.; Delgado-Moreno, L.; Rodríguez-Liébana, J. A. A Review of the Impact of Wastewater on the Fate of Pesticides in Soils: Effect of Some Soil and Solution Properties. *Sci. Total Environ.* **2020,** *718,* Online; https://doi.org/10.1016/j.scitotenv.2019.134468
91. Petousi, I.; Daskalakis, G.; Fountoulakis, M. S.; Lydakis, D.; Fletcher, L.; Stentiford, E. I.; Manios, T. Effects of Treated Wastewater Irrigation on the Establishment of Young Grapevines. *Sci. Total Environ.* **2019,** *658,* 485–492.
92. Petrie, B.; Barden, R.; Kasprzyk-Hordern, B. A Review on Emerging Contaminants in Wastewaters and the Environment: Current Knowledge, Understudied Areas and Recommendations for Future Monitoring. *Water Res.* **2015,** *72,* 3–27.
93. Phonphoton, N.; Pharino, C. Multi-Criteria Decision Analysis to Mitigate the Impact of Municipal Solid Waste Management Services During Floods. *Resourc. Conserv. Recycl.* **2019,** *146,* 106–113.
94. Qi, Z.; Bartling, P. N. S.; Ahuja, L. R.; Derner, J. D.; Dunn, G. H.; Ma, L. Development and Evaluation of the Carbon-Nitrogen Cycle Module for the GPFARM-Range Model. *Comput. Electron. Agric.* **2012,** *83,* 1–10.
95. Raschid-Sally, L. The Role and Place of Global Surveys for Assessing Wastewater Irrigation. *Irrigat. Drainage Syst.* **2010,** *24* (1–2), 5–21.
96. Richardson, S. D. Water Analysis: Emerging Contaminants and Current Issues. *Anal. Chem.* **2003,** *75* (12), 2831–2857.
97. Ruiz-Rosa, I.; García-Rodríguez, F. J.; Mendoza-Jiménez, J. Development and Application of a Cost Management Model for Wastewater Treatment and Reuse Processes. *J. Cleaner Prod.* **2016,** *113,* 299–310.
98. Saini, G.; Deepak, P. S. Wastewater-Based Epidemiology for Novel Coronavirus Detection in Wastewater. *Glob. J. Environ. Sci. Manag.* **2021,** *7* (4), 643–658.
99. Samaras, V. G.; Stasinakis, A. S.; Mamais, D.; Thomaidis, N. S.; Lekkas, T. D. Fate of Selected Pharmaceuticals and Synthetic Endocrine Disrupting Compounds During Wastewater Treatment and Sludge Anaerobic Digestion. *J. Hazard. Mater.* **2013,** *244–245,* 259–267.
100. Sato, T.; Qadir, M.; Yamamoto, S.; Endo, T.; Zahoor, A. Global, Regional, and Country Level Need for Data on Wastewater Generation, Treatment, and Use. *Agric. Water Manag.* **2013,** *130,* 1–13.

101. Seiler, R. L.; Zaugg, S. D.; Thomas, J. M.; Howcroft, D. L. Caffeine and Pharmaceuticals as Indicators of Waste Water Contamination in Wells. *Ground Water* **1999,** *37* (3), 405–410.
102. Shaffer, M. J.; Halvorson, A. D.; Pierce, F. J. Nitrate Leaching and Economic Analysis Package (NLEAP): Model Description and Application. *Manag. Nitrogen Groundwater Q. Farm Profitability* **2015,** 285–322.
103. Sharma, P.; Goel, G.; Ashekuzzaman, S. M.; Saini, G.; Singh, R. Groundwater Arsenic in South-East Asia: Extent, Effects and Solutions. *Asian J. Water, Environ. Pollut.* **2014,** *11* (2), 1–11.
104. Singer, A. C.; Nunn, M. A.; Gould, E. A.; Johnson, A. C. Potential Risks Associated with the Proposed Widespread Use of Tamiflu. *Environ. Health Perspect.* **2007,** *115* (1), 102–106.
105. Singh, A. an Overview of Drainage and Salinization Problems of Irrigated Lands. *Irrigat. Drainage* **2019,** *68* (3), 551–558.
106. Singh, A. A Review of Wastewater Irrigation: Environmental Implications. *Resourc., Conserv. Recycl.* **2021a,** *168*, Online; https://doi.org/10.1016/j.resconrec.2021.105454
107. Singh, A. Soil Salinization Management for Sustainable Development: A Review. *J. Environ. Manag.* **2021b,** *277*, Online; https://doi.org/10.1016/j.jenvman.2020.111383
108. Song, P.; Feng, G.; Brooks, J.; Zhou, B.; Zhou, H.; Zhao, Z.; Li, Y. Environmental Risk of Chlorine-Controlled Clogging in Drip Irrigation System Using Reclaimed Water: the Perspective of Soil Health. *J. Cleaner Prod.* **2019,** *232,* 1452–1464.
109. Sun, Q.; Lv, M.; Hu, A.; Yang, X.; Yu, C. P. Seasonal Variation in the Occurrence and Removal of Pharmaceuticals and Personal Care Products in a Wastewater Treatment Plant in Xiamen, China. *J. Hazard. Mater.* **2014,** *277,* 69–75.
110. Teijon, G.; Candela, L.; Tamoh, K.; Molina-Díaz, A.; Fernández-Alba, A. R. Occurrence of Emerging Contaminants, Priority Substances (2008/105/CE) and Heavy Metals in Treated Wastewater and Groundwater at Depurbaix Facility (Barcelona, Spain). *Sci. Total Environ.* **2010,** *408* (17), 3584–3595.
111. Thuptimdang, P.; Limpiyakorn, T.; McEvoy, J.; Prüß, B. M.; Khan, E. Effect of Silver Nanoparticles on Pseudomonas Putida Biofilms at Different Stages of Maturity. *J. Hazard. Mater.* **2015,** *290,* 127–133.
112. Tian, B. B.; Zhou, J. H.; Xie, F.; Guo, Q. N.; Zhang, A. P.; Wang, X. Q.; Yu, Q. Q.; Li, N.; Yang, H. Impact of Surfactant and Dissolved Organic Matter on Uptake of Atrazine in Maize and Its Mobility in Soil. *J. Soils Sediments* **2019,** *19* (2), 599–608.
113. Urbano, V. R.; Mendonça, T. G.; Bastos, R. G.; Souza, C. F. Effects of Treated Wastewater Irrigation on Soil Properties and Lettuce Yield. *Agric. Water Manag.* **2017,** *181,* 108–115.
114. Van Stempvoort, D. R.; Roy, J. W.; Brown, S. J.; Bickerton, G. Artificial Sweeteners as Potential Tracers in Groundwater in Urban Environments. *J. Hydrol.* **2011,** *401* (1–2), 126–133.
115. Vance, M. E.; Kuiken, T.; Vejerano, E. P.; McGinnis, S. P.; Hochella, M. F.; Hull, D. R. Nanotechnology in the Real World: Redeveloping the Nanomaterial Consumer Products Inventory. *Beilstein J. Nanotechnol.* **2015,** *6* (1), 1769–1780.
116. Vassiljev, A.; Kaur, K.; Annus, I. Modelling of Nitrogen Leaching from Watersheds With Large Drained Peat Areas. *Adv. Eng. Softw.* **2018,** *125,* 94–100.
117. Wang, F.; Yu, C.; Xiong, L.; Chang, Y. How Can Agricultural Water Use Efficiency be Promoted in China? A Spatial-Temporal Analysis. *Resourc., Conserv. Recycl.* **2019,** *145,* 411–418.

118. Wang, H. C.; Cui, D.; Han, J. L.; Cheng, H. Y.; Liu, W. Z.; Peng, Y. Z.; Chen, Z. B.; Wang, A. J. A2O-MBR as an Efficient and Profitable Unconventional Water Treatment and Reuse Technology: A Practical Study in a Green Building Residential Community. *Resourc., Conserv. Recycl.* **2019,** *150,* Online; https://doi.org/10.1016/j.resconrec.2019.104418
119. Wang, X.; Huang, G. Evaluation on the Irrigation and Fertilization Management Practices Under the Application of Treated Sewage Water in Beijing, China. *Agric. Water Manag.* **2008,** *95* (9), 1011–1027.
120. Wang, Y.; Westerhoff, P.; Hristovski, K. D. Fate and Biological Effects of Silver, Titanium Dioxide, and C 60 (Fullerene) Nanomaterials During Simulated Wastewater Treatment Processes. *J. Hazard. Mater.* **2012,** *201–202,* 16–22.
121. Yang, G. C. C.; Yen, C. H.; Wang, C. L. Monitoring and Removal of Residual Phthalate Esters and Pharmaceuticals in the Drinking Water of Kaohsiung City, Taiwan. *J. Hazard. Mater.* **2014,** *277,* 53–61.
122. Ye, S.; Zeng, G.; Wu, H.; Liang, J.; Zhang, C.; Dai, J.; Xiong, W.; Song, B.; Wu, S.; Yu, J. The Effects of Activated Biochar Addition on Remediation Efficiency of Co-Composting with Contaminated Wetland Soil. *Resourc., Conserv. Recycl.* **2019,** *140,* 278–285.
123. Zaibel, I.; Zilberg, D.; Groisman, L.; Arnon, S. Impact of Treated Wastewater Reuse and Floods on Water Quality and Fish Health Within a Water Reservoir in an Arid Climate. *Sci. Total Environ.* **2016,** *559,* 268–281.
124. Zuccato, E.; Castiglioni, S.; Bagnati, R.; Chiabrando, C.; Grassi, P.; Fanelli, R. Illicit Drugs, a Novel Group of Environmental Contaminants. *Water Res.* **2008,** *42* (4–5), 961–968.

PART III

METHODS AND APPLICATIONS

CHAPTER 8

EVALUATION OF GROUNDWATER VULNERABILITY TO CONTAMINATION THROUGH WATER QUALITY INDEX

ABDUL GANI, SHRAY PATHAK, and ATHAR HUSSAIN

ABSTRACT

Groundwater is an essential source of drinking water for the existence of mankind. Due to anthropogenic actions, this source is getting depleted and contaminated daily. This leads to the importance of estimating the groundwater vulnerability to pollution and mitigating its impact. A water quality index (WQI) is an instrument for comprehensively reporting to policymakers and the public with a lot of data about the quality of the water in one term. It makes it easier to interpret data than numerical values. A WQI is obtained using existing models, and the validation is performed to understand its applicability in the study region. The estimation of the vulnerability index involves the assignment of weights to the parameters involved using the overlay and index method. Further, the assigned weights will be validated specifically for the study region.

8.1 INTRODUCTION

Water naturally occurs below the Earth's surface and fills all or a portion of the voids in soils is called groundwater. The term "subsurface water" is also used to distinguish it from "surface water," which is water that is found in large bodies of water like lakes or oceans or that flows through streams. Through the hydrologic cycle, surface water and underground water are

Sustainable Use of Wastewater in Agriculture: Applications, Methods, and Technology. Megh R. Goyal & Gaurav Saini (Eds.)

connected (water is constantly moving around in the Earth's environment). Groundwater primarily originates from precipitation. Rainfall seeps into the soil zone below the surface of the ground. Water percolates downward when the soil zone becomes saturated. All the interstices fill with water in a zone of saturation. There is also a zone of aeration, where air and water alternately occupy the interstices. Groundwater keeps sinking until, at a certain depth, it combines with a region of dense rock. Such rocks retain water within their pores, but because the pores are not connected, the water cannot travel through them. Recharge is the process by which precipitation replenishes the groundwater supply. In temperate or tropical climates, recharge typically only takes place in the colder months of the year. The amount of precipitation that enters aquifers—water-bearing strata—usually ranges between 10 and 20%.

The movement of groundwater is continuous. It travels much more slowly than surface water does, with the actual speed depending on the aquifer's transmissivity and storage capacity. When the groundwater pressure is greater than the atmospheric pressure closes to the Earth's surface, natural groundwater discharges occur through springs and riverbeds. Internal circulation is difficult to measure; however, it can take a year or less for water to cycle on average near the water table and thousands of years in deep aquifers. In dry and semiarid regions, groundwater is crucial to the development of enormous industrial and agricultural operations that might not otherwise be possible. The fact that aquifers that predate the development of deserts have not been impacted by the aridity gradient over time is especially fortunate. However, withdrawal would exhaust even the most significant groundwater basins, making any development predicated on the presence of aquifers at best transitory.

There is a tremendous amount of groundwater dispersed over the globe, and many groundwater reserves are yet unexplored or underdeveloped. The upper 2 km of the Earth's surface is thought to contain about 5.97 quintillion gallons of groundwater. Unconsolidated carbonate hard rock types found in coastal plains and alluvial valleys under temperate or desert environments make up the majority of the examined or exploited groundwater reservoirs.

Most groundwater is free of microbial pathogens; therefore, treatment for home or industrial use is not necessary, even if some water absorbs materials from rocks and may contain residues of old seawater. Additionally, groundwater supplies are accessible in many regions without reliable surface water supplies and are not significantly impacted by brief droughts. However, chemical pollution from fracturing, fertilizers, leaking or unsafe landfills, and septic tanks, as well as other point and uncertain information sources of pollution, poses a threat to aquifers as well as other groundwater supplies. It can be costly and challenging to clear up such contamination, which might render groundwater unsafe for human consumption.

8.2 GROUNDWATER QUALITY

There are numerous and varied sources of contaminated groundwater. Groundwater can become contaminated in different ways.[5,39] Once pollutants have entered the water table, they flow slowly because groundwater velocities are typically modest to local surface water discharge locations or deeper regions of the groundwater system.[5] Similarly, once an aquifer has some areas that are contaminated, it might take a while for natural processes to restore a better quality of water, even after the original causes of pollution have stopped being a problem.[16]

Remediation efforts to improve the quality of groundwater are typically very expensive and only partially successful. In some situations, the pace at which pollutants travel to deeper groundwater can be significantly accelerated by steep slopes brought on by groundwater pumping.[5] Surface water bodies are frequently locations that lead to wells, and surface water is receiving more attention as a potential source of pollution for wells.[5] One of the situations of most concern for the contamination of ground water by streams is karst terrains. Because groundwater frequently makes up a significant portion of streamflow, the caliber of groundwater discharge may have an impact on the caliber of the receiving stream. Pumping-related decreases in groundwater discharge to a stream may potentially have an effect if surface runoff and point source contamination levels are significantly diluted by this discharge.

Significant changes in organic matter content, chemical oxidation conditions, and microbial activity frequently take place in many aquifers within a relatively confined area or interface where groundwater and surface water meet (on the order of cm to m). Therefore, the transit and destiny of metals, nutrients, organic compounds, and other pollutants between the two sources can be strongly impacted by circumstances close to the interface between surface water and groundwater. Saline water surrounds the fresh groundwater resource laterally and below. It is particularly noticeable along the coastline where the fresh groundwater interacts with the sea; however, the opposite is also true in interior regions where the freshwater is surrounded by deep salty water. Where freshwater and saltwater part ways and keep a steady equilibrium when things are natural. Usually, the boundary involves a change in water's salinity from fresh. The boundary between saltwater and freshwater will shift in response to water pumping from an aquifer that is close to or contains saline groundwater. The water supply could become contaminated if the boundary shifts far enough.[7]

8.2.1 WATER QUALITY CHARACTERISTICS

It is based on its physical, chemical, and microbial qualities. These characteristics of groundwater resources vary widely across the world. Therefore, the quality of sources of water utilized for several purposes should be assessed in terms of the water characteristics that most influence the potential use of water. Water quality characteristics can be classified as follows:

- Physical
- Chemical
- Microbiological
- Biological

8.2.2 STANDARDS FOR DRINKING WATER

Quality criteria established for water to drink are referred to as drinking water standards. Governments and international organizations must establish standards for drinking water in order to guarantee the safety and quality of water intended for human consumption. It is a fundamental human right to have access to clean, safe drinking water, and these requirements are essential for protecting the general public's health and preventing waterborne illnesses. Establishing maximum permissible levels for various contaminants and ensuring that water satisfies strict quality requirements are the main goals of drinking water standards. These requirements were created using scientific data, health risk analyses, and suggestions from professionals in a variety of disciplines, including toxicology, epidemiology, and water quality management. Some of the key components covered by drinking water standards include physical characteristics, chemical contaminants, microbial contaminants, radiological contaminants, disinfection byproducts, etc.

Standards for drinking water might change from nation to nation and may be set by various governmental entities or international organizations. For instance, under the Safe Drinking Water Act in the United States, the Environmental Protection Agency (EPA) established and enforces the National Primary Drinking Water Regulations. These laws define maximum contaminant levels and maximum contaminant level goals for each chemical and address a variety of pollutants. The World Health Organization (WHO) is crucial in establishing global standards for the quality of drinking water. They offer guidelines and best practices in their Guidelines for Drinking Water Quality to guarantee clean, safe drinking water around the world. It is crucial for water suppliers and authorities to periodically test and monitor water quality to ensure compliance with drinking water standards.

8.2.2.1 BUREAU OF INDIAN STANDARDS (BIS)

The drinking water standards are listed in Table 8.1 of the Indian drinking water standard.[9]

TABLE 8.1 Indian Drinking Water Standards as per IS 10500:2000.[9]

Parameter*	Acceptable limit	Cause of rejection
Aluminum	0.03	0.2
Ammonia	0.5	No relaxation
Arsenic	0.01	0.05
Barium	0.7	No relaxation
Boron	0.5	1.0
Cadmium	0.003	No relaxation
Calcium	75.0	200.0
Chloramines	4.0	No relaxation
Chloride	250.0	1000.0
Chromium	0.05	No relaxation
Color	5.0	15.0
Copper	0.05	1.5
Cyanide	0.05	No relaxation
Fluoride	1.0	1.5
Free residual chlorine	0.2	1.0
Iron	0.3	No relaxation
Lead	0.01	No relaxation
Magnesium	30.0	100.0
Manganese	0.1	0.3
Mercury	0.001	No relaxation
Nickel	0.02	No relaxation
Nitrate	45.0	<45.0

TABLE 8.1 *(Continued)*

Parameter*	Acceptable limit	Cause of rejection
pH	6.5–8.5	>6.5 and <8.5
Silver	0.1	No relaxation
Sulphate	200.0	400.0
Sulfide	0.05	No relaxation
TDS	500.0	2000.0
Total Alkalinity	200.0	600.0
Total Hardness	200.0	600.0
Turbidity	1.0	5.0
Zinc	5.0	15.0

**With the exception of pH, color, and turbidity, all units are in mg/L, while color has a unit of Hazen unit and turbidity has a unit of nephelometric turbidity unit (NTU).*

8.2.2.2 WORLD HEALTH ORGANIZATION (WHO)

Guidelines created by the WHO on the quality of water and health impacts provide the framework for international standards setting and regulation. Guidelines created by the WHO on the quality of water and health impacts serve as the foundation for international regulation and standard setting in both developing and developed nations. Drinking water quality is a significant environmental factor that affects health. The basis for the control and prevention of waterborne diseases is the assurance of the safety of drinking water. The WHO specifies the drinking water standards in Table 8.2.

8.3 GROUNDWATER QUALITY INDEX (WQI)

The water quality of various groundwater systems has been affected because of anthropogenic sources directly or indirectly. Many times, this water is contaminated because of the presence of several quality parameters. In such situations, the interpretation of large sets of data makes it difficult for policymakers to classify water quality in a single term (e.g., very good, good, fair, poor, etc.).[23] The WQI is an instrument for comprehensively reporting to policymakers and the public.[27] It makes it easier to interpret data than numerical values.[28]

TABLE 8.2 WHO Drinking Water Standards.[41]

Parameters*	Acceptable
Arsenic	0.01
Boron	0.5
Calcium	75.0
Chloride	250.0
Chlorine	5.0
Chromium	0.05
Fluoride	1.5
Iron	–
Lead	0.01
Magnesium	50.0
Manganese	0.4
Nitrate	45.0
Nitrite	3.0
pH	6.5–8.5
Sulphate	250.0
TDS	–
Total Hardness	100.0
Zinc	–

**With the exception of pH, color, and turbidity, all units are in mg/L.*

Ongoing groundwater surveillance becomes compulsory to reduce groundwater contamination and to manage pollution.[12] The groundwater's quality relies on inland surface water, recharged water quality, atmospheric precipitation, and geochemical procedures subsurface.[33] In addition to affecting water quality, water pollution also threatens economic development, human health, and social prosperity.[40] The evolution of the amount and quality of groundwater and the establishment of a database is therefore essential for the growth of further civilization and future policies for the growth of water resources.

Horten first developed the idea of the WQI to describe gradations in water quality. WQI shows a single figure, such as a grade that expresses the general quality of water for a given region, and it is based on several parameters of water quality. WQI represents a composite impact of water quality contributing variables for any water system.[34] It is a well-known technique and the most prevalent efficient instrument to describe the quality of the water, offering a straightforward, stable, consistent unit of measurement and communicating water quality data to policymakers and people involved. It, therefore, becomes a significant parameter for groundwater evaluation and management.[36]

The WQI is useful and unique to illustrate the general status of the quality of water in a single word that is useful in selecting the suitable method to address the problems involved. WQI was introduced by Horton. It shows a single amount, like a grade, that expresses the general quality of water at a given region and moment based on several parameters of water quality. WQI represents a composite impact of water quality contributing variables for any water system. WQI is a well-known technique and one of the utmost efficient instruments to describe water quality, offering a straightforward, stable, consistent unit of measurement and communicating water quality data to policymakers and people involved.[36] It, therefore, becomes a significant parameter for groundwater evaluation and management. Different sources of water quality were conveyed based on calculated indices of water quality.[14]

For government reporting, WQI is a method of regularly condensing data on water quality. It merely informs us of the water quality for drinking from a supply of drinking water, much such as the air quality index or UV index.[22] WQI is a precious and distinctive rating to illustrate the general status of the quality of water in a single word that helps to select the suitable therapy method to address the problems involved.[28] WQI, on the opposite side, portrays the composite impact of distinct indicators of water quality and communicating water quality data to policymakers and people involved.[2, 4] The WQI measures the excess water quality range, frequency, and amplitude, and the three measurements are then combined into one score. The calculation yields a score from 0 to 100. The better the rating, the better the water quality.[13] The results are then classified into one of the following six categories:[26]

- **Excellent:** if WQI Value lies between 95–100;
- **Very Good:** if WQI Value lies between 89–94;
- **Good:** if WQI Value lies between 80–88;
- **Fair:** if WQI Value lies between 65–79;
- **Marginal:** if WQI Value lies between 45–64;
- **Poor:** if WQI Value lies between 0–44.

Initially, Horton[18] created WQI in the United States by choosing 10 most frequently used water quality parameters that are universally used and adopted in European, African, and Asian nations. The weight allocated shows a parameter importance for its specific use and has a significant effect on the index. A new WQI, like Horton's index, was also developed by the Brown group in 1970. Recently, numerous researchers and professionals have regarded many changes to the WQI idea. The three phases that follow outline the most prevalent factors that form the foundation of a general WQI strategy:[38]

- **Parameter selection:** Judgment of professional specialists, organizations, or governmental organizations determined in the legislative field shall be executed. It is suggested to select the variables from the five groups, namely, the amount of oxygen, physical features, health aspects, eutrophication, and dissolved substances, which have a significant effect on water quality.
- **Quality Function (curve) Calculation for Every selected Parameter:** From the factors of their unique units, subindices transform into non-dimensional scale values.
- **Aggregation of Subindices Using a Logical Function:** This is typically used with geometrical or arithmetic averages.

Several domestic and international organizations, namely, the Canadian Council of Ministers for the Environmental WQI (CCME WQI), Oregon WQI (O WQI), Weighted Arithmetic WQI (WA WQI), National Sanitation Foundation WQI (NSF WQI), etc. have developed an enormous number of indicators of water quality.[17,31] A specific area's water quality has been assessed using this WQI. Furthermore, these indices frequently depend on different quantities and types of water quality measures in comparison to the relevant area norms.[30] Water quality indices are accredited in an effective and timely way to show yearly cycles, temporal and spatial differences in water quality, and water quality trends, even at small levels. Based on the literature reviewed, accessible indices have many variants and constraints based on the use of and lack of universal acceptance for the basis of quality of water.[38] Therefore, with a variable amount of water quality factors, it requires global acceptability.[3] The various water quality indices developed globally are as follows:

- NSF WQI
- O WQI
- CCME WQI
- WA WQI

8.3.1 NSF WQI

This method includes the following water quality parameters, namely, dissolved oxygen (DO), turbidity, pH, phosphates, biochemical oxygen demand (BOD_5), Coliforms, temperature, nitrates, and total dissolved solid (TDS).[15,19,32,35] As individual parameter contributes differently to the alteration of the water quality, a particular weight was given to each of them in the development of the NSF-WQI, as shown in Table 8.3.[8,15]

TABLE 8.3 Assigned Weight Scores to the Following Parameters.[8,15]

Parameter	Weight
BOD_5	0.11
Coliforms	0.16
DO	0.17
Nitrate	0.10
pH	0.11
Temperature	0.10
Total dissolved solids	0.07
Total Phosphates	0.10
Turbidity	0.08

This index is determined in the following steps:

- Water quality scores (q_i),
- Find out the WQI.

a) Water Quality Score (q_i):

The score (q_i) for the individual parameter's water quality was calculated using the equation 8.1.

$$q_i = \frac{C_i - S_o}{S_i - S_o} * 100 \tag{8.1}$$

where C_i is the physiological–chemical parameter's concentration, S_i is the recommended physiological–chemical parameter for drinking water according to Indian norms (BIS:10500), and S_o is the ideal concentration of *i*th physiological—chemical parameter.

The ideal concentration value for each parameter is taken as zero except pH; for pH, its value is 7.

b) WQI

WQI was determined using equation 8.2.

$$NSF\,WQI = \sum_{i=1}^{n} W_i * q_i \tag{8.2}$$

where W_i is assigned weight for each parameter; q_i is the subindex of the *i*th water quality parameter.

The result placed the examined body of water in one of the five categories depicted in Table 8.4 for that category.

TABLE 8.4 Water Quality Value (NFS–WQI).[8,15]

NSF-WQI value	Water Quality
0–25	Very bad
25–50	Bad
50–70	Medium
70–90	Good
90–100	Very good

Advantages

- It quickly and objectively summarizes the data pertaining to the evaluated parameters into a single number that is also replicable.
- The assessment of regional variations in water quality.
- The index value shows the possible consumption of water.

Disadvantages

- A complicated scale of water quality parameters is not used.
- Certain data are lost while being altered.

8.3.2 O WQI

This approach is used to determine the quality of surface water bodies from Oregon and other regions. In this descriptive approach to water quality, eight physiological, chemical, and biological parameters, namely, DO, temperature, BOD_5, total nitrogen, pH, total phosphorus, TDS, and Coliforms, are used.[10,38] This index was proposed using the NSF-WQI model as a starting point. The difference is in the mathematical approach and the parameter weights, which, in this instance, are not even considered. Equation 8.3 provides the following mathematical formula for this strategy, which makes use of the idea of an arithmetical average:[38]

$$OWQI = \sqrt{\frac{n}{\sum_{i=1}^{n} \frac{1}{SI_i^2}}} \tag{8.3}$$

where S_i is the subindex of the *i*th parameter.[11,37,38]

In Cude's article from 2001, the calculus approach of subindex for individual parameters is described in great depth.

The analyzed water quality is indicated by the acquired value following the index computation of the water using the O WQI technique, as given in Table 8.5.

TABLE 8.5 Values that Correspond to Water Quality in Accordance with O WQI.[11,38]

Oregon WQI	Water quality
0–60	Very poor
60–70	Poor
70–85	Fair
85–90	Good
90–100	Excellent

Advantages

- The most impacted factors can have the most impact on the O WQI by combining subindexes using a weighted harmonic.
- The equation is susceptible to changes in the environment and significant effects on water quality.

Disadvantages

- It is unable to offer conclusive evidence about modifications to habitat, biology, or toxic concentrations.
- It is unable to assess all health-harming substances, such as metals, germs, and toxins.

8.3.3 CCME WQI

This method has been used in a variety of surface water all over the world.[1] Due to its simplicity of use and ease in selecting water quality factors to be included in the model, it is quite extensively utilized.[21,24] Three primary factors F_1, F_2, and F_3 are used to construct the WQI. These factors can be easily estimated by using simple formulas.

- The percentage of variables (failed variables) that do not satisfy the targets (reference norms) at least once over the period is represented by the factor F_1 (how many), and it is often calculated using equation 8.4:

$$F_1 = \frac{\text{Number of failed variables}}{\text{Total number of variables}} \times 100 \quad (8.4)$$

- The following equation yields the factor F_2 (how often), which shows the proportion of specific tests that completely fail to achieve objectives (eq 8.5):

$$F_2 = \frac{\text{Number of failed Tests}}{\text{Total number of Tests}} \times 100 \quad (8.5)$$

- The amount by which test values deviated from their reference values is represented by the factor F_3 (How much). Three steps are involved in the calculation:

a. When the test value cannot be greater than the goal

$$\text{Excursion}_i = \frac{\text{Failed test value}}{\text{Objective}} - 1 \quad (8.6)$$

b. When the test value cannot be lower than the goal

$$\text{Excursion}_i = \frac{\text{Objective}}{\text{Failed test value}} - 1 \quad (8.7)$$

- The total number of excursions that violate regulations is determined using the following formula:

$$\text{nse} = \frac{\sum_{i=1}^{n} \text{Excursion}_i}{\text{Number of tests}} \quad (8.8)$$

$$F_3 = \frac{\text{nse}}{0.01\ \text{nse} + 0.01} \quad (8.9)$$

The CCME WQI can be produced using the following formula once all of these variables have been determined.

$$\text{CCME WQI} = 100 - \frac{\sqrt{F_1^2 + F_2^2 + F_3^2}}{1.732} \quad (8.10)$$

The analysis of the water may place it in one of the feature groups listed in Table 8.6, depending on the value achieved by computing the index in accordance with the methods previously presented.

Advantages

- It is highly adaptable to various uses of water.
- Simple to calculate.

- It is not very sensitive to situations of missing data.
- It is suitable for assessing information obtained from automatic sampling.

TABLE 8.6 The Matching Water Quality in Accordance with CCME WQI.[24,25,30]

CCME–WQI value	Water quality
0–45	Poor
45–60	Marginal
60–80	Fair
80–95	Good
95–100	Excellent

Disadvantages

- The relevance of each variable in determining the index is equal.
- It cannot be combined with some other indicators or biological facts.
- When too few variables are considered, F_1 does not function properly.

8.3.4 WA WQI

This technique gives information on how well a body of water is rated. This technique takes advantage of the most widely used water quality.[20,29] WQI was developed in five steps:

a. **Assigning weights to each parameter (w_i)**
   Every parameter has assigned a particular weight (w_i) based upon their importance in drinking water quality. The weights were assigned in a range of 1–5.[6]
b. **Relative weight of each parameter (W_i)**
   The relative weight (W_i) for individual parameters is determined as given in equation 8.11.

$$W_i = \frac{w_i}{\sum_{i=1}^{n} w_i} \tag{8.11}$$

c. **Water Quality Score (q_i)**
   The score (q_i) for the individual parameter's water quality was calculated using equation 8.12.

$$q_i = \frac{C_i - S_o}{S_i - S_o} * 100 \tag{8.12}$$

The ideal concentration value for each parameter is taken as zero except pH; for pH, its value is 7.

d. **Subindex (SI_i) calculation**
SI_i is determined using equation 8.13.

$$SI_i = W_i * q_i \tag{8.13}$$

e. **WQI**
WQI was determined using equation 8.14.

$$WQI = \sum_{i=1}^{n} SI_i \tag{8.14}$$

The value obtained after the WQI was evaluated using the WA WQI method shows the water's quality, as given in Table 8.7.

TABLE 8.7 The Matching Water Quality in Accordance with WA WQI Method.[29,30,38]

WQI Range	Classification
>300	Water unsuitable for drinking use
200–300	Very poor water
100–200	Poor water
50–100	Good water
<50	Excellent water

Advantages

- It incorporates the values of numerous physiochemical water quality criteria into a mathematical model, indicating the ecological status of water.
- This method illustrates the significance of individual parameters in the assessment and control of water quality.
- It can be utilized to indicate if a water source is suitable for human intake on the surface or underneath.

Disadvantages

- This indicator might not offer sufficient details regarding the current condition of the water quality.
- Factors that can be used to describe the water quality are not all incorporated into this index.
- Only the direct impact of pollution on a water body is measured by this index.[30,38]

8.4 SUMMARY

The chemical pollution from fracturing, fertilizers, leaking or unsafe landfills and septic tanks, as well as other point and uncertain information sources of pollution, poses a threat to aquifers as well as other groundwater supplies. It can be costly and challenging to clear up such contamination, which might render groundwater unsafe for human consumption. Also, WHO water quality standards are followed worldwide, while Indian drinking water standards are followed only in India. However, there are certain parameters that are not covered in WHO standards, but those are covered in Indian drinking water standards, including TDS, zinc, and iron The Indian drinking water standards suggest two values of which one is the acceptable limit and another is the rejection limit, while WHO standards provide only acceptable value. Among the four methods of assessing WQI include CCME WQI and WA WQI, which are the first two methods that provide the freedom to choose the different water quality parameters—the WA WQI method depends on the opinion of the expert to give the different weights to the different water quality parameters, and, hence, the results obtained from WA WQI varies from person to person. However, in CCME WQI, there is no weightage to different water quality parameters; hence, the results procured from the CCME WQI do not deviate as it depends on the opinion of a person. In the context of the WQI; therefore, there is a need to investigate more methods/models, which should not be opinion based but it should depend on the variation of the data under different water quality standards.

KEYWORDS

- **biochemical oxygen demand**
- **dissolved oxygen**
- **groundwater**
- **National Sanitation Foundation**
- **pollution**
- **total dissolved solids**
- **Water Quality Index**

REFERENCES

1. Abbasi, T.; Abbasi, S. Water Quality Indices. In *Water Quality Indices*; Elsevier, 2012. https://doi.org/10.1016/C2010-0-69472-7
2. Acharya, S.; Sharma, S. K.; Khandegar, V. Assessment of Groundwater Quality by Water Quality Indices for Irrigation and Drinking in South West Delhi, India. *Data Brief* **2018,** *18,* 2019–2028. https://doi.org/10.1016/j.dib.2018.04.120
3. Akkoyunlu, A.; Akiner, M. E. Pollution Evaluation in Streams Using Water Quality Indices: A Case Study from Turkey's Sapanca Lake Basin. *Ecol. Indic.* **2012,** *18,* 501–511. https://doi.org/10.1016/j.ecolind.2011.12.018
4. Ali, S. Determination of the Pot Ability of Subsurface Water by using Water Quality Index (WQI) Method of Different Wells- A Case Study of East Wollega Zone. *Western Ethiopia.* **2018,** *24,* 10–17.
5. Alley, W. M. Ground Water. In *Encyclopedia of Inland Waters*;Elsevier, 2009; pp 684–690. https://doi.org/10.1016/B978-012370626-3.00015-6
6. Banoeng-Yakubo, B.; Yidana, S. M.; Emmanuel, N.; Akabzaa, T.; Asiedu, D. Analysis of Groundwater Quality Using Water Quality Index and Conventional Graphical Methods: The Volta Region, Ghana. *Environ. Earth Sci.* **2009,** *59* (4), 867–879.
7. Basu, N. B.; Van Meter, K. Sustainability of Groundwater Resources. In *Comprehensive Water Quality and Purification*; 2014; Vol. 4, pp 57–75.
8. Behmanesh, A. Assessment the Water Quality of Babolrood River Based on the nsf Water Quality Index. *Int. J. Biol., Pharm. Allied Sci.* **2015,** *4* (5), 342–352.
9. BIS, 2012. *Indian Standard Drinking Water Specifications IS 10500:2012*; 2012.
10. Cude2, C. G. Oregon Water Quality Index A Tool for Evaluating Water Quality Management Effectiveness. *J. Am. Water Resour. Assoc.* **2001,** *37* (1), 125–137.
11. Darvishi, G.; Kootenaei, F. G.; Ramezani, M.; Lotfi, E.; Asgharnia, H. Comparative Investigation of River Water Quality by OWQI, NSFWQI and Wilcox Indexes (Case study: The Talar River - Iran). *Arch. Environ. Prot.* **2016,** *42* (1), 41–48.
12. Deshmukh, K. K.; Aher, S. P. Assessment of the Impact of Municipal Solid Waste on Groundwater Quality Near the Sangamner City using GIS Approach. *Water Resour. Manag.* **2016,** *30* (7), 2425–2443.
13. Dojlido, J.; Raniszewski, J.; Woyciechowska, J. Water Quality Index Applied to Rivers in the Vistula River Basin in Poland. *Environ. Monit. Assess.* **1994,** *33* (1), 33–42.
14. Effendi, H. River Water Quality Preliminary Rapid Assessment Using Pollution Index. *Proc. Environ. Sci.* **2016,** *33,* 562–567.
15. Ewaid, S. H. Water Quality Evaluation of Al-Gharraf River by Two Water Quality Indices. *Appl. Water Sci.* **2017,** *7* (7), 3759–3765.
16. Huang, G. Semantics Oriented Spatial Temporal Data Mining for Water Resource Decision Support. *J. Controll. Release* **2011,** *11* (2), 430–439.
17. Gupta, N.; Pandey, P.; Hussain, J. Effect of Physicochemical and Biological Parameters on the Quality of River Water of Narmada, Madhya Pradesh, India. *Water Sci.* **2017,** *31* (1), 11–23.
18. Horton, R. K. An Index Number System for Rating Water Quality. *J. Water Pollu. Fed.* **1965,** 37 (3). 300–305.
19. Ichwana, I.; Syahrul, S.; Nelly, W. Water Quality Index by Using National Sanitation Foundation-water Quality Index (NSF-WQI) method at Krueng Tamiang Aceh. *Int. Conf. Technol., Innov., Soc.* **2016,** 110–117. https://doi.org/10.21063/ictis.2016.1019

20. Iticescu, C.; Georgescu, L. P.; Topa, C. M. Assessing the Danube Water Quality Index in the City of Galati-Romania. *Carpathian J. Earth Environ. Sci.* **2013,** *8* (4), 155–164.
21. Jakovljevic, D. Serbian and Canadian Water Quality Index of Danube River in Serbia in 2010. *J. Geogr. Inst. Jovan C., SASA* **2012,** *62* (3), 1–18.
22. Khot, I. M. IoT Assisted Drinkable Water Quality Analysis System Using Machine Learning Techniques. *Int. J. Res. Appl. Sci. Eng. Technol.* **2020,** *8* (9), 228–236.
23. Kumar, M.; Munoz-Arriola, F.; Furumai, H.; Chaminda, T. (Eds.). *Resilience, Response, and Risk in Water Systems*; Springer: Singapore, 2020.
24. Lumb, A.; Sharma, T. C.; Bibeault, J.-F. A Review of Genesis and Evolution of Water Quality Index (WQI) and Some Future Directions. In *Water Quality, Exposure and Health*; 2011, Vol. 3 (1), pp 11–24.
25. Mahagamage, M. G. Y. L.; Manage, P. M. In *Water Quality Index (CCME - WQI) Based Assessment Study of Water Quality in Kelani River Basin, Sri Lanka*, The 1st Environment and Natural Resources International Conference, November 2014, 199–204.
26. Mirrasooli, E.; Ghorbani, R.; Molaei, M. Water Quality Assessment in Terms of Water Quality Index (WQI): Case Study-Gorganroud River, Golestan Province, Iran. *Open J. Ecol.* **2017,** *07* (13), 640–649.
27. Mohammed, S. E.; Adel Abdulrazzaq, K. Developing Water Quality Index to Assess the Quality of the Drinking Water. *Civil Eng. J.* **2018,** *4* (10), 2345–2350.
28. Nayak, S. K.; Mohanty, C. R. Influence of Physicochemical Parameters on Surface Water Quality: A Case Study of the Brahmani River, India. *Arab. J. Geosci.* **2018,** *11* (17), 514–520.
29. Oni, O.; Fasakin, O. The use of Water Quality Index Method to Determine the Potability of Surface Water and Groundwater in the Vicinity of a Municipal Solid Waste Dumpsite in Nigeria. *Am. J. Eng. Res. (AJER)* **2016,** *5* (10), 96–101.
30. Paun, I.; Cruceru, L. V.; Chiriac, F. L.; Niculescu, M.; Vasile, G. G.; Marin, N. M. Water Quality Indices - Methods for Evaluating the Quality of Drinking Water. *Cude* **2016,** *2001,* 395–402. https://doi.org/10.21698/simi.2016.0055
31. Rahman, M. In *Quality Assessment of Water Using Water Quality Index Method : A Case Study on the old Brahmaputra River, Jamalpur Segment Method - case Study on the Old Brahmaputra River*, 5th International Conference on Civil Engineering for Sustainable Development, February 2021, p 8.
32. Samadi, M. T.; Sadeghi, S.; Rahmani, A.; Saghi, M. H. Survey of Water Quality in Moradbeik River Basis on WQI Index by GIS. *Environ. Health Eng. Manag. J.* **2015,** *2* (1), 7–11.
33. Singaraja, C.; Chidambaram, S.; Anandhan, P.; Prasanna, M. V.; Thivya, C.; Thilagavathi, R.; Sarathidasan, J. Determination of the Utility of Groundwater with Respect to the Geochemical Parameters: A Case Study from Tuticorin District of Tamil Nadu (India). *Environ., Dev. Sustain.* **2014,** *16* (3), 689–721.
34. Singh, P. K.; Panigrahy, B. P.; Verma, P.; Kumar, B. *Evaluation of the Surface Water Quality Index of Jharia Coal Mining Region and Its Management of Surface Water Resources*; 2018, pp 429–437. https://doi.org/10.1007/978-981-10-5792-2_34
35. Singh, S.; Ghosh, N. C.; Krishan, G.; Galkate, R.; Thomas, T.; Jaiswal, R. K. Development of an Overall Water Quality Index (OWQI) for Surface Water in Indian Context. *Curr. World Environ.* **2015,** *10* (3), 813–822.
36. Singh, V. P.; Yadav, S.; Yadava, R. N. (Eds.). *Environmental Pollution*; Springer: Singapore, 2018; Vol. 77.

37. Tirkey, P.; Bhattacharya, T.; Chakraborty, S. Water Quality Indices- Important Tools for Water Quality Assessment. *Int. J. Adv. Chem.* **2013,** *1* (1), 15–28.
38. Tyagi, S.; Sharma, B.; Singh, P.; Dobhal, R. Water Quality Assessment in Terms of Water Quality Index. *Am. J. Water Resour.* **2013,** *1* (3), 34–38.
39. UNESCO. *Resources of the World and Their Use*. IHP-VI, Series on Groundwater No. 6. In Ihp-Vi, Series on Groundwater No. (Vol. 6), 2004.
40. Vasanthavigar, M.; Srinivasamoorthy, K.; Vijayaragavan, K.; Rajiv Ganthi, R.; Chidambaram, S.; Anandhan, P.; Manivannan, R.; Vasudevan, S. Application of Water Quality Index for Groundwater Quality Assessment: Thirumanimuttar Sub-basin, Tamilnadu, India. *Environ. Monit. Assess.* **2010,** *171* (1–4), 595–609.
41. World Health Organization. *Guidelines for Drinking-Water Quality*, 4th Ed; Incorporating the First And Second Addenda, 2022.

CHAPTER 9

APPLICATION OF SLUDGE FROM SEQUENCING BATCH REACTOR—SEWAGE TREATMENT PLANT: A CASE STUDY

AKASH GARG, PRIYA AGARWAL, and GAURAV SAINI

ABSTRACT

Chemical and synthetic fertilizers are commonly used in agricultural fields worldwide, especially in developing and underdeveloped countries, due to resource and alternative scarcity and an ever-increasing pressure on limited land resources due to the burgeoning population. This results in damage to different environmental components, including (but not limited to) water (both surface and groundwater), soil, crops, animals, and other living beings. Sustainable alternatives need to be developed and implemented to minimize environmental damage while ensuring an adequate supply of food and materials generated from agriculture. The sludge generated from the sewage treatment plants is another nuisance. The application of sewage sludge as fertilizer can serve as a solution to both these problems. In this study, sludge generated by a sequencing batch reactor (SBR)-based sewage treatment plant (STP) was used as a fertilizer for tomato crops, and growth of the plants was studied for a duration of 12 weeks. It was found that the sludge resulted in good growth (characterized by plant height, wet weight, dry weight, etc.), and 15% sludge was found to yield the most optimal results. This provides a sustainable means for the simultaneous reuse of sludge while reducing the farmers' dependence on chemical fertilizers.

Sustainable Use of Wastewater in Agriculture: Applications, Methods, and Technology. Megh R. Goyal & Gaurav Saini (Eds.)

9.1 INTRODUCTION

The increasing population with expanded urbanization and the standard of living, produces a large amount of city sewage.[13] Thus, the proper treatment and management of sewage is an essential step for not only a healthy and good quality of human life but for the environment as well. Nowadays, sewage needs to be properly treated before being disposed of in waterways to make it safe for the receiving water bodies. Sewage treatment generally includes the removal of organic matter, microorganisms, and other contaminants from the raw wastewater and making it fit for disposal and/or reuse.[15]

Sewage sludge, a semisolid residue, is obtained from the wastewater treatment plant after the process of sedimentation. The conventional wastewater treatment process generates two types of sewage sludge: primary sludge, which includes suspended solids and organics through gravitational sedimentation (from the primary sedimentation tanks or clarifiers), and secondary sludge generated from the secondary settling tank/s (after the secondary treatment processes) including microorganisms.[24]

The concentration and quantity of sludge are a function of the characteristics of the received wastewater and the treatment process followed.[42] The generated sewage sludge is generally disposed of in a number of ways including, sanitary landfills, incineration, or reused through its applications on land for structural soil improvement, soil amendment, and as a soil buffer.[5] The reuse of the sewage sludge costs less than incineration or sanitary landfill.[24] Firstly, landfill requires huge space and the effluent leaching out has the potential to contaminate the groundwater, especially in regions with high water tables.[8,43]

Similarly, the process of incineration can potentially result in 70% mass reduction and 90% volume reduction of sludge.[14,25] Unfortunately, the same comes at a high energy cost as well as concomitant air pollution from the generated gases. The main sources of ecological degradation are the residues from incineration and the air pollutants that are released into the atmosphere.[1,25] Moreover, the treated sludge used to cover landfills is considered transitional and unsustainable due to phosphorus and energy loss. Given the high cost and environmental issues related to the disposal of sewage sludge, a number of approaches to reduce sludge production have been tried, including the use of metabolic uncouplers.[35,36]

For these reasons, the practice of adding sludge to agricultural soil is considered the most ecological and economical as it provides nutrients, refines the physical properties of soil, and reduces human health risks.[13,44]

It can serve as a source of both the macro as well as the micronutrients that are required for the growth of plants.[22,31] It also contains significantly high levels of organic matter, which aids soil fertility.[26] This method of management of soil is very effective and economical as it enables nutrient recycling from organic wastes[3,30] and enhances the structure, permeability, porosity, and water retention capacity of soil which further improves crop yields.[21,37] The usage of treated municipal sewage sludge as a land amendment tool accounts for two key nutrients (nitrogen and phosphorus), which are essential for several crops.[24]

In most countries, the application of sludge in agriculture is considered as a final treatment. Several countries actively recommend the use of sewage sludge as a fertilizer. This is very common throughout the underdeveloped and developing world. According to Eurostat's official data, it is mostly practiced in Bulgaria, Cyprus, the Czech Republic, Hungary, Spain, Poland, Ireland, Greece, Portugal, Latvia, and Lithuania.[14] The sludge is added as a fertilizer/manure and conditioner to soil, in place of the commonly used chemical fertilizers. The sludge provides a significant amount of organic matter and nutrients to the soil/crop and enables sustainable disposal of the sludge. There are certain issues related to the application of sludge as a soil conditioner. The leaching of heavy metals (and their ability to bioaccumulate and enter the food chain) as well as the presence of pathogens and volatile organic compounds are causes of concern.[32] Arsenic in drinking water causes cancer, and also cardiovascular and neurological diseases.[38]

Therefore, sewage sludge can serve as an effective alternative to conventional fertilizers, with certain limitations.[9] Sewage sludge has an enduring impact on the microbial activity in the top soil layer (which is the key to its fertility).[40] It not only maintains the soil humus content but also improves the enzymatic activity therein.[4] As a source of nutrients, the combined application of chemical fertilizer and sewage sludge enhances soil fertility and farm profitability.[21,25] Therefore, it is encouraged for maintaining the soil fertility for sustainable agriculture. The integrated application of chemical fertilizers and sewage sludge improves crop yield and causes significant improvements in mineralizable nitrogen and microbial biomass.[34] Given the difficulties and costs associated with its storage, it is recommended that the sewage sludge is reused (rather than disposed of or treated) in a widespread activity such as agriculture.[20]

In recent years, sequential batch reactor (SBR) systems have been used widely to treat industrial effluents and municipal sewage due to their low cost and high pollutant removal efficiency.[27] SBR is a modification to the conventional activated sludge process (ASP) with several flexibilities and alternatives in both the process as well as the design which allows it to achieve the applicable wastewater discharge standards, as per the regulatory agencies.[13] The term SBR was first introduced by Prof. Robert L. Irvine at the University of Notre Dame, United States.[17] It consists of five stages, which work sequentially (and hence the name SBR): (i) filling (of sewage), (ii) reaction, (iii) settling, (iv) effluent removal, and (v) idle.

An SBR can efficiently remove key contaminants including biological oxygen demand (BOD), total suspended solids (TSS), and nutrients such as phosphorus, nitrogen, as well as ammonia.[15] SBR is extremely effective in nutrient removal for irrigation purposes and sludge production for fertilizers.[29]

The chemical differences between sludge obtained from SBR-STP and other conventional processes are mentioned in Table 9.1.

TABLE 9.1 Differences in the Chemical Composition of the SBR-STP Sludge and the Conventional STP Sludge.

	SBR-STP sludge	Conventional sludge
Al_2O_3	6.87%	8.90%
Br	–	0.50%
CaO	7.46%	7.30%
Fe_2O_3	5.10	0.90%
K_2O	1.83%	1.80%
MgO	3.37%	2.10%
MnO	0.11%	–
Na_2O	0.42%	0.40%
P_2O_3	–	1.30%
P_2O_5	3.43%	–
SiO_2	24.00%	54.50%
SO_3	2.22%	–
SrO	0.09%	–
TiO_2	1.10%	–
ZnO	0.35%	0.20%
Other	–	0.80%
Loss on ignition	43.65%	21.03%

Sewage sludge has been used to grow many crops such as rice,[6,19,42] wheat,[6,18,19] tomato,[16] sweet basil,[7] spinach (*Beta vulgaris*),[12,33] maize (*Zea mays L.*),[19] lettuce, and cucumber plants.[2] Not only the sludge but the wastewater from SBR can also be reused in agriculture for the purpose of irrigation.[23,29] This leads to the use of better nutrient value of wastewater, eliminating the application of fertilizer. Further, it assures the availability of wastewater to grow multiple crops.[23]

This chapter focuses on the application of SBR-STP sewage sludge for the growth of a native crop in the Delhi region, India. SBR technology is highly recommended for India and there are a number of STPs based on this technique. Yet there are hardly any studies that investigated the possibility of using SBR-STP sludge for plant growth. Different amounts of sludge were added to the crop and changes in key plant properties (including height, mass, root length, and so on) were studied. The results show a significant improvement in plant characteristics with the addition of SBR-STP's sludge. If taken up at a large scale, such sludge can be a potential game-changer, in terms of easier sludge disposal, reduction in chemical fertilizer usage, and overall, a sustainable solution to soils and their amendment.

9.2 LITERATURE REVIEW

While the use of sewage sludge as a soil conditioner is common in various states of the European Union, the actual consumption varies from state to state. Countries such as Denmark, France, Ireland, Spain, and the United Kingdom use more than half of the STP sludge as a soil conditioner for agricultural activities.[28] It is to be noted that sludge can be used either alone or in conjunction with chemical fertilizers.

In northern parts of India, especially the alluvial plains of the river Ganges, a dual cropping system is commonly practiced. In this system, farmers grow the two chief food crops: wheat and rice, as part of the Rabi-Kharif cropping cycle. This practice is safe, sustainable, and profitable, given the climatic, and soil conditions. In a study conducted to investigate the potential use of sewage sludge for such a cropping pattern, interesting observations were made.[19] When the chemical fertilizer recommended dose was augmented with 30 Mg/ha STP sludge, the mean plant height increased by 14.85% for rice and 13.90% for wheat, as compared to the control case with no augmentation by the sludge. The grain yield was also found to increase by 8.10% for rice and 18.90% for wheat. When only 70% of the recommended chemical fertilizer dose was used, along with 20 Mg/ha sludge, the productivity was comparable to that obtained by the control case of recommended chemical fertilizer dose only (no sludge addition).[19] In another study, the grain yield of maize increased by 9%, when sewage sludge was used at a dose of 20 Mg/ha. It was also reported that an application of 30 Mg/ha sludge can result in sustainability of the soil organic matter.[21]

In another field study, an addition of 15 tons of farm yard manure to the 100% recommended fertilizer dose significantly increased the yields of rice and wheat. In addition, it was found that this mix resulted in a higher number count of soil microbes, including bacteria, fungi and *actinomycetes*.[6]

In a field study on wheat, in calcareous soils, different doses (ranging from 10–100 tons per hectare) of STP sewage sludge were used, in addition to different doses of chemical fertilizer (ranging from 60–120 kg per hectare). It was observed that a sewage dose of 40 tons per hectare resulted in the highest levels of spike length, grain yield per plant, grain number per spike, and number of productive tillers per plant. In comparison, other parameters such as height of the plant, number of tillers per plant, and straw yield per plant achieved maximum values at a dosage of 80 tons per hectare of the STP sewage sludge.[18]

In a study investigating the potential of using sludge as a soil amendment tool, rice was used as a test crop and its growth, biomass, and yield were measured. It was found that sludge led to increased shoot length, leaf numbers, leaf area, and total plant biomass, along with a reduction in the root length. Most significantly, as much as a 137% increase in the crop yield was reported for a sludge dose of 12 kg/m^2. Unfortunately, the study also reported an increase in the heavy metal's availability in the soil, and the same were also found to have accumulated in different parts of the plants.[42]

A small-scale, indoor study was conducted using cherry tomatoes in pots. The temperature was controlled and these plants were grown in soils amended with a combination of fertilizer and biochar. A 64% increase in the tomato yield was observed at a normalized dose of 10 tons of biochar per hectare, applied in conjunction with chemical fertilizer, due to a potential increase in the availability of key nutrients (such as phosphorus and nitrogen) as well as better soil chemistry.[16]

9.3 MATERIALS AND METHODS

The soil samples, for this study, were collected from the agricultural land in the Dadri area of district Gautam Buddha Nagar in the state of Uttar Pradesh, India. These soils were dried using a standard laboratory oven at a temperature of 80 °C for 24-h duration. The dried soil was sieved using a 150-micron standard sieve. The sieved soil samples were characterized for different physicochemical parameters and heavy metals. The physicochemical parameters quantified in this study included pH, total organic carbon (TOC), total nitrogen (TN), phosphorus (P), potassium (K), and electrical conductivity (EC). The heavy metals characterized included iron (Fe), cadmium (Cd), manganese (Mn), lead (Pb), copper (Cu), and zinc (Zn). The physicochemical, as well as metal analysis, were conducted as per the standard analytical protocols.

The sewage sludge was collected from the 137 MLD (million liters per day) sewage treatment plant located in the Kasna region of the district Gautam Buddha Nagar in the state of Uttar Pradesh, India. This plant operates on the sequencing batch reactor (SBR) technology, a relatively new technique that is gaining significant traction in India, especially, for municipal wastewater treatment. The air-dried sludge was passed through a 150-micron standard sieve. The dried and sieved sludge was analyzed for various physicochemical parameters and also for heavy metal concentration, the same as the soil, using standard laboratory quantification protocols.

9.3.1 POTS PREPARATION

An easy-to-grow and native crop, tomato (*Solanum lycopersicum*), was used for studying the possibility of using SBR-STP's sludge as manure. The tomato plant seeds were grown in the polybags. To carry out the study, 150 such polybag pots were prepared, and an investigation was carried out. The tap water was used for irrigating the plants, using the same water source that is used for irrigating the crops in that region. Pots were further classified into five sets according to the percentage of sludge used. Specifically, five different percentages of sludge, varying from 0% to 15% were used for this study. Pots of size 12 cm diameter were filled with a total mass of 1 kg (consisting of soil and/or sludge, as detailed later), obtained from local agricultural fields. Tomatoes were grown under five different conditions, such as:

1. 0% sludge + 100% farmyard soil
2. 2% sludge + 98% farmyard soil
3. 5% sludge + 95 % farmyard soil
4. 10% sludge + 90% farmyard soil
5. 15% sludge + 85% farmyard soil

The sludge and soil were well mixed in a larger container before preparing the polybag pots. For adequate plant growth, the pots were kept in an open space to ensure adequate sunlight and air movement. The plant growth was studied promptly throughout this study. The plants were allowed to grow for a total duration of 12 weeks, before harvesting the plants for characterization. The growth of the plants was measured every week in terms of the plant height.

9.3.2 DIGESTION OF SOIL, SLUDGE, AND PLANT SAMPLES FOR HEAVY METAL ANALYSIS

9.3.2.1 SOIL DIGESTION

First, a conical flask is filled with 0.5 g of the soil sample. Nitric acid (HNO3) of 10 mL is then added to the conical flask and the mixture is heated using a hot plate maintained at 35 °C. After 1 h of heating, 3 mL of perchloric acid ($HClO_4$) is added to the conical flask, mixed and heated again using the hot plate for sample digestion. Once the digestion process is completed, the digested sample is mixed with 100 mL of distilled water were combined and then passed through a Whatman 42 filter paper. The resulting filtered soil sample is stored in plastic bottles for further analysis.

9.3.2.2 SLUDGE DIGESTION

The sludge digestion protocol is similar to that used for soil digestion, with minor differences. As a first step in the process, a conical flask is filled with 0.5 g of sludge sample. Then, 15 mL of nitric acid (HNO_3) is added to the flask and the mixture is heated using a hot plate maintained at 35 °C. After 1 h of heating duration, 5 mL of perchloric acid ($HClO_4$) is added to the flask, mixed and the resulting mixture is heated using a hot plate for the digestion. After the digestion of the sample, 100 mL of distilled water is added to it, and the resulting mixture is filtered through a Whatman 42 filter paper. The resulting filtered sample is stored in the plastic bottles for further analysis.

9.3.2.3 PLANT DIGESTION

Once the tomato plants attained a significant amount of growth (through visual inspection and height measurements), they were taken out from the polybag pots and washed with tap water to remove the soil particles attached to the plant & roots. Parameters such as the total plant height, root length, and biomass accumulation were measured for each plant. After that, plants were oven-dried at 80°C for 24 h (shown in Fig. 3.3) and the dry weight of the plants was taken. Then, the leaf, stem, and roots were separated for analysis of heavy metals and converted into the powder form which was passed through a 150-micron standard sieve; 0.5 g of the sample was taken in separate conical flasks for different leaves, stems, and roots for plants grown in different concentrations of sludge. Then, 10 mL of nitric acid (HNO_3) was added to the conical flask and heated using a hot plate, maintained at 35 °C. After 1 h of heating period, 3 mL of perchloric acid ($HClO_4$) was added to the sample in the flask, mixed, and the flask was heated again using the hot plate for digestion.

After the completion of the digestion step, 100 mL of distilled water was added to the conical flask and the mixture was filtered through a Whatman 42 filter paper. Then, the filtered sample was stored in the plastic bottles for further analysis. After the digestion, the samples were taken for heavy metals analysis using an atomic absorption spectroscope (AAS).

9.4 RESULTS AND DISCUSSION

The study was conducted to investigate the possibility of using SBR STP's sludge as a potential manure/fertilizer for agriculture and thus reduce the dependency on chemical fertilizers, while also reducing the disposal cost of the generated sludge. A native crop, tomato, was grown by using different amounts of sludge and the resultant growth pattern and related parameters were examined.

The characterization of the various physicochemical parameters of soil, sludge, and plants was carried out in the laboratory using a standard method and the results are summarized in Tables 9.1–9.3.

TABLE 9.2 Physicochemical Characteristics of Soil and SBR STP Sludge.

Parameters	Unit	Soil	Sewage sludge
EC	(moh/cm)	380	568
K	Kg/ha	267.68	1565
N	Kg/ha	167.2	1100
P	Mg/L	0.04	0.12
pH	–	8.56	6.49
TOC	(%)	3.50	6.35

The physicochemical characteristics of the soil and sludge (pH, EC, N, P, K, and TOC) were determined and included in Table 9.2. It is noticeable that soil and sludge differ in terms of characteristics. For example, the SBR-STP's sludge has a significantly high content of potassium and nitrogen as compared to the soil used for plant growth. The soil, on the other hand, was determined to be significantly more alkaline as compared to the sludge.

TABLE 9.3 Heavy Metal Concentration in Soil and Sludge Samples.

Parameter	Unit	Soil	Sewage sludge
Cd	ppm	ND*	ND*
Cu	ppm	0.138	0.446
Fe	ppm	45.375	24.482
Mn	ppm	2.081	1.211
Pb	ppm	0.520	1.192
Zn	ppm	0.696	3.277

*ND: Not detected

TABLE 9.4 Summary of Plant Growth Parameters.

% Sludge	Wet weight (gm)	Dry weight (gm)	Avg. height (cm)	Avg. root length (cm)
0%	10.00	1.10	26.20	10.00
2%	10.05	0.97	25.90	10.30
5%	16.60	1.25	29.40	11.36
10%	17.25	1.83	33.60	11.90
15%	23.32	3.12	37.37	16.06

Table 9.3 indicates the concentration of various heavy metals present in the soil and sludge which were analyzed in the laboratory using standard method. The heavy metals characterization was done through AAS. Similar to the physical characteristics, the soil and the STP sludge differed in terms of the heavy metal concentrations. Cadmium was not detected in any of the soil and sludge samples. The sludge derived from SBR-STP had a significantly higher concentration of zinc, lead, and copper. Soil, on the other hand, was found to be richer in terms of iron and manganese content, as compared to sewage sludge.

Table 9.4 indicates various plant growth parameters measured in this study. The growth of plants was determined with respect to the total weight of the plant, root length, plant height, plant weight, and dry weight. For this, the seeds of the tomato plant are sown in pots with different mixes of sludge and soil at different percentages, i.e., 0, 2, 5, 10, and 15%. From Table 9.4, it can be seen that average plant height increased as a function of increasing sludge content in the soil. The only exception was the plants grown in 2% sludge, wherein the plant height reduced slightly from the control samples (plants grown without sludge). For all other cases, the increasing sludge concentration resulted in a taller plant, with the maximum plant height of 37.37 cm attained with 15% sludge content in the soil.

A similar trend was observed with the wet weight of the plant, wherein successive increases in sludge concentrations resulted in increased wet weight. For the control sample (containing 0% sludge), the wet weight was measured to be 10 g, whereas for the maximum sludge concentration of 15%, the corresponding value was 23.32 g (Table 9.4). This shows that 15% sludge concentration was the most optimal from the perspective of wet weight and plant height.

Similar results were found for dry plant weights, and (with the exception of 2% sludge concentration), a rise in sludge concentration was followed by an increase in the dry weight of the plant. For the control samples, the average dry weight was measured as 1.1 g, which dropped to 0.97 g for 2% sludge concentration samples. For other sludge concentrations (5%, 10%, and 15%), the dry weight increased successively with the highest dry weight of 3.12 g for 15% sludge samples (Table 9.4). These results are in line with the observations for wet plant weight and plant heights.

The plant growth and average plant root length were measured from the uprooted plants at the end of the growth cycle. The results, for different sludge concentrations, are mentioned in Table 9.4. It can be observed that an increase in sludge concentrations resulted in longer plant roots. For the control sample, the average plant root length was measured as 10 cm. This increased to 16.06 cm for the plants grown in 15% SBR STP sludge. These results confirm that 15% SBR STP sludge was the most optimal sludge content for plant growth.

The plant growth measured in terms of plant height, plant weight, dry weight, and root length was presented graphically.

Figure 9.1 shows the average weekly growth of plants calculated when no (0%) sludge fertilizer was added to the soil. Average weekly plant growth is calculated by recording the plant height over a week and the average is calculated using the growth rate formula. It was seen that the average height of the plants increased over a period of time but the average height of the 0% sludge plants was comparatively less than that of sludge plants.

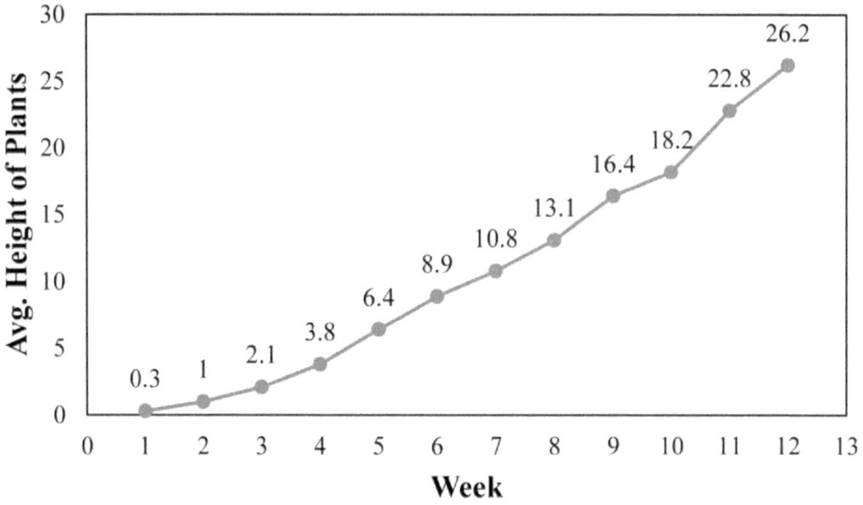

FIGURE 9.1 Weekly average variation of plant height with 0% sludge addition.

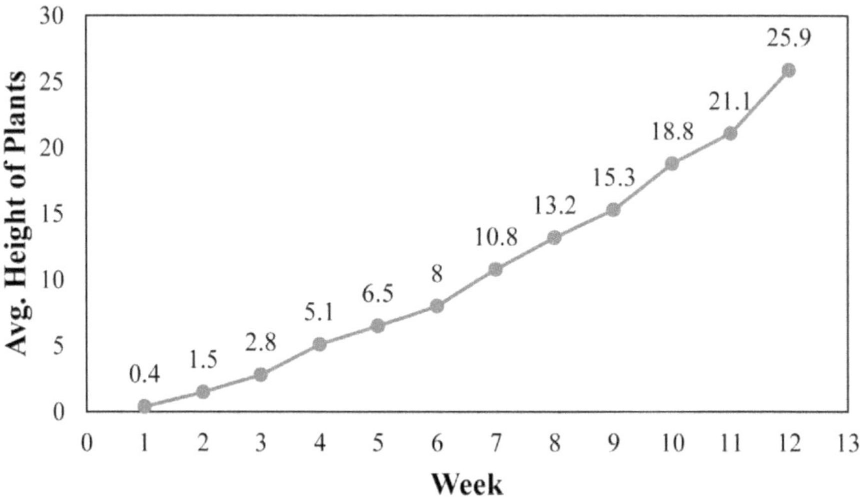

FIGURE 9.2 Weekly average variation of plant heights with 2% sludge addition.

Figure 9.2 shows the average weekly growth of plants when 2% of SBR sludge fertilizer is mixed with soil. It is evident from the graph that the average weekly growth rate of plants improved with the addition of sludge in the soil. Over time, it was observed that the plants' average height increased. It was noticed that the plants' average height was almost the same as that of 0% sludge plants.

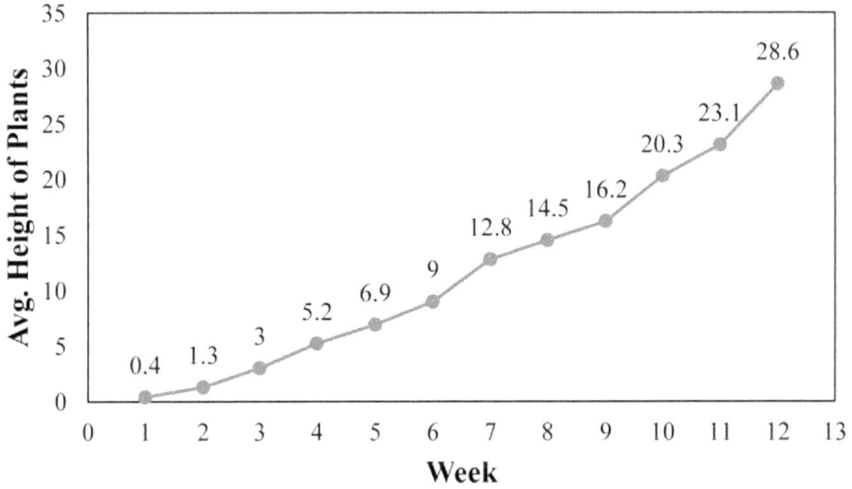

FIGURE 9.3 Weekly average variation of plant heights with 5% sludge addition.

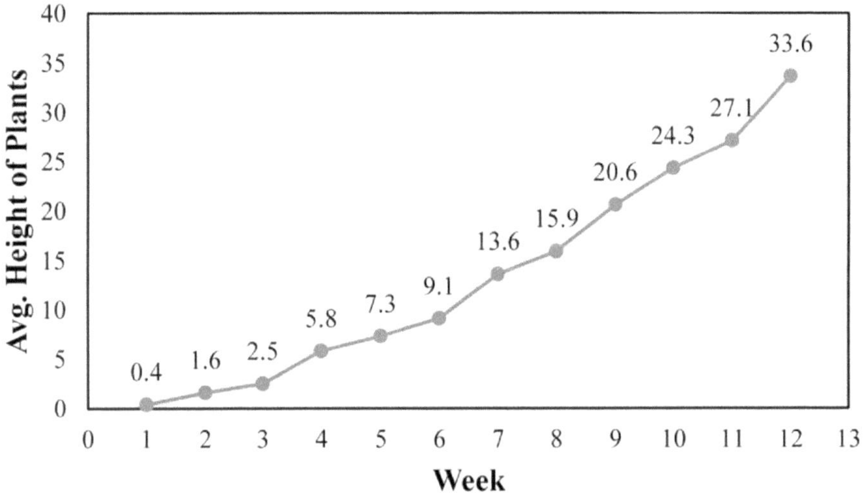

FIGURE 9.4 Weekly average variation of plant heights with 10% sludge addition.

Figure 9.3 gives an account of the average weekly growth of plants when 5% sludge is added to soil; an account of the growth of plants is taken over a week and the average growth of the plants is calculated using the growth formula. Over time, it was observed that the plants' average height increased. It was noticed that the average height of the plant was more than that of 0% and 2% sludge plants.

Figure 9.4 includes an analytical observation of the average weekly growth of plants calculated when 10% SBR-based sludge fertilizer is added to the soil. The growth rate has improved when compared to the previously used sludge soil ratio. It shows the weekly growth of 10% of sludge plants. Over time, it was observed that the plants' average height increased. It was observed that the plant's average height was more than that of 0%, 2%, and 5% sludge plants. The growth of the plants increased gradually.

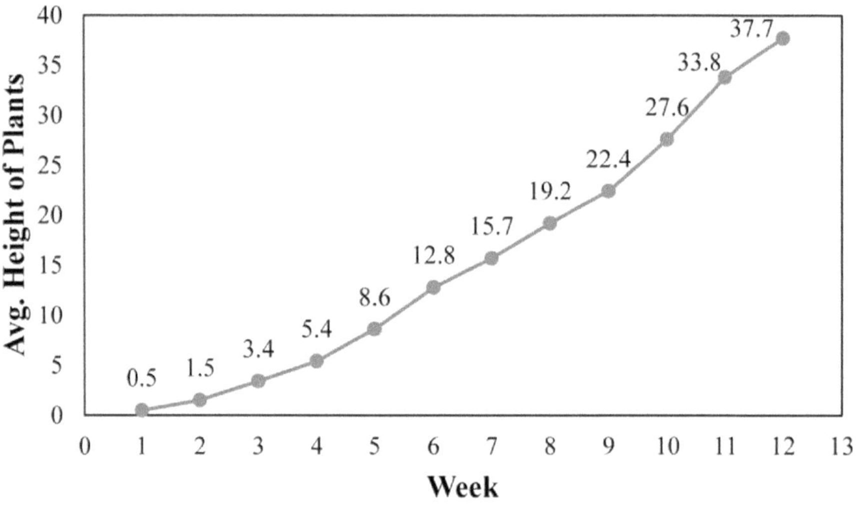

FIGURE 9.5 Weekly average variation of plant heights with 15% sludge addition.

Figure 9.5 shows the average weekly growth of plants when 15% sludge-based fertilizer is added to the soil. It can be observed that the growth rate is maximum for the plants. It was seen that the average height of the plants increased over a period of time and was found more than that of 0%, 2%, 5%, and 10% sludge plants. The growth of the plants increased gradually.

In other studies, the yield of ladyfinger (*Abelmoschus esculentus*) was increased significantly by 75% and 135% at 20% and 40%, respectively, on sewage sludge application in comparison with plants grown in non-amended soils.[41] The amendment rates of 10, 20, 30, and 40 g per kg of soil were applied to agricultural soil to improve the growth of *Corchorus*. The highest enhanced growth resulted in an amendment rate of 20 g/kg.[11]

According to another study, applying SS at rates of 0.5%, 1.0%, 1.5%, and 2.5% improved rice grain yield by 114.4%, 267.3%, 537.1%, and 811.7%, respectively, in comparison to the control. Comparing the rice plants at

2.5% SS to the control, the total root length, total volume, root surface area, and root diameter all increased by 467.0%, 393.9%, 588.5%, and 25.2%, respectively.[37]

The study in this chapter shows that SBR STP's sludge can be successfully used as manure, in place of chemical, and synthetic fertilizers and can help farmers gain a higher yield. This can provide a viable, sustainable alternative to the farmers while minimizing the amount of sludge that the STPs need to dispose of. This is a clear example of converting industrial waste to a useful product. This is of special relevance to the farmers in developing and under-developed countries wherein the farmers lack resources and are suffering from utter poverty. It is to be noted that sludge usage for agricultural activities is accompanied in many cases by heavy metal leaching into soil and possibly groundwater. This can also cause the uptake of contaminants (present in the sludge) by the plants. Since SBR-STPs are somewhat of a novelty technique, hardly any studies are available in such directions. Hopefully, future research efforts, including long-term studies on the effect of SBR STP's sludge usage as a soil conditioner on the soil and plants can further complement our work and provide a clear road map for its adoption.

9.5 SUMMARY

Experimentations were carried out to study the possibility of the use of the SBR-type STP sludge as a potential fertilizer in the NCR area and five different concentrations of sludge were used and mixed with the soil and 30 tomato seeds were seeded for each concentration. The growth of plants was studied over time. Based on the results it is concluded that the sludge from SBR-type STP at Kasna, Greater Noida can be effectively used as a fertilizer. For the tomato plants planted in this study, it was found that 15% of STP sludge was the most optimal percentage in terms of average plant height. It was also found that the length of the roots of 15% sludge plants was the longest and their wet and dry weights were also more than as compared to the lower percentages of the sludge. One can observe that the use of 15% sludge application as fertilizer turned out to be the best results in terms of the wet weight, dry weight, average plant height, and average root length of the tomato plants, after 12 weeks of the growth period. The dry weight was calculated after drying the harvested plants at 80°C for 24 h. It was found that the sludge contains heavy metals such as Pb (1.192 ppm), Cu (0.446 ppm), Fe (24.482 ppm), Zn (3.277 ppm), and Mn (1.211 ppm). It was found that branches contain less Cu, Fe, Zn, Mn, and more Pb than leaves. All the heavy metals found in the soil were under the saturation level. The utilization of sewage sludge as a possible source of nutrients for tomato seedlings can meet the need for environmentally safe disposal.

KEYWORDS

- **heavy metals**
- **nutrients**
- **plant growth**
- **sequencing batch reactor (SBR)**
- **sustainable**
- **tomato**

REFERENCES

1. Abuşoğlu, A.; Özahi, E.; Kutlar, A. I.; Al-jaf, H. Life Cycle Assessment (LCA) of Digested Sewage Sludge Incineration for Heat and Power Production. *J. Clean. Prod.* **2017,** *142* (4), 1684–1692.
2. Al-Saikhan, M. S.; Badr, E. A.; Babeker, M. Y. Study of Sewage Sludge use for the Cultivation of Plants and its Effects on Soil Properties in Al Ahsa. *Sci. J. King Faisal Univ.* **2020,** *21* (2), 88–95.
3. Antonkiewicz, J.; Kołodziej, B.; Bielinska, E. J.; Glen-Karolczyk, K. The use of Macro Elements from Municipal Sewage Sludge by the Multiflora Rose and the *Virginia fanpetals*. *J. Ecol. Eng.* **2018,** *19,* 1–13.
4. Armenta, R.; Vaca, R.; Lugo, J.; del Aguila, P. Microbiological and Biochemical Properties of an Agricultural Mexican Soil Amended with Sewage Sludge. Revista Brasileira *de* Ciência *do Solo* (Brazilian J. of Soil Science) **2011,** *36,* 1646–1655.
5. Babatunde, A. O.; Zhao, Y. Q. Constructive approaches Toward Water Treatment Works Sludge Management: An International Review of Beneficial Reuses. *Crit. Rev. Environ. Sci. Technol.* **2007,** *37,* 129–164.
6. Bhatt, B.; Chandra, R.; Ram, S.; Pareek, N. Long-term Effects of Fertilization and Manuring on Productivity and Soil Biological Properties Under Rice (*Oryza sativa*)-Wheat (*Triticum aestivum*) Sequence in Mollisols. *Arch. Agron. Soil Sci.* **2016,** *62* (8), 1109–1122.
7. Burducea, M.; Lobiuc, A.; Asandulesa, M.; Zaltariov, M. F.; Burducea, I.; Popescu, S. M.; Zheljazkov, V. D. Effects of Sewage Sludge Amendments on the Growth and Physiology of Sweet Basil. *Agronomy* **2019,** *9,* 548–555.
8. Costa, A. M.; Alfaia, R. G. S. M.; Campos, J. C. Landfill Leachate Treatment in Brazil –An Overview. *J. Environ. Manag.* **2019,** *232,* 110–116.

9. Deeks, L. K.; Chaney, K.; Murray, C.; Sakrabani, R.; Gedara, S.; Le, M.; Tyrrel, S.; Pawlett, M.; Read, R.; Smith, G. H. A New Sludge-derived Organo-Mineral Fertilizer Gives Similar Crop Yields as Conventional Fertilizers. *Agron. Sustain. Dev.* **2013,** *33,* 539–549.
10. Dutta, A.; Sarkar, S. Sequencing Batch Reactor for Wastewater Treatment: Recent Advances. *Curr. Pollut. Rep.* **2015,** *1,* 177–190.
11. Eid, E. M.; Hussain, A. A.; Taher, M. A.; Galal, T. M.; Shaltout, K. H.; Sewelam, N. Sewage Sludge Application Enhances the Growth of *Corchorus olitorius* Plants and Provides a Sustainable Practice for Nutrient Recirculation in Agricultural Soils. *J. Soil Sci. Plant Nutrit.* **2019,** *20,* 149–159.
12. Eid, E. M.; El-Bebany, A. F.; Alrumman, S. A.; Abd El-latif, H.; Taher, M. A.; Fawy K. F. Effects of Different Sewage Sludge Applications on Heavy Metal Accumulation, Growth and Yield of Spinach (*Spinacia oleracea L.*). *Int. J. Phytoremediation* **2017,** *19,* 340–349.
13. Elmi, A.; Al-Khaldy, A.; Al-Olayan, M. Sewage Sludge Land Application: Balancing Act between Agronomic Benefits and Environmental Concerns. *J. Clean. Prod.* **2020,** *250*, E-article ID 119512.
14. Eurostat Sewage Sludge Production and Disposal [Online]. 2020; http://appsso.eurostat.ec.europa.eu/nui/show.do?lang=en&dataset=env_ww_spd; (accessed Mar 31, 2023).
15. Gupta, D.; Singh, S. K. Energy use and Greenhouse Gas Emissions from Waste Water Treatment Plants. *Int. J. Environ. Eng.* **2015,** *7* (1), 1–10.
16. Hossain, M. K.; Strezov, V.; Chan, K. Y.; Nelson, P. F. Agronomic Properties of Wastewater Sludge Biochar and Bioavailability of Metals in Production of Cherry Tomato (*Lycopersicon esculentum*). *Chemosphere* **2010,** *78* (9), 1167–1171.
17. Irvine, R. L.; Davis, W. B. In *Use of Sequencing Batch Reactor for Wastewater Treatment.* CPC International, Corpus Christi, TX. Presented at the 26th Annual Industrial Waste Conference, West Lafayette - IN: Purdue University, 1971.
18. Jamil, M.; Qasim, M.; Umar, M.; Rehman, K. Impact of Organic Wastes (sewage sludge) on the Yield of Wheat (*Triticum aestivum L.*) in a Calcareous Soil. *Int. J. Agric. Biol.* **2004,** *6* (3), 465–467
19. Jatav, S. S.; Singh, S. K.; Parihar, M.; Alsuhaibani, A. M.; Gaber, A.; Hossain, A. Application of Sewage Sludge in a Rice (*Oryza sativa* L.) & Wheat (*Triticum aestivum* L.) System Influences the Growth, Yield, Quality and Heavy Metals Accumulation of Rice and Wheat in the Northern Gangetic Alluvial Plain. *Life (Basel)* **2022,** *12* (4), 484–495.
20. Kacprzak, M.; Neczaj, E.; Fijałkowski, K.; Grobelak, A.; Grosser A.; Worwag, M.; Rorat, A.; Brattebo, H.; Almås, A.; Singh, B. R. Sewage sludge Disposal Strategies for Sustainable Development. *Environ. Res.* **2017,** *156,* 39–46.
21. Kayikcioglu, H. H.; Yener, H.; Ongun, A. R.; Okur, B. Evaluation of Soil and Plant Health Associated with Successive Three-year sewage Sludge Field Applications Under Semi-arid Biodegradation Condition. *Arch. Agron. Soil Sci.* **2019,** *65,* 1659–1676.
22. Kominko, H.; Gorazda, K.; Wzorek, Z. Potentiality of Sewage Sludge-based Organo-mineral Fertilizer Production in Poland Considering Nutrient Value, Heavy Metal Content and Phytotoxicity for Rapeseed Crops. *J. Environ. Manag.* **2019,** *248*, E-article ID 109283.
23. Kurian, M.; Reddy, V. R.; Dietz, T.; Brdjanovic D. Wastewater Re-use for Peri-urban Agriculture: A Viable Option for Adaptive Water Management? *Sustain. Sci.* **2013,** *8,* 47–59.

24. Lamastra, L.; Suciu, N. A.; Trevisan, M. Sewage Sludge for Sustainable Agriculture: Contaminants' Contents and Potential use as Fertilizer. Chem. *Biol. Technol. Agric.* **2018,** *5* (10), 156–162.
25. Lynn, C. J.; Dhir, R. K.; Ghataora, G. S. Environmental Impacts of Sewage Sludge Ash in Construction: Leaching Assessment. *Resour., Conserv. Recycl.* **2018,** *136,* 306–314.
26. Malinowska, E.; Jankowski, K.; Wiśniewska-Kadźa, B.; Sosnowski, J.; Kolczarek, R. Effect of Different Methods of Treatment of Municipal Sewage Sludge on Their Physicochemical Properties and Their Agricultural Utilization. *J. Ecol. Eng.* **2015,** *16* (2), 76–81.
27. Metcalf & Eddy, *Wastewater Engineering Treatment: Disposal and Reuse.*McGraw Hill Book Co.: New York, USA, 2003.
28. Milieu Ltd., WRc, Risk and Policy Analysts (RPA) Ltd. Environmental, Economic and Social Impacts of the Use of Sewage Sludge on Land. Final Report, Part III: Project Interim Reports 2010 [Online]; http://ec.europa.eu/environment/archives/waste/sludge/pdf/part_iii_report.pdf; (Accessed 20 Mar 2023).
29. More, R. P.; Thackeray, S. B. Review Paper on Use of Sequencing Batch Reactor Treated Wastewater on Urban Agriculture. *Int. J. Sci. Res.* **2015,** *6* (3), 2346–2348.
30. Nahm, M.; Morhart, C. Virginia mallow (*Sida hermaphrodita* (L.)) as Perennial Multipurpose Crop: Biomass Yields, Energetic Valorization, Utilization Potentials, and Management Perspectives. *Glob. Change Biol.: Bioenerg.* **2018,** *10,* 393–404.
31. Ociepa-Kubicka, A.; Pachura, P. The Use of Sewage Sludge and Compost for Fertilization of Energy Crops on the example of *Miscanthus* and Virginia Mallow. *Rocznik Ochrona* Srodowiska **2013,** *15,* 2267–2278.
32. Official Gazette, *The Ordinance on the Protection of the Agricultural Land from Pollution* (No. 71/2019), **2019,** 1–4; online; https://narodne-novine.nn.hr/clanci/sluzbeni/2019_07_71_1507.html (accessed 31 Mar 2023).
33. Roy, T.; Singh, R. D.; Biswas, D. R.; Patra, A. K. Effect of Sewage Sludge and Inorganic Fertilizers on Productivity and Micronutrients Accumulation by Palak (Beta vulgaris) and Their Availability in a *Typic Haplustept. J. Indian Soc. Soil Sci.* **2013,** *61* (3), 207–218.
34. Roy, T.; Biswas, D. R.; Ghosh, A.; Patra, A. K.; Singh, R. D.; Sarkar, A.; Biswas, S. S. Dynamics of Culturable Microbial Fraction in an Inceptisol under Short-term Amendment with Municipal Sludge from Different Sources. *Appl. Soil Ecol.* **2019,** *136,* 116–121.
35. Saini, G. Metabolic Uncoupling: Biomass Control Strategy in Microbial Processes. *Journal* of *Microbial & Biochemical Technology* **2014,** *6* (5), 185–192.
36. Saini, G.; Wood, B. D. Metabolic Uncoupling of *Shewanella oneidensis* MR-1, under the Influence of Excess Substrate and 3, 3′, 4′, 5-Tetrachlorosalicylanilide (TCS). *Biotechnol. Bioeng.* **2007,** *99* (6), 1352–1360.
37. Shan, Y.; Lv, M.; Zuo, W.; Tang, Z.; Ding, C.; Yu, Z.; Shen, Z.; Gu, C.; Bai, Y. Sewage Sludge Application Enhances Soil Properties and Rice Growth in a Salt-affected Mudflat Soil. Sci. Rep. 2021, *11,* 282–291.
38. Sharma, P.; Goel, G.; Ashekuzzaman, S. M.; Saini, G.; Singh, R. Groundwater Arsenic in South-East Asia: Extent, Effects and Solutions. *Asian J. Water*, *Environ. Pollut.* **2014,** *11* (2), 1–11.
39. Seleiman, M. F.; Santanen, A.; Jaakkola, S.; Ekholm, P.; Hartikainen, H.; Stoddard, F. L.; Mäkelä, P. S. Biomass Yield and Quality Of Bioenergy Crops Grown with Synthetic and Organic Fertilizers. *Biomass Bioenerg.* **2013,** *59,* 477–485.

40. Siebielec, G.; Siebielec, S.; Lipski, D. Long-term Impact of Sewage Sludge, Digestate and Mineral Fertilizers on Plant Yield and Soil Biological Activity. *J. Clean. Prod.* **2018,** *87,* 372–379.
41. Singh, R. P.; Agrawal, M. Use of Sewage Sludge as Fertilizer Supplement for *Abelmoschus esculentus* Plants: Physiological, Biochemical and Growth Responses. *Int. J. Environ. Waste Manag.* **2009,** *3,* 91–106.
42. Singh, R. P.; Agrawal, M. Variations in Heavy Metal Accumulation, Growth and Yield of Rice Plants Grown at different Sewage Sludge Amendment Rates. *Ecotoxicol. Environ. Saf.* **2010,** *73* (4), 632–641.
43. Ziyang, L.; Youcai, Z.; Tao, Y.; Yu, S.; Huili, C.; Nanwen, Z.; Renhua, H. Natural Attenuation and Characterization of Contaminants Composition in Landfill Leachate under Different Disposing Ages. *Sci. Total Environ.* 2009, *407* (10), 3385–3391.
44. Zhou, G.; Gu, Y.; Yuan, H.; Gong, Y.; Wu, Y. Selecting Sustainable Technologies for Disposal of Municipal Sewage Sludge Using a Multi-criterion Decision-making Method: A Case Study from China. *Resour., Conserv. Recycl.* **2020,** *161,* E-article ID 104881.

CHAPTER 10

SCOPE OF RAINWATER HARVESTING FOR FOOD SECURITY AND FLOOD MITIGATION: A REVIEW

USMAN USMAN ALIYU, AUWAL ALHASSAN MUSA, ABDULHAYAT MUHD JIBRIN, LASMAR GARBA, AWAISU SHAFIU IBRAHIM, YUSUF BASHIR, IBRAHIM ABDULLAHI IBRAHIM, and MUSTAPHA NUHU GARKO

ABSTRACT

As an important and integral natural resource, water has considerable influence on health, wealth, food creation, and every other useful production for us. Freshwater scarcity is one of the few scenarios and among the top topics of discussion due to its usage to the exponentially increasing populace in some regions across the world. This scarcity has been caused mainly because of inconsistent geographical arrangements and temporal dispersion of precipitation intensity in arid and semi-arid regions. Also, spontaneous imbalances between demand and supply of fresh water are the main challenges faced by many communities across the world, especially those living in African communities. Likewise, reliance and over-dependence on groundwater as the main source of fresh water for irrigation may lead to its depletion and result in difficulties in recharging groundwater aquifers. Again, exposure of pollutants due to industrial and municipal wastes to some of the surface water reservoirs is posing a threat to the community resulting in an inability to tap the full potential of water from these sources. Hence, it is necessary to explore other sources of water supply for small-scale irrigation systems in the African continent. This chapter encompasses the integrated approach of discussing the impacts of small-scale irrigation

Sustainable Use of Wastewater in Agriculture: Applications, Methods, and Technology. Megh R. Goyal & Gaurav Saini (Eds.)

schemes in rural Nigerian communities. This chapter focuses on the scope of rainwater harvesting on small-scale irrigation systems for food security and flood mitigation.

10.1 INTRODUCTION

Water, energy, and food security are key concerns as growing populations and rapid urbanization imposed increased pressure on available natural resources.[10] In some of the tropical developing countries, rural households are still faced with the burden of having access to good quality water; and this affects the health of people and crippled agricultural production.[55] As an important and integral unprocessed resource, water has considerable influence on health, wealth, food production, and every other production, which may be natural or artificial. Thus, the importance of water cannot be overemphasized.[29]

In general, water covers about 70% of the earth's surface while only 2.5% is freshwater, leaving 97.5% as salty, which mainly belongs to seas and oceans.[47] The main sources of fresh water are basically categorized into underground, surface, and atmospheric sources, and rainwater constitutes almost the entire atmospheric source.

Freshwater scarcity is one of the few scenarios and among the top topics of discussion across the world due to its threatening effects on the continued and exponentially rising human population. More than 750 million people in the world in 2012 were having difficulties in accessing safe and portable water;[54] and this is increasing as a result of the exponential rise of population, which reached 1.8 billion people in the year 2015. In most developing cities, there is unevenness in water supply and demand, which has resulted from the over-dependence on groundwater as the major resource of fresh water.[7] This has led to such consistent challenge of misbalancing the supply and demand. Also, over-reliance on this source of water may be associated with the depletion of groundwater reservoirs, leading to difficulties in recharging the groundwater aquifers.[35]

Other hazardous events associated with such depletion of water tables include landslides, decreasing the level of moisture content of natural soil, which in turn affects the level of vegetation and brings about general drought.[44] Another challenge faced by those conventional sources of water for human use is potential huge pollution due to industrial and municipal waste dumps in most surface waters. Additionally, inadequate agricultural input, lack of public authority intervention, and lack of access facilities (such as roads) are also one of the challenges of under-utilization of available resources.[50]

Therefore, to reduce such intense usage of groundwater and safeguard inhabitants from the aforementioned menaces, it is good to explore other sustainable water supply systems capable of meeting the demand apart from underground sources, and one of the most promising alternatives means to surface water is the rainwater.[51] Also, to have a better understanding of the challenges of balancing the demand and supply of water, it has been mandatory to know the expected water demand and amount of water capable of being generated by these conventional sources under normal circumstances so that an estimated deficit can be obtained.[33]

In recent years, there have been drawbacks in adapting multiscale analysis in achieving the growth and changes of sustainable food networks due to spatial and temporal changes, especially in low- and median-income countries.[39] However, there is a need to identify major factors for illustrating the continuous technology diffusion and the significant elements that could affect the embracing of the technology to fully understand the role of modern multiscale analysis, which involves imperative measures to maintain agricultural activities and climate condition through the provision of efficient water supply system due to alarming danger caused by human activities on the ecosystems in some regions.[46,52]

This chapter analyzes the available data from relatively verifiable research articles, which consider the origin of rainwater harvesting, methods appropriate to rainwater harvesting (RWH), parameters influencing rainwater harvesting projects, rainwater harvesting's associated challenges, energy savings and improvement of greener environment, stormwater management, financial implications, and recommendations to enhance the irrigation systems.

10.2 CONCEPT OF RAINWATER HARVESTING (RWH) SYSTEMS

The RWH is one of the earliest methods of collecting water to balance the water supply and it has been in existence for the last four millenniums.[28] The extent of its current practice is considerably changing around the globe. From a universal perspective, relevant pragmatic, hypothetical, and social factors of the Center for Economic Performance (CEP) are considered to determine the modern status of RWH.[14] Being eco-friendly, it could be achieved by taking up traditional methods of directly collecting water from roofs to storage tanks for domestic use, or by collecting through pipe networks and channeling to specific reservoirs.[13] After collection, rainwater could undergo some chemical treatment due to its contact with the surface or collecting systems, which may temper its compounded properties conforming to regulations by

State Environmental Protection Agencies. Although the process of collection and storage of rainwater is associated with challenges of contamination of some chemicals, these are prone to health issues. However, if mitigated through proper treatment, it could serve as a multipurpose use of water for public, commercial, agricultural, and industrial usage, especially in water-scarce areas. Therefore, the most important aspect is to ensure excellent attributes of water for various uses by conducting certain physical/chemical tests, such as physiochemical parameters, organic pollutants, and microbial contamination tests, depending on the location and natural condition of the designed area.

To determine the volume to harvest rainwater, it is necessary to understand the rainfall pattern of the region, although this will vary depending on the desired purpose of the irrigation systems, storage with high efficiency, saving capacity of the storage, and the sustainability of the storage.[20] These considerations have mandated the designers to focus on a certain set of parameters leading to achieving their objectives ranging from the catchment area, conveyance system, final storage, and water demand that is affected by rainfall patterns in the region.[26]

This indicates that the rainwater harvesting system (RWHS) is a system that will contain and store relatively large portions of the total runoff by not allowing it to infiltrate into the ground, moving to surface water or contaminated pollutants by ensuring that physical, chemical, and other characteristics are preserved in relation to certain water standards. The RWHS would be more beneficial to the region where annual rainfall is minimal and threatened with drought and food insecurity. To maintain a consistent supply of fresh water and to minimize these challenges, the volume of rainwater harvested[44] is determined as below:

$$V = A \times R \times C \tag{10.1}$$

where V is the volume of harvestable rainwater, A is the catchment area, R is the total amount of rainfall, and C is the runoff coefficient.

In addition to the quantity of harvested rainwater, roofs commonly serve as an ideal rainwater catchment area. However, only a certain portion of rainwater can be harvested from the roof; and this amount can be represented theoretically[33] as follows:

$$V_t = \sum_{i=1}^{n} \sum_{j=1}^{k} [0.001 \times (R_{IJ} \times A_j \times C)] \tag{10.2}$$

where V_t is the nonempirical volume of water collected (m^3), R_{IJ} is the average annual rainfall in two zones i and j (mm), A_j is the roof area in a region of i

and j (m^2), C is the average runoff coefficient, and n and j are total numbers of respective regions and zones of the roof.

However, the estimation of water demand by the individual household is difficult to predict due to the variability of the amount of water usage by members of the household based on gender/age/attitude, season of using the water (summer or winter), time spent by the members of household in the house, cultural/social/religious activities, and local irrigation purpose. It is estimated that the daily consumption of drinking water in the southern coastal belt of Bangladesh is about 6 liters per person.[27] However, without considering small-scale irrigation schemes, the sustainability aims of the water harvesting system may not be achieved.[8]

10.3 CONCEPT OF SMALL-SCALE IRRIGATION

Personalized maintenance of water for agricultural purposes has been going on in South Asia, where almost all irrigation is recently based on private wells. Currently, mini private irrigation systems are also operated in sub-Saharan Africa.[22] The majority of rural dwellers (i.e., in sub-Saharan regions) depend directly or indirectly on farming. Despite the availability of irrigable land and water resources in sub-Saharan countries, only a small percentage is being utilized for irrigation due to its support in the economy of the region; and it is insignificant in comparison with rain-fed agriculture.[50]

Also, previous studies on large-scale irrigation systems around the world have focused on internal and external performance. Irrigation produces direct benefits by increasing crop yield and intensification.[37] The massive expansion of irrigated agriculture[15] was an essential component of past green revolutions; but it does not get much attention in sub-Saharan Africa. However, irrigation research and investment efforts have mainly focused on the inefficient general irrigation field, with the least focus on the small-scale private irrigation systems.

The variation of rainfall pattern in Nigeria leads to a high risk of agricultural production (hinders the development of agriculture) thereby bringing down the farmer's livelihood because a large percentage of crops were produced from rain-fed agriculture and only a few farms benefitted from the irrigated farming.[12] Hence, the availability of artificial means of irrigation helps farmers to (a) grow more crops in a year; (b) improve crop production and family income; and (c) optimize their food scarcity and dependency on government. Adopting a small-scale irrigation scheme helps farmers to be safe against rainfall shortage by providing a sustainable supply of water for tillage and rearing of livestock, which also leads to an increase in food security and improvement of human nutrition.[45,50]

Small-scale irrigation systems can make significant contributions to agricultural productivity growth. However; the full potential to expand small-scale irrigation with alternative irrigation technologies is still unknown in most of the sub-Saharan countries.[41] The adoption of irrigation will surely smoothen the zigzags of RWH agriculture[12] and agricultural produce, and it will also promote food security. If the farmers can have access to groundwater through several small-scale irrigation systems,[48] then this can help to cultivate several crops in a year. Several facts determine the adoption of small-scale irrigation either negatively or positively, such as: (a) lack of access to good roads, lack of nearby water resources, and distance of marketing points from the household residence negate the acceptance of irrigation; and (b) land size, farmer's knowledge, and means to fund will affect the adoption of irrigation positively. Additionally, the generation of income, reduction of poverty, increased food security, creation of employment, and agricultural productivity can also have an impact on the adoption of irrigation.[45]

In the Sahel savanna areas of Nigeria, pasture lands are not readily available during the dry season; therefore, irrigated farming would contribute toward the enhancement of pasture lands in such areas. Hence, the migration/conflict between pastoralists and crop farmers would be reduced.[42] Some farmers in Africa prefer to adopt an individual water source for irrigation purposes than grouping for a large water source (where many farmers will irrigate their farms from systems. This is due to rainwater/surface runoff that can be stored; the use of surface water; and pumping out from subsurface water. Also, the provision of private water sources would be of advantage to farmers and owners of such systems because the rainwater/surface runoff can be stored for later use.[56]

10.4 ENERGY SAVING AND IMPROVEMENT OF GREENER ENVIRONMENT

Among of the earliest methods of balancing water supply is rainwater harvesting systems (RWHSs), e.g., 2.3–23 million m^3 of water can be generated through rainwater harvesting in Sydney, Australia.[11] The current system of urban water supply systems is entirely energy-driven, which causes the accumulation of resources especially in fast-growing cities of Asia, where the population is increasing geometrically. However, RWHS will relieve some portion of the energy generated for water supply due to its cost-effectiveness scheme for small-scale water needs especially in local communities in Nigeria. Other challenges faced by the water treatment systems are their limited capabilities of treatment processes[44] to have a more sustainable water supply system capable of meeting the demand and increasing the efficiency of the conventional water treatments and irrigation systems; and RWH systems should be given special attention for proper implementation, as these could drastically reduce the crisis for water demand and relieve the excessive burden on the conventional sources of water while assisting in balancing the climate changes, flood control, stormwater management and helping in activities related to small irrigation schemes.

The entire runoff volume from the roof is minimal when compared to the total runoff in a particular catchment area. However, RWHS is capable of saving an average of 44% of the total runoff volume from the roof, which is a reasonable quantity for effective usage.[31] Geographic Information System (GIS)-Simulation-Based Design System (GSBDS) was employed to explore viable usage in terms of water balance model, energy saving,[19] and its economic feasibility as an alternative to conventional water supply systems. These results indicate that 21.6% of water can be saved and 138.6 kWh/year could also be saved by each household; thus, making the system a reliable means of renewable energy approach to water supply.

Good farming location results in a rise in profits and the decrease of environmental pollution from the pesticides; by other means, high-tech techniques could aggravate social imbalances; and in some instances, can alter power linkages if dependent on the high-tech[36] Therefore, modifications include human–environment relationships to attain environmentally sustainable food security.[25]

There is a need to impose imperative measures to maintain agricultural activities and climate conditions through the provision of efficient water supply systems.[46] However, climatic change, improving food safety, income and nutrition, and other developmental purposes (that require small-scale farmers to gain admittance to prevalent climate-smart technologies and other sustainable productivity growth means) can be achieved. Ecosystem stabilization is a serious challenge due to environmental effects, and the survival of humanity in arid and semi-arid regions is in alarming danger due to the negative effects of human activities on ecosystems.[46] This has posed a serious threat to climate change, which includes severe weather, rainfall instabilities, and unusual temperature variations leading to extreme risks to the agricultural economy. Therefore, due to the significance of irrigation farming to the Nigerian and other African nation's economies, the nations must adopt climate change techniques to avoid susceptibility of these farming methods, where small-scale systems can contribute to agriculture production.[41]

Policy interventions (such as providing agricultural credit and dispersing access to irrigation systems) are crucial in reducing the negative influence of climate change on poor farmers, and capital constraints. Crop modeling is the primary tool available to assess the impact of climate change and other factors on agricultural produce. In addition, climate change also creates chances for farmers (all sellers to goods founders) through income from rising food prices.[49] Also, the massive expansion of irrigated agriculture was an essential component of past green revolutions; however, little or no attention was given to the implementation of such policies in sub-Saharan Africa.[16] However, in the Sahel savanna areas of Nigeria, pasture lands are not readily available during the dry season, and irrigated farming would contribute toward the enhancement of pasture lands in such areas. Hence, the migration conflict between pastoralists and crop farmers would be reduced. Also, Greenland would hinder the development of the greenhouse effect.[12]

10.5 FINANCIAL IMPLICATION AND FOOD SECURITY

Most RWHS analyses were focused on the quantity and quality of the harvested water; besides, there is a need to establish an economic basis for the reliable implementation of RWHS. Such economic analysis includes environmental benefits, comparable cost to the conventional water supply systems, energy and water saving potentialities, as well as maintenance costs.[9] Economically, it has been proven that a rainwater harvesting system is better compared to groundwater extraction; and its implementation will lead to savings of >11% of general water supply cost.[30] Although it was arguably stated that the cost reduction would be just 4.5%,[31] which is viable to most households in comparison with high water bills. The household can save a certain amount of tax by utilizing public sources of water, and therefore, it can be concluded that rainwater harvesting is a sustainable and cost-effective technology, and it also contributes in saving energy consumed by the water treatment plants.

One strategy considered for low-income small-scale farmers to unravel money deficiency is the usage of small-scale irrigation techniques.[48] The availability of artificial means of irrigation helps the farmer to enhance crop production, and to gain and promote their food scarcity level as well as dependency on government. Different studies and analyses indicate that there is a crucial impact on participants' earnings in small-scale irrigation methods.

Nonetheless, the poverty analysis indicates that non-irrigating households are poorer than households with irrigation system.[23]

Most of the households faced food insecurity and the level of small farmholders' mean efficiency was 85.9%.[3] Disparities among the household characteristics influence the deviation in earnings of households and food security. Furthermore, a means of irrigation has the ability to improve household income, and multiple crops, hence improving the standard of living of small-scale farmers.[43] It will also lead to dietary diversity; thus, irrigation helps to optimize agricultural income and multiform in crop production. Small-scale irrigation helps in addressing irregular rainfall pattern issues. This helps in reducing food insecurity, and rural poverty, as well as upgrading the entire support of agriculture to the country's economy.[17] The associated higher-income and middle-income farmers are usually better and can generate more income than the low-income farmers because the latter use commonly furrow irrigation or manual treadle pumps and the former use motorized pumps.[8]

To expand the farming systems in Africa through the international research sector, the Consultative Group for International Agricultural Research (CGIAR) may have a significant impact in achieving sustainability and food scarcity in the district, as in the context of the existing governmental wealth toward the promotion of farming. Nevertheless, the activities of CGIAR are not fully-funded and are exaggerated in Africa.[4] Incompatible government strategies, lack of knowledge among the farmers about farming practices, financial burden, and inadequate provisions by the government are the main drawbacks that can hinder the progress of achieving a sustainable irrigation system in Nigeria.[2]

The development of the agricultural field, which has been predicted as subjective to a future economic existence, can only be achieved if the challenges facing the irrigation systems are being addressed.[2] Sustainable agricultural growth could increase per capita consumption and can lead to poverty alleviation.[7] Some impacts of the adoption of irrigation to household livelihood[45]include: the generation of income, reduction of poverty, increased food security, creation of employment, and agricultural productivity.

Many irrigation techniques can be used in promoting household gains and reducing food scarcity; and these involve enlarging farm products to help poorer households meet their primary demands, provide guidance on agricultural risks by scarce rainwater supplies, and optimize the use of crops affecting agricultural inputs and provision of entrepreneurship.[6] The irrigation helps to increase the household income and multiform crops; thus improving the standard of living of small-scale farmers. It also leads to dietary diversity. Thus, irrigation provides increased agricultural gains and multiform crops.[43]

10.6 PARAMETERS AFFECTING SMALL-SCALE IRRIGATION SYSTEMS WITH RAINWATER HARVESTING

The components of roofing characteristic are major factors to influence the quality and quantity of water; and these characteristics include roof types, roofing materials, roof location, and weather conditions to which it is exposed. In the 21st century, most of the newly developed or existing irrigation systems that need modifications must be integrated by considering sustainable parameters (social, economic, and environmental).[53] In the analysis, several parameters (such as roof area, storage properties, catchment area, quantity and intensity of the rainfall, and runoff coefficient of the region) should be taken into account. The availability of fresh water and an increasing human population are dependent on each other; thus, balancing and sustaining the demand and supply of water from reliable sources for human needs is of paramount importance. Also, social and economic implications are highly associated with water scarcity, which directly affects the general households in rural areas.[18,40]

A recommendation to increase the catchment area for the rainfall will increase the corresponding water quantity harvested which could relieve the burden of water shortage to the vulnerable regions.[40] This approach may help in relieving higher dependence on other sources of water for both portable and non-portable human purposes especially in mega cities where the water demand is very high.

Climate change due to atmospheric concentrations of greenhouse gases, political shifts, global demand, agricultural policies, and shortage of rainfall have negatively affected the crop production systems in Nigeria and especially in the northern region with high drought.[2] Climate has a significant effect on RWH agriculture and also has a significant impact toward the level of production of products.[12] Thus, agricultural production can effectively decrease in food-scarce sectors, if problems associated with climate change are not properly managed.[21]

Also, the rapid spread of knowledge and innovative technologies has aided in achieving maximum benefits from farmland, water, and man power; and these improvements can enhance income and food security. Selection of the right crop and proper soil management practices would also enhance crop yield and income even when adopting rain-fed agriculture.[32]

10.7 CHALLENGES ASSOCIATED WITH THE RAIN WATER HARVESTING AND SMALL-SCALE IRRIGATION SYSYEMS

The main challenge is the uneven distribution of the precipitation, which makes it harder to predict the actual quantity of water for a given period in a particular catchment area; this has endangered the rain-fed agriculture. Also, efforts should be made in sub-Saharan Africa to secure food security that is impacted by soil fertility, poverty, increasing population, uneven global trade relations, and weak infrastructure, as well as climate change and future uncertainties.[49]

There are many challenges faced by small-scale farmers in developing nations that limit the chances of achieving quantitative agricultural produce and improved standards of living.[24] Increased irrigation of fishponds has also caused multiple needs for irrigation water, which has resulted a local contestation between residents within the farming system.[8] Incompatible government strategies, lack of knowledge among the farmers about farming practices, financial burden, inadequate government, inadequate participation of individual actors, and limited concern for the operation and maintenance of irrigation networks are main drawbacks that hinder the progress of achieving a sustainable irrigation system. To attain food security and socio-economic improvement for irrigation implementation, the government should support an appropriate framework, and proper technology, and help farmers in decision-making policy.[2]

Lack of technical knowhow is among the major problems faced by small-scale irrigation schemes especially in sub-Saharan countries. Frequent breakdown and unreliable water supplies may have been resulted,[16] when the farmers are not provided with regular and efficient maintenance of the irrigation equipments. Also, the lack of infrastructural facilities (such as water reservoirs and conveyance systems) lead to the siltation of canals, leakage, and irrigation water losses. The main drawbacks that hinder the progress of achieving a sustainable irrigation system include incompatible government strategies, lack of knowledge among the farmers about farming practices, financial burden, and inadequate government strategies.[2]

The small-scale-irrigation is faced with different challenges, such as inadequate market for the produces, connection between micro finance institutions and smallholder farmers, lack of access to water, and low-technical knowhow on the effective use of irrigation. To increase the participation of stallholder households in irrigation, these challenges can be solved by teaching irrigation technology to stakeholders.[1]

10.8 STORMWATER MANAGEMENT

Some techniques for stormwater management should be employed to bridge the problem of high water demand due to uneven spatial and temporal distribution of rainfall intensity.[5] It is among the most effective solution to urban flood mitigation affecting many cities.[22] As mentioned earlier, RWHSs can function as standards in mitigating floods, especially in an area experiencing heavy rainfall in most cases and the existing drainage is incapable of collecting the rainfall. In view of this, some regions in Taiwan provide other means to reduce the impact of such menace on the environment termed low-impact development (LID), which is included and can greatly impact water management.[28]

The reduction of runoff is advantageous in stormwater management, and increasing the catchment area for the rainfall will increase the corresponding quantity of water harvested.[40] Despite its limitation in stormwater management, it is capable of saving 44% of the total runoff volume from the roof, but the entire runoff volume from the roof is minimal compared to the total runoff in the catchment area.[31] However, it contributes to stormwater management through the reduction of pollutants except dissolved pollutants before reaching any spillage, while assisting in balancing climate changes, flood control, and help in stormwater management. It is recommended that reliable stormwater management systems must provide sustainable drainage systems in conjunction with RWS so that the efficiency can be increased.[31]

10.9 RECOMMENDATIONS

- To improve agricultural transformation in localities, agricultural innovations should be improved, integrated with interventions that boost rural infrastructure, and advanced use of technology. Thus, proper development of small-scale rainwater will help in addressing irregular rainfall pattern issues, which will reduce food insecurity and rural poverty, and promote general agricultural contribution to the economy of developing states. Hence, it has been indispensable to provide other supportive technologies that will reclaim, conserve, and recharge the underground water.
- There have been inadequate participation of individual actors and limited concern for the performance and sustenance of small hierarchy irrigation networks by the relevant policymakers. Therefore, it is

necessary to explore other means/ tools that will fill the gap of high-water demand in water stressed regions.

- The need for minimum international standard water supply (per capita of 1000 m^3) is of paramount importance, where most of the countries have failed to attend most urban areas especially in developing countries. Thus, implementation of modern irrigation systems can reduce high dependence of other sources of water for both portable and non-portable human purposes especially in mega cities with high-water demand.

10.10 SUMMARY

Having vigorously attempted to discuss the concept, sustainability, challenges, financials implication, and flood management of the two systems, the following conclusions can be summarized:

- Rainwater harvesting offers significant contribution to water-energy-food nexus especially in remote localities. Also, the nature and size of the catchment area has an influence on small-scale irrigation systems; and increasing the catchment area for the rainfall will increase the corresponding quantity of water harvested. Rainfalls have greater impact on recharging groundwater reservoirs; therefore, the harvesting of rainwater should be optimal in drought-hit area.
- The challenges of irregular spatio-temporal dispersion of rainfall intensity are mostly experienced in desert and semi-desert areas.
- Numerous small holder farmers can increase their income, enhance crop yield and reinforce food security, when they carefully invest in private irrigation systems. Farmers that practice irrigation farming can regulate the soil moisture at critical conditions, can efficiently regulate crop growth, and can also have high yield of farm product, more than those adopting rain-fed agriculture.
- In semi-arid regions, rainwater harvesting system has been demonstrated to be successful; however, their dependability cannot be guaranteed in arid regions. Because the irrigation data for managing the problems of small reservoirs are not readily available for arid regions.
- The individuation of irrigation and its random as well as rampant spill-over makes it difficult to balance admittance and sustainable surveillance of water resources. While, some disasters caused by excessive runoff (such as flooding and soil erosions) can be neutralized through harnessing the importance of modern irrigation technology.

KEYWORDS

- **flood**
- **food**
- **harvesting**
- **irrigation**
- **rainwater**
- **wastewater**

REFERENCES

1. Abanega, A.; Mohammed A. Impact of Small-scale Irrigation Schemes on Small Holder's Household Income and Food Security in Ethiopia. *J. Agric. Food Secur.* **2021,** *11* (1). https://doi.org/10.1186/s40066-021-00294-w.
2. Adelodun, B.; Choi, K. A Review of the Evaluation of Irrigation Practice in Nigeria : Past, Present and Future Prospects. *Afr. J. Agric. Res.* **2018,** *13* (40), 2087–2097.
3. Adeniyi, D. A.; Mulugeta F. D. Efficiency, Food Security and Differentiation in Small-scale Irrigation Agriculture: Evidence from Northwest Nigeria. *Cogent Soc. Sci.* **2020,** *6*, 1749508.
4. Adenle, A. A.; Wedig, K.; Azadi, H. Technology in Society Sustainable Agriculture and Food Security in Africa: The role of Innovative Technologies and International Organizations. *Technol. Soc.* **2019,** *58* (April), 101143. https://doi.org/10.1016/j.techsoc.2019.05.007.
5. Adham, A.; Riksen, M.; Ouessar, M.; Ritsema, C. J. A Methodology to Assess and Evaluate Rainwater Harvesting Techniques in Semi-arid Regions. *Water* **2016,** *8,* 198–205.
6. Agidew, A. The Determinants of Small-scale Irrigation Practice and its Contribution on Household Farm Income: The case of Arba Minch Zuria Woreda, Southern Ethiopia. *Afr. J. Agric. Res.* **2017,** *12,* 1136–1143.
7. Amare, M.; Nathaniel, D. J.; Shiferaw, B.; Cissé, J. D. Rainfall Shocks and Agricultural Productivity: Implication for rural Household Consumption. *Agric. Syst.* **2018,** *166* (July), 79–89.
8. Amede, T.; Desta, T.; Harris, L. T.; Kizito, D.; Xueliang, F.; Chinyanja, C. Triangle in the Zambezi River Basin, Southern Africa: Status of, and Prospects for, Agriculture, *Natural Resources Management and Rural Development*. WLE Research for Development (R4D) Learning Series 1; 2014 [Online]; http://www. iwmi.cgiar.org/Publications/wle/r4d/wle_research_for_development-learning_series-1.pdf_(accessed Mar 20, 2023).
9. Amos, C. C.; Rahman, A.; Gathenya. J. M. Economic Analysis and Feasibility of in Urban and Peri-urban Environments: A Review of the Global Situation with a Special Focus on Australia and Kenya. *Water* **2016,** *8,* 149–154.
10. Arani, K.; Siddiqi, A.; Diaz, L. Food Security Amidst Water Scarcity: Insight on Sustainable Food Production from Saudi Arabia. *Sustain. Prod. Consump.* **2015,** *2,* 67–78.

11. Ataur, R.; Joseph, K.; Monzur, A. I. Rainwater Harvesting in a Greater Sydney Water Saving, Reliability and Economic Benefits. *Resour. Conserv. Recycl.* **2012,** *61,* 16–21.
12. Bielmyer, G.; Arnold, W. R.; Tomasso, J. R.; Isely, J. J.; Klaine, S. J. Effects of Roof and Rainwater Characteristics on Copper Concentrations in Roof Runoff. *Environ. Monit. Assess.* **2012,** *184* (5), 2797–2804.
13. Biswas, B. K.; Mandal, B. H. Construction and evaluation of rainwater Harvesting System for Domestic use in a Remote and Rural Area of Khulna, Bangladesh. *Int. Sch. Res. Notices* 2014**,** *2014*, 751952. http://dx.doi.org/10.1155/2014/751952.
14. Campisano, A.; Butler, D.; Ward, S.; Burns, M. J.; Friedler, E.; Debusk, K.; Fisher-jeffes, L. N.; Ghisi, E.; Rahman, A.; Furumai, H. Urban : Research, Implementation and Future Perspectives. *Water Res.* **2017,** *115,* 195–209.
15. Carracelas, G.; Guilpart, N.; Cassman, K. G.; Grassini, P. Distinguishing Between yield plateaus and Yield Ceilings: A Case Study of Rice in Uruguay. *Field Crops Res.* **2023,** *292*, 108808. https://doi.org/10.1016/j.fcr.2023.108808.
16. Cassman, K. G.; Grassini, P. "Can there be a Green Revolution in Sub-Saharan Africa without Large Expansion of Irrigated Crop Production? *Glob. Food Secur.* **2013,** *2* (3), 203–209. https://doi.org/10.1016/j.gfs.2013.08.004.
17. Chauhan, G. S.; Musa, A. A. Attenuating the Downstream Flooding of Tagwai Dam from the Main Channel using Lateral Weir. *J. Eng. Sci.* **2020,** *11,* 326–330.
18. Chauhan, G. S.; Mustapha, T.; Musa, A. A.; Abubakar Tahir Garba, A. T.; Danish Rashid, D. Assessment of Rajasthan Surface Water Quality During Covid19 Pandemic Lockdown. *Int. J. Eng. Res. Technol.* **2020,** *6* (9), 400–408.
19. Chiu, Y. R.; Tsai, Y. L.; Chiang, Y. C. Designing Cost-effectively in a Urban Water-Energy Saving Scheme by using a GIS Simulation Based Design System. *Water* **2015,** *7,* 6285–6300.
20. Czarny, J.; Präbst, A.; Spinnler, M.; Biek, K.; Sattelmayer, T. Development and Simulation of Decentralized Water and Energy Supply Concepts–a case Study of Rainwater Harvesting at the Angkor Centre for Conservation of Biodiversity in Cambodia. *J. Sustain. Dev. Energy, Water, Environ. Syst.* **2017,** *5* (4), 626 644.
21. Dahiru, T. M.; Tanko, H. The Effects of Climate Change on Food Crop Production in Northern Nigeria. *Int. J. Res.–Granthaalayah* **2018,** *6* (9), 458–469.
22. De Fraiture, C.; Giordano, M. Small Private Irrigation: A Thriving but Overlooked Sector. *Agric. Water Manag.* **2014,** *131,* 167–174.
23. Emmanuel, O.; Kow, I.; Popoola, L. Differential Impacts of Rainfall and Irrigation on Agricultural Production in Nigeria: Any Lessons for Climate-Smart Agriculture? *Agric. Water Manag.* **2016,** *133,* 30–36.
24. Garbero, A.; Songsermsawas, T. Impact of Modern Irrigation on Household Production and Welfare Outcomes. In *International Fund for Agricultural Development (IFAD) Research Series 31*; Rome–Italy, 2018; p 48.
25. Hadjikakou, M.; Ritchie, E. G.; Watermeyer, K. E.; Bryan, B. A. Improving the Assessment of Food System Sustainability. *Lancet Planetary Health* **2019,** *3,* 62–63.
26. Hanson, L. S.; Vogel, R. M. Generalized Storage–Reliability–Yield Relationships for. *Environ. Res. Lett.* **2014,** *9*, 075997. https://doi.org/1088/1748-9326/9/7/075007.
27. Harun, M. A. Y. A.; Kabir, G. M. M. Evaluating Pond sand Filter as Sustainable Drinking Water Supplier in the Southwest Coastal Region of Bangladesh. *Appl. Water Sci.* **2013,** *3* (1), 161–166.
28. Huang, C. L.; Hsu, N. S.; Wei, C. C.; Luo, W. J. Optimal Spatial Design of Capacity and Quantity of for Urban Flood Mitigation. *Water* **2015,** *7,* 5173–5202.

29. Indhumathi, G.; Suresh, C.; Sangeetha, M. A study on Modern Techniques used in Irrigation for Farming in Coimbatore city. *Int. J. Interdiscip. Res. Arts Hum.* **2017,** *2* (2), 166–170.
30. Islam, Md. T.; Ullah, Md. M.; Amin, M. G. M.; Sahadat Hossain, S. Rainwater Harvesting Potential for Farming System Development in a Hilly Watershed of Bangladesh. *Appl. Water Sci.* **2017,** *7,* 2523–2532.
31. Jeffes, L. N. F.; Armitage, **N. P.**; Carden, K. The Viability of Domestic Rainwater Harvesting in the Residential Areas of the Liesbeek River Catchment, Cape Town. *Water SA* **2017,** *43,* (1), 81–90. http://dx.doi.org/10.4314/wsa.v43i1.11.
32. Karpouzoglou, T.; Barron, J. A global and Regional Perspective of Rainwater Harvesting in Sub-Saharan Africa's Rainfed Farming Systems. *Phys. Chem. Earth* **2014,** *75* (72), 43–53.
33. Kartolo, J.; Kusumawati, E. In Feasibility Study of Rainwater Harvesting for Domestic use (case study: West Jakarta Rainfall Data). Proceedings of the 3rd International Conference on Construction and Building Engineering (ICONBUILD) **2017,** *1903* (1), 100008. https://doi.org/10.1063/1.5011618.
34. Liaw, C. H.; Chiang, Y. C. Framework for Assessing the Rainwater Harvesting Potential of Residential Buildings at a National Level as an Alternative Water Resource for Domestic Water Supply in Taiwan. *Water* **2014,** *6,* 3224–3246.
35. Maina, C. W.; Raude, J. M. Assessing Land Suitability for Rainwater Harvesting Using Geospatial Techniques: A Case Study of Njoro Catchment - Kenya. *Appl. Environ. Soil Sci.* **2016,** *10,* 467635. http://dx.doi.org/10.1155/2016/4676435>
36. Maru, A.; Berne, D.; De Beer, J. E. A. Digital and Data-driven Agriculture: Harnessing the Power of Data for Smallholders in Rome. *Glob. Forum Agric. Res. Innov. Res.* **2018,** *7,* 525–534.
37. Mateos, L.; Borgia, C.; García-bola, M. Patterns of variability in large-scale irrigation schemes in Mauritania. *Agric. Water Manag.* **2012,** *112,* 1–12.
38. Mbonigaba, J.; Muhoyi, E. Stakeholder Consultations on Small-scale Irrigation Schemes' Constraints in Zimbabwe. *Environ. Dev. Sustain.* **2021,** *24,* 13198–13217.
39. Müller, B.; Hoffmann, F.; Heckelei, T.; Müller, C.; Hertel, T. W.; Polhill, J. G.; Wijk, M. V.;Achterbosch, T.; Alexander, P.; Brown, C.; Kreuer, D.; Ewert, F.; Ge, J.; Millington, J. D. A.; Seppelt, R.; Verburg, P. H.; Webber, H. Modelling Food Security: Bridging the Gap Between the Micro- and the Macro- Scale. *Glob. Environ. Change* **2020,** *63* (March), 102085. https://doi.org/10.1016/j.gloenvcha.2020.102085.
40. Mwamila, T. B.; Katambara, Z.; Han, M. Y. Strategies for Household Water Supply Improvement with Rainwater Harvesting. *J. Geosci. Environ. Prot.* **2016,** *4,* 146–158.
41. Ojo, T. O.; Baiyegunhi, L. J. S. Climate Change Perception and its Impact on net Farm Income of Smallholder Rice Farmers in Southwest-Nigeria. *J. Clean. Prod.* **2021,** *310* (June), 127373. https://doi.org/10.1016/j.jclepro.2021.127373
42. Olawale, E.; Kow, I.; Popoola, L. Differential Impacts of Rainfall and Irrigation on Agricultural Production in Nigeria: Any Lessons for Climate-Smart Agriculture? *Agric. Water Manag.* **2016,** *178,* 30–36.
43. Passarelli, S.; Dawit, M.; Elizabeth, B.; Ringler, C. Evaluating the Pathways from Small-scale Irrigation to Dietary Diversity: Evidence from Ethiopia and Tanzania. In *The Science, Sociology and Economics of Food Production and Access to Food. Published in Cooperation with The International Society for Plant Pathology (ISPP)*; 2018, Vol. 10, pp 981–997.

44. Rahman, S.; Khan, M. T. R.; Akib, S.; Din, N. B. C. D.; Biswas, S. K.; Shirazi, S. M. Sustainability of Rainwater Harvesting System in Terms of Water Quality. *Sci. World J.* **2014,** *10,* 721357. http://dx.doi.org/10.1155/2014/721357.
45. Rasa, B. B. Review on adoption of irrigation and its impact on household livelihood in Ethiopia. *Adv. Sci. Human.* **2021,** *7* (3), 52–58.
46. Shereif, Muhmoud, H.; Adamowski, J.; Alazba, A. A.; El-Gindy, A. M. Rainwater Harvesting for the Management of Agricultural Draughts in arid and Semi-arid Regions. *Puddy Water Environ.* **2016,** *14,* 231–246.
47. Shittu, O. I.; Okareh, O. T.; Coker, A. O. Design and Construction of Rainwater Harvesting System for Domestic Water Supply in Ibadan–Nigeria. *J. Res. Environ. Sci. Toxicol.* **2012,** *1* (6), 153–160.
48. Temesgen, H.; Mengistu, K.; Fekadu, B. Evaluating the Impact of Small-Scale Irrigation Practice on Household Income in Abay Chomen District of Oromia National Regional State-Ethiopia. *J. Dev. Agric. Econ.* **2018,** *10,* 384–393.
49. Tesfamicheal, W.; Berger, T.; Swamikannu, N.; Ramila, T. Climate Variability, Consumption Risk and Poverty in Semi-arid Northern Ghana: Adaptation Options for Poor Farm Households. *Environ. Dev.* **2014,** *12,* 2–15.
50. Tesgera, W. D.; Guluma, W. The Role and Significance of Small-scale Irrigation in Improving Household Income in Ethiopia. *Int. J. Res. Bus. Manag.* **2020,** *7* (3), 20–35.
51. Tizita, D. The Effect of Small-scale Irrigation on Household Food Security in Bona-Zuria Woreda, Sidama Zone, Southern Ethiopia. Master's Thesis, The School of Environmental, Gender and Development Studies, Hawassa University, Hawassa—Ethiopia, 2017; p 127.
52. Ugochukwu, A. I.; Phillips, P. W. B. Technology Adoption by Agricultural Producers: A Review of the Literature. *Innov., Technol. Knowledge Manag.* **2018,** *2018,* 361–377.
53. UN Water Report 48. 2014 p 87 [Online]. http://www.unwater.org/ (accessed March 20, 2023).
54. WaterAid, *Water Security Framework*; WaterAid: London, UK, 2012; p 85.
55. Wen, Y.; Schoups, G.; Van de Giesen, N. Organic Pollution of Rivers: Combined threats of Urbanization, Livestock Farming and Global Climate Change. *Sci. Rep.* **2017,** *7,* 43289. https://doi.org/10.1038/srep4 3289.
56. Wichelns, D. Investing in Small, Private Irrigation to Increase Production and Enhance Livelihoods. *Agric. Water Manag.* **2013,** *131,* 163–166.

CHAPTER 11

DESIGN AND CONSTRUCTION OF MODIFIED LIFTING DEVICE FOR WATER AND WASTEWATER IN NIGERIA

MOHAMMED RABIU MUSA, MUHAMMAD ISA ABDURRAHMAN, and GAURAV SAINI

ABSTRACT

Water-lifting devices (driven by animal and/or human beings) have been used for supplying water to agricultural fields for centuries. This is particularly important from the perspective of small-scale farmers, especially in the developing and underdeveloped world. Riparian farming in Nigeria is limited by the ability of farmers to irrigate their fields. Conventionally and historically, *shadoufs* have been used for such purposes. However, shadoufs have their drawbacks (such as low efficiency and non-ergonomic design/operation); thus, requiring a modified water-lifting system. This chapter details the design and construction of a water-lifting system that is cheap and easy to construct, is ergonomic, and has significantly better output than the conventional *shadouf.* The new device was tested extensively in the Kano River basin, Nigeria, using multiple operators (including local farmers). The statistical analysis revealed that it has a higher water-lifting potential than the conventionally used devices. Furthermore, the device can be customized to each farmer. Such devices can significantly enhance the livelihoods of riparian farmers with small land holdings.

11.1 INTRODUCTION

Water-lifting devices have been used throughout history to supply water for both human consumption and agricultural purposes.[13] It has been estimated

Sustainable Use of Wastewater in Agriculture: Applications, Methods, and Technology. Megh R. Goyal & Gaurav Saini (Eds.)

that the earliest example might have appeared 5500 years ago, the numbers involved today are enormous, and indeed several varieties of them in the order of millions are currently in use globally.[4] Many of these devices have been used extensively for irrigation purposes, especially in China and Egypt.[13] Irrigation is the practice of supplying water to crops, orchards, grass, or any other cultivated plants during dry conditions or erratic rainfall periods.[5] As in some other parts of the world where irrigation is essential if crops are to flourish, several large dams have been built and water distribution systems based on canals have been established; it is usual to transfer water from these canals to the fields to sustain the plants.[1,7]

In Nigeria, there are areas with no electricity and where the capital and support services required to run solar or engine-driven pumps are absent; therefore; hand, wind, water, or animal-powered lifting systems are all that can be used.[6] Along the Kano River, this usually means that the shadouf is employed by many farmers. The *shadouf* is a simple, almost primitive, manually operated device that is used by humans to lift water from streams, rivers, ponds, and shallow wells.[8] It originated along the Nile and was recorded in tomb drawings at Thebes dating from 1250 B.C. It has been used in the Middle East, Asia, and Africa. In northern Nigeria, it is most frequently used by farmers to irrigate vegetables during the dry season (October to March).[1]

Toward the bottom of the farming hierarchy are the riparian farmers (who practice farming during the dry season along river banks and/or their tributaries). The lands are irrigated from the river directly or from shallow wells close by.[3] Such farmers do not have access to a pump and have to rely on the shadouf. The output of their land will be limited by the amount of water they can physically lift through shadouf and the area over which this water can be distributed (which will be to some extent limited by the local topography). In some cases, the area of land they hold is very limited, they may be able to irrigate and attend to their crops with much time to spare so that work can be done for others to supplement income. In other cases, it will not be practical to irrigate parts of the land at their disposal because it is too distant from the source of water or too elevated to supply water at a realistic rate.[3]

The farmers with larger landholdings are more likely to have access to resources such as electricity, pumps, and so on, and in comparison, the riparian farmers are on the other extreme. Their plight can be significantly improved by enhancing the efficiency of water-lifting devices and making them more user-friendly and ergonomic.

This chapter details the construction and operation of a novel water-lifting device that is ergonomic, cheap, and more efficient as compared to the traditional shadouf. Two hypotheses were formulated based upon which this study was conducted. The first hypothesis was focused on the possibility of using (new) manufactured materials to improve the traditional method of water lifting. The second hypothesis was to explore the possibility of a modified design of the lifting devices from ergonomic considerations, making them more efficient than traditional devices. Extensive design and testing were conducted to find the most optimal design for the water lifter device. The results of this work are likely to be of significant interest to small-scale farmers and policymakers world over.

11.2 STUDY AREA

The study area is comprised of a catchment along the River Kano below the Tiga dam to the point where it meets River Challawa (Figure 11.1) of Kano state in Nigeria.[1] The River Kano flows in a northerly direction between the Tiga dam and the Challawa extending eventually to Lake Chad; the distance between the two (Tiga and Challawa dams) is 42.5 km and about 70 km from Kano city.[7] The catchment, which is some 13 km wide in the northerly section and widens out to approximately 34 km near the dam, covers some 79,000 hectares with the Tiga dam alone sustaining irrigated agricultural farms with an area of 22,000 ha.[7] Within this area, there are essentially two types of farmers; those who are fortunate enough to be able to use water supplied by gravity from canals fed by the dam and those who have to lift water, if they are to irrigate their lands.[12]

The terrain through which the Kano River flows is generally higher than that found in the south of the country, the altitude varying from 300 to 1500 m. It rises in the Jos plateau, which is centrally situated in Nigeria and has extensive plains and broad valleys.[9] This plateau and the high ground around its northwest edge form the escarpment of the Kano River.[2]

11.3 MATERIALS AND METHODS

11.3.1 ROWING SEAT

An adjustable moving seat rig was constructed. The frame on which it was mounted was made from 5.0 × 5.0 × 0.4 cm angle iron, 5 cm diameter steel pipe, and a 3.0 mm thick steel sheet. A padded stackable chair was mounted on it. The frame sloped from back to front by 14 cm over a distance of 136 cm, so that when the leg muscles were relaxed it moved forward under the operator's weight. This movement was possible because two rollers (lengths of 50 mm pipe) were located between two sheet metal troughs, one welded to the frame, and the other to the underside of the chair. Two stops on the base of the lower trough prevented the rollers from moving from their desired position.

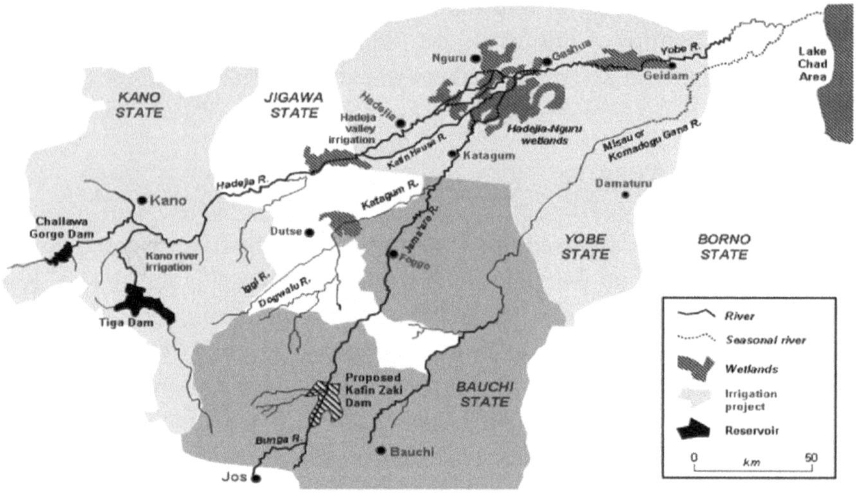

FIGURE 11.1 Tiga dam and Kano River (on the left) flowing to Lake Chad.

Source: Reprinted from Ref. [1]. Copyright © 2018 by the authors. Licensee MDPI, Basel, Switzerland. http://creativecommons.org/licenses/by/4.0/

A timber footrest was clamped in such a position on the frame that when the seat was in its forward position, the bent legs of the operator were comfortable. The seat was 34 cm above the ground level and the length of the troughs gave enough movement to allow the knees to be straightened.

In the first instance, the operator pulled on a horizontal piece of pipe pivoted at one end. The length of the pull exceeded the chair movement by the distance the operator was prepared to bend forward and stretch out his arms plus the distance between the outstretched arms position and that when the hands were adjacent to the chest. A frame equipped with two pulleys was placed in front of the sliding seat and various steel weights were lifted and lowered repeatedly to indicate the load that might be imposed on the operator. A weight of 25–30 kg could be lifted and lowered continuously, as determined by preliminary investigations.

11.3.2 DESIGN OF A SINGLE BUCKET LIFTER

The sliding seat arrangement had many attractive features from an ergonomic point of view; so it was decided to examine the possibility of applying it to some more conventional lifting devices. The force that the operator could exert directly had been established as being around 25–30 kg, so the logical follow-up step was to investigate whether a container of water weighing this much could be lifted to the required height in a series of 60–70 cm steps. It was then a relatively simple step to decide that a 25-L bucket suspended on a rope that could be hauled in steps (and held in some form of a gripping device between steps) might be worth further consideration.

A tripod arrangement was constructed from 10 cm 5 cm rough-sawn timber and equipped with a plastic bucket; the hauling rope was taken over pulleys to the sliding seat frame. The method of attaching the bucket to the ropes in such a way that it would always fill when dropped onto water, and then tip when required, caused some problems. These were eventually overcome by making a 30 × 3 mm steel hoop, which was bolted to the bucket. The hoop had welded to it two short diametrically opposed rods to which the ropes could be tied, a 0.5 kg steel weight to tip it over when it hits the water, and a loop to which a third rope could be attached. The purpose of this latter rope was to tip the raised bucket to discharge the contents, the lower end of the rope being attached to a point near the ground. An image of this version of the device set near the water source is shown in Figure 11.2.

FIGURE 11.2 The tripod/bucket lifting set near the brook, with the "rowing" seat arrangement.

A collecting trough was attached to the main pole using a steel strap. The volume of this trough was able to accommodate all the water from the bucket and it was fixed so that it emptied into a discharge pipe which in Figure 11.3 is shown set to discharge onto the ground, but which would normally run into a distribution system. The aim was to hoist the bucket to the level shown in Figure 11.3, at which point the tipping rope would become tight and tilt the bucket so that it discharged its contents into the trough.

FIGURE 11.3 The operator sitting on the swing and positioned in such a way so that he can see the bucket in the water source.

In practice, the position of the discharge point would be chosen with the topography of the area to be irrigated, but for this study, it was fixed at 4.3 m above the water surface (2.8 m above the bank).

Although the system was then working well, the operator was not close enough to the water source and could not see the bucket before it rose above the bank. A thought was given as to how the operator could best be moved forward, toward the position traditionally occupied by the shadouf operator. It was realized that the ideal position would be near the legs supporting the main pole and another major benefit of putting him there was that in this position the sliding seat could be replaced with a swing seat (Figure 11.3).

The construction of a swing had much in common with that of the existing support legs: by choosing the dimensions carefully, it was possible to provide as much movement on the swing seat as on the sliding seat assembly and this arrangement allowed the main pole to serve a dual purpose, by also acting as a stay for the swing. Any additional construction cost was more than offset by the elimination of the sliding seat frame (a structure that had to be welded).

FIGURE 11.4 The operator's feet placed on footrests and under a horizontal bar.

FIGURE 11.5 The six buckets in the output test ranged in volume from 4.5 to 35 L. The one that was used in most of the development work is shown second from left.

A close-up of the front end of the system is shown in Figure 11.4, where one can observe that the operator's feet have to be placed beneath a horizontal bar as well as on the footrests; and if this is not done, the operator is liable to fall-off the swing backward as he is not able to hold onto the vertical ropes.

11.3.3 DETERMINATION OF THE APPROPRIATE LOAD

Once a satisfactory design for the bucket lifting assembly was reached, it was decided to test it for output when fitted with different sizes of the bucket. It was thought to be necessary to find the most optimum load to be lifted so that two operators (A and B) of different weights (80 and 65 kg), heights (1.7 and 1.75 m), and ages (45 and 33 years) could use it. Six buckets of various sizes were also used, and these are shown in Figure 11.5.

The instructions to the operators, who had both worked the lifter during its development, were to fill and return the empty bucket 18 times, working at a rate they thought they could maintain for at least an hour.

11.3.4 THE OPERATION OF BUCKET LIFTER

After completing the work described earlier, the modified bucket lifter was put into operation to attempt to gauge the reaction of local farmers to this approach and then to test the equipment in the environment for which it was designed, using local operators.

The farm on which the lifter was first installed was located on the banks of the Jakara River within Kano City where labor is relatively expensive and where farmers can often afford to buy or hire an engine-powered pump. In Kano City, comparatively high profits can normally be made by growing vegetables because of the proximity of a large market. The city also provides supplies of fuel, spare parts, and competent mechanics. Three lifters were constructed within the city; the first single bucket lifter was constructed on the Jakara river, within the Kano City boundary, the second single bucket lifter was installed 20 km south of Kano City, and the third was constructed on a site some 20 km south of Kano, on the river bank, and adjacent to a shadouf.

Eleven men, all of whom had used a shadouf, were asked if they would take part in an evaluation of the bucket lifter. A demonstration was given, and each person was allowed about 10 min to become familiar with its operation. Their ages ranged from 23 to 70 years and their weights from 48 to 75 kg. All persons were reasonably fit and involved in agriculture in some way. They were asked to work for 15 min each at a rate that could be kept up for at least an hour. Instructions were given that the bucket should be allowed to become completely submerged between lifts so that it was full each time (the operation was supervised); the number of buckets lifted was counted every 15 min and the details of the operators and their performances were recorded.

The test was run with the lift set at 2.85 m: the main pole was then adjusted so that the lift height corresponded (approximately) with that of the neighboring shadouf. A second test was then carried out, each person working both the lifter and the shadouf for 15 min. Arrangements were made to collect and measure the volumes of water on each occasion and the results were recorded.

11.4 RESULTS

11.4.1 APPROPRIATE LOAD DETERMINATION OF A SINGLE BUCKET LIFTER

The observations consisted of the time required per lift and the total amount of water lifted over 4.3 m. From this information, the work output in watts was calculated (Table 11.1). Both operators were able to do most of the work when using the second largest bucket.

TABLE 11.1 Summary of Test Results for Water Lifting Undertaken by Two Operators (Designated A and B) and Working with a Range of Six Buckets of Varying Sizes.

		Operator A or B					
Bucket ID	—	**One**	**Two**	**Three**	**Four**	**Five**	**Six**
Size (liters)	—	35	30	25	16	13	4.5
Average time to lift and empty the bucket (sec)	**A**	51.6	27	28.3	29.4	24.6	25.7
	B	55.7	31.6	34.9	30	28.6	25.6
Average Volume of water lifted (Liters/min)	**A**	40.7	42.4	48.8	28.6	31.7	10.5
	B	37.7	49.8	39.5	28	27.3	10.5
Average output (Watts)	**A**	28.6	35	34.3	20.1	22.3	7.4
	B	26.5	29.8	27.7	19.7	19.2	7.4

11.4.2 OPERATIONAL PERFORMANCE TEST: BUCKET LIFTER VERSUS SHADOUF

The general reaction of the eleven operators was encouraging; they were able to observe that they could lift water at a greater rate than with the *shadouf*. In addition, they were attracted by the novelty of being able to work while in a sitting position. The 11 operators were selected to lift water using both the bucket lifter and the *shadouf* to evaluate their performances.

Table 11.2 shows that the performance results obtained were in favor of the bucket lifter; the mean output using the shadouf was some 4.3 Watts lower than that of the bucket lifter, i.e., a difference of 19%. Similarly, the shadouf lifted lower volume with a mean difference of 9.9 L/min (~15%).

TABLE 11.2 Summary of Results of Lifting Test Carried out for *Shadouf* and Bucket Lifter.

Operator #	Age (years)	Weight (kg)	Bucket lifter		Shadouf	
			Volume of water lifted per minute (liter)	Output (watts)	Volume of water lifted per minute (liter)	Output (watts)
1	70	63	79.2	27.8	70.2	23.8
2	62	50	64.6'	22.7	59.8	20.2
3	24	53	80	28.1	68.6	23.2
4	37	64	80.9	23.4	71.1	24.1
5	37	75	93.1	32.7	72	24.4
6	42	66	94.8	33.3	70.9	24
7	23	50	66	23.2	58.8	19.9
8	31	48	63.5	22.3	64	21.7
9	30	58	67.3	23.7	62.2	21.1
10	35	53	62.9	22.1	67.3	22.8
11	33	70	90.2	31.7	68.9	23.3
Average	38.54	59.1	76.6	26.9	66.7	22.6

11.4.3 STATISTICAL ANALYSIS

11.4.3.1 PAIRWISE COMPARISON BETWEEN THE METHODS OF LIFTING WATER BY SHADOUF AND BUCKET LIFTER

Let (s_i, m_i), $i = 1, 2......, 50$ be pairs of the observations of the water lifted by the *shadouf* (s_i) and modified method (m_i). It is assumed that these pairs are independent. The difference $(D_i) = m_i - s_i$ for all *i*'s is shown in Table 11.3; hence, the confidence interval for the mean difference is obtained as follows:

The average amount of water (in liters) lifted using shadouf is $s = 6177.10$ L, the amount lifted using the bucket method is $m = 6872.50$ L, and the standard errors are 1029.01 and 1228.49 L, respectively.

Assuming that the difference D_i is normally distributed with mean μ and finite variance ʃɣ², the confidence interval for the mean difference D_i at $a = 0.01$ is 506.86 and 883.95.

To test the hypothesis Ho: $\mu_D = 0$ against $\mu_D > 0$ gives $t_o = 10.62$ and, hence, reject Ho at $\alpha = 0.01$, so that the difference between the bucket and *shadouf* methods is the same. This shows that the amount of water lifted by the two methods is not the same and that the bucket method is the most effective.

TABLE 11.3 Differences between the Volume of Water per Hour Lifted by *Shadouf* and Bucket Lifter.

S. No.	Shadouf (s)	Modified (bucket) (m)	D_i
1	6210	6699	489
2	7335	6649	414
3	7540	7791	251
4	6670	7098	428
5	7260	7182	78
–	–	–	–
–	–	–	–
49	5910	6699	789
50	5430	6069	639

TABLE 11.4 ANOVA for the Interaction between Head, Operators, and Water Lifted by Modified Lifter.

Source	Sum of squares (SS)	df	MSE	F_0
Head (H)	106,857,281	4	26,714,320.25	1292.72(a)
Operator (0)	14,936,433	9	1,659,603.67	80.31(a)
Water lifted (W)	12,089,528	1	12,089,528.00	585.07(a)
HxO	1,358,514	36	37,736.50	1.83(b)
HxW	2,015,491	4	503,872.75	24.38(a)
OxW	2,490,801	9	276,756.11	13.39(a)
Residual	743,948	36	20,665.22	–
Total	140,492,000	99	–	–

(a) Significant at 1%; (b) Significant at 5%

11.4.3.2 ANALYSIS OF VARIANCE (ANOVA) FOR INTERACTION BETWEEN HEAD, OPERATORS, AND WATER LIFTED BY THE MODIFIED LIFTER

The analysis of variance is used here to find out if there were any significant interactions between water lifted (W) per hour, Head (H), and Operator (O).

Table 11.4 provides this analysis of variance (ANOVA). This table indicates that (as expected) there are highly significant interactions between the rate of water lifted, and both the head and operator; while the interaction between the Head and the Operator is significant at 95% confidence interval ($\alpha = 0.05$).

TABLE 11.5 Correlations between Age, Weight of the Operators, and the Volume of Water Lifted Using Modified Lifter.

Operators, O, estimate	Age (years) (A)	Weight, W (kg)	Average volume of water lifted (L)	
1	36	48	12,205.00	12,046.58
2	45	63	13,351.80	13,452.68
3	42	65	13,517.00	13,640.16
4	25	55	12,561.20	12,702.76
5	35	54	13,105.80	12,609.2
6	62	56	12,396.60	12,796.50
7	23	49	12,025.80	12,140.32
8	38	75	14,601.60	14,577.56
9	48	64	13,925.40	13,546.42
10	52	58	12,805.80	12,983.98
Total	406	587	13,0496	130,496

It was found that "$\gamma_{AL} = 0.1934$, $\gamma_{WL} = 0.9426$, and $\gamma_{AW} = 0.2993$"; where, γ_{AL} is the correlation between operators' age and the quantity of water lifted, γ_{WL} is the correlation between operators' weight and the quantity of water lifted, and γ_{AW} is the correlation between operators' age and weight. For each correlation coefficient *y*, the hypothesis Ho: $\gamma = 0$ against H_1: $\gamma \neq 0$ was tested. It shows that γ_{WL} is significant at 1% ($\alpha = 0.01$). However, γ_{AL} and γ_{AW} are not significant at neither at 1% nor at 5%.

11.4.3.3 REGRESSION ANALYSIS FOR THE RELATIONSHIP BETWEEN OPERATORS' WEIGHT AND THE QUANTITY OF WATER LIFTED

It was assumed that there is a linear relationship between weight (W) and average water lifted (L). Therefore, the following linear regression equation was obtained:

$$L = 7547.0750 + 93.7398\,W \tag{11.1}$$

The hypothesis tested was that there is no linear relationship between *L* and *W*. The ANOVA results from the hypothesis test are shown in Table 11.6.

TABLE 11.6 Analysis of Variance for the Relationship between Operators' Weight (*W*) and the Quantity of Water Lifted (*L*).

Source	SS	Df	MSE	F
Regression	5,308,314.9070	1	5,308,314.9070	604.9258
Residual	66,258.0930	8	8282.2616	
Total	5,974,573	9		

Based on these results, the hypothesis Ho can be rejected at $\alpha = 0.01$ and it can be concluded that there is a significant statistical evidence that the model is linear. Furthermore, about 89% of the variability in *L* is explained by linear regression and the linear model is satisfactory in this respect.

11.5 DISCUSSIONS

In Northern Nigeria, all *shadoufs* are of the same basic design that consists of a counterweighted pivoting pole, which is tied to a light vertical pole (sometimes the dried stalk of guinea corn) that carries at its lower end a calabash.[8] The latter is sometimes suspended on four short ropes, but a slightly better arrangement is to use two split flexible branches, which are formed around the calabash to provide four handles. The farmer may grow his calabash; if he does not, they are cheap to buy and the operators can fill them almost to the brim with ease. Although both plastic and metal containers have been used to replace the calabash,[5] the advantages to be gained from this seem extremely limited. The poles themselves are readily available, cheap and reasonably easy to grip, and the counterweight is again cheap to install.

Apart from that provided by the short pieces of rope, the only movement that takes place between adjacent members of the *shadouf* either between the main pole and its support or between the latter and its two supports. In a few cases, the main pole is pegged to fixed horizontal support; this could be greased to reduce friction or the peg and the whole arrangement could be replaced by a hinge. The latter would seem preferable but the improvement likely to follow would certainly be small. In general, the horizontal member is supported by two vertical posts that are set into the ground, each having a V-shaped branch at the top. Once again, this could be greased, or the horizontal member could be permitted to roll backwards and forwards along a short horizontal track. Although this arrangement would almost eliminate friction, yet the increased efficiency would again be relatively small and the modified supports might be difficult to construct from naturally available materials. In summary, careful consideration of all aspects of *shadouf* construction and operation did not bring any particular promising departures from current practice, so that the first hypothesis almost led to a dead-end with only minor recommendations emerging.

The manual lifting operation involves a man standing while pulling down on a pole, scooping, lifting the pole then grabbing the calabash to lift and tip it (and getting wet in the process). Practically all the work is done with the arms and the trunk has to twist from side to side. Similarly, a container filled from a deep well can be lifted either by a rope pulled by hand or by a hand-powered windlass; and a single-piston pump can be moved directly as in the Rower design, or a twin pump has two positions linked together by a rope as in the treadle pump. All involve standing position; only the last use the leg muscles and in this case only the legs.

Having decided to first examine how to work produced by the rowing action could be used to lift water, the first device attached to a rowing rig was deemed unsuccessful. It was too complicated, too slow to refill after discharging and the forces involved caused the components to stretch, absorbing some of the energy for no useful purpose. It was only when it was decided to match the force developed to the weight of water carried in a container (i.e., to make it a batch-type operation) so that the design of the bucket lifter emerged as described in this chapter.

When the design was completed, the two people were actively involved in the fabrication and modification was then done in a simple test procedure and the results was encouraging. The two operators performed the work at the rate of 29.8 and 35.0 Watts, respectively. Some caution must be expressed at this point: both operators were well-motivated, they were fit, they did not work for a long period (although they were instructed to work at a rate they could maintain), and the work was done in a comfortable environment (all factors were expected to enhance the performance).

During the time spent carrying out the experimental work of this study, several points have emerged that warrant more attention. A major expenditure of time and money would be required to make any significant progress, while others would require relatively little investment. One of them is the ineffective control of the amount of water released from the Tiga Dam for both irrigation and human consumption. Despite the modest amount of money invested in irrigation research and development,[11] much more effort is required and additional technical and financial resources are needed to improve the water efficiency with which high-quality crops are grown. The loss of water through seepage and evaporation from the canals and wetted soil surfaces is another area that requires immediate attention. Pipes or even envelopes could reduce seepage, but possibly a greater contribution could be made by changing the application method from one that is based on flooding basins to alternate approaches, such as furrow irrigation (as it would facilitate the mechanization of the crop production, even at a very basic level).[6] Finally, there are places, within the study area, that have the potential to produce crops limited by a high water table. Within the Kano River irrigation scheme, areas exist, where the groundwater level is only about 40 cm below the ground surface.[10] In such areas, the possibility of drainage should be investigated; and if the water released is suitable, this might be used for irrigation or other purposes downstream, where water scarcity is becoming a serious problem.

11.6 SUMMARY

The *shadouf* is unlikely to be improved substantially by the use of "modern" materials. Minor improvements could be made by lubrication or redesign at the point(s), where the main pole pivots; but the most obvious place for improvement is in the conditions, where the operator stands. This should be flat, not slippery, free draining, and capable of being shaded.

A new method of manual lifting water was developed with new features, such as the operator is seated and can use his/her leg, back, arm, and shoulder muscles by pulling against a load in the direction in which they are most efficient; the operator then relaxes each set of muscles and returns to the starting position of the lifting cycle while the water container falls by gravity and refills. This bucket lifter has shown better efficiency than the shadouf, even though the operators had only recently been introduced to it. It is easy to adapt the lifter to suit each operator. Although slightly more expensive than a shadouf, yet it would appear to be cheaper than most other lifters that have been investigated.

The statistical analysis shows that the bucket lifter is substantially more efficient in lifting water than the conventional *shadouf* and that there exists a strong relationship between the rate of water lifting and both the head and operator. These results show that substantial improvements can be made in the lives of small-scale riparian farmers by facilitating water-lifting in an ergonomic manner coupled with higher efficiency. Such modified water-lifting devices are likely to find attention from farmers and policy-makers alike in Africa.

KEYWORDS

- **ergonomic**
- **irrigation**
- **Nigeria**
- **riparian farmers**
- **shadoufs**
- **water-lifting device**

REFERENCES

1. Ahmad, M. T.; Haie, N. Assessing the Impacts of Population Growth and Climate Change on Performance of Water use Systems and Water Allocation in Kano River Basin, Nigeria. *Water* **2018,** *10* (12), 1766–1772.
2. Appleton, J. B.; Belden, W. Northern Nigeria: A Study in Political Geography. *Econ. Geogr.* **1936,** *12* (4), 325–331.
3. Buckley, C.; Hynes, S.; Mechan, S. Supply of an Ecosystem Service—Farmers' Willingness to Adopt Riparian Buffer Zones in Agricultural Catchments. *Environ. Sci. Policy* **2012,** *24,* 101–109.
4. Chen, S. H. Introduction. In *Hydraulic Structures*; Chen, S. H., Ed.; Springer: Berlin - Heidelberg, 2015; pp 1–39.
5. Dauda, T.; Akinbode, S.; Saka, J. An Assessment of the Roles of Irrigation Farming in the Millennium Development Goals. *Afr. J. Agric. Res.* **2009,** *4* (5), 445–450.
6. Koech, R.; Langat, P. Improving Irrigation Water use Efficiency: A Review of Advances, Challenges and Opportunities in the Australian Context. *Water* **2018,** *10* (12), 1771–1778.
7. Maina, M. M.; Amin, M. S. M.; Aimrun, W.; Sani, I. *Soil Salinity Assessment of Kadawa Irrigation of the Kano River Irrigation Project (KRIP). J. Food Agric. Environ.* **2012,** *10* (3), 1028–1034.
8. Mirti, T. H.; Wallender, W. W.; Chancellor, W. J.; Grismer, M. E. Performance Characteristics of the *shadouf*: A Manual Water-lifting Device. *Appl. Eng. Agric.* **1999,** *15* (3), 225–231.
9. Odunuga, S.; Badru, G. Landcover Change, Land Surface Temperature, Surface Albedo and Topography in the Plateau Region of North-Central Nigeria. *Land* **2015,** *4* (2), 300–324.

10. Sangari, D. An Evaluation of Water and Land Uses in the Kano River Project, Phase I, Kano State. *J. Appl. Sci. Environ. Manag.* 2010 *11* (2), online; https://doi.org/10.4314/jasem.v11i2.55002
11. Shawai, S. Status of Water used for Drinking and Irrigation in Kano: A Critical Review on Physicochemical and Heavy Metals Concentration. *Am. J. Biomed. Sci. Res.* (2019) *5* (3), 235–242.
12. Ugalahi, U. B.; Adeoye, S. O.; Agbonlahor, M. U. Irrigation Potentials and Rice Self-sufficiency in Nigeria: A Review. *Afr. J. Agric. Res.* **2016,** *11* (5), 298–309.
13. Yannopoulos, S.; Lyberatos, G.; Theodossiou, N.; Li, W.; Valipour, M.; Tamburrino, A.; Angelakis, A. Evolution of Water Lifting Devices (pumps) Over the Centuries Worldwide. *Water* **2015,** *7* (12), 5031–5060.

PART IV

RECYCLED WATER USE VERSUS IRRIGATION

CHAPTER 12

INTERNET OF AGRIBOT THINGS (IOAT) FOR IRRIGATION AND WASTEWATER SYSTEMS

SANUSI MOHAMMED SADIQ, INVINDER PAUL SINGH, and MUHAMMAD MAKARFI AHMAD

ABSTRACT

In the 21st century, automation has a big impact on how people live. Appliances can be controlled automatically thanks to automation. In addition to comfort, it also saves time, energy, and resources. The global lack of clean water resources has resulted in a need for their best usage. The data collection and advanced computation of application-specific sensors form the foundation of IoT solutions, which link the physical and digital worlds. By using various technologies to improve agriculture's functioning, nations are attempting to make it more sustainable. By making improvements to irrigation systems, which are crucial for water use efficiency, it is possible to meet the "Sustainable Development Goals (SDGs) of the United Nations," in particular Goal 6 and Target 6. In the context of precision agriculture, IoT-based smart irrigation management systems can aid in reaching the highest level of water resource efficiency. These cutting-edge systems provide efficient water management as well as excellent soil and weather monitoring.

12.1 INTRODUCTION

As the backbone of the economy and one of the most vital societal sectors that encourages social advancements, agriculture is a significant industry.[7]

Sustainable Use of Wastewater in Agriculture: Applications, Methods, and Technology. Megh R. Goyal & Gaurav Saini (Eds.)

Unprecedented strain is being placed on agricultural operations by factors such as population growth, resource scarcity, the impact of pandemics on the workforce, financial instability, and unpredictability of the weather.[5] Today's farmers must contend with worsening water shortages or floods, dwindling land availability, and shifting costs as they work to feed the world. Farmers would need to increase food output by 70% to feed everyone by 2050 given the expected rapid growth of the global population.[5,52]

All nations have serious concerns about the developing problem of agriculture automation. As the world's population grows, which swiftly is happening, so does the demand for food. It has been highly challenging for the agriculture business to find methods and procedures that would enable them to fully satisfy the rising expectations and needs due to the expanding need for food as well as shifting customer preferences.[18]

To enhance the industry's overall outcomes and results, it is crucial to make sure that improvements are made. To meet consumers' evolving expectations, food production technology needs to be greatly improved and innovated. Since the majority of nations rely on the agricultural sector, it is imperative to utilize agricultural resources.

Due to the growing global population and the increasing need for freshwater, there is currently a water shortage in some areas of the world, and the problem is only going to get worse. By 2050, the 7.2 billion people that currently inhabit the planet are projected to number 9 billion.[19] Policies relating to climate change and water management have been linked. The demands for water from different industries or the consequences of different levels of global warming on hydrological resources are only two examples of the many factors that may have an impact on water management. Research papers on water resources and agriculture frequently tackle climate change and its implications. To ensure that there is enough water for both human consumption and food production as well as ecosystem maintenance, water adaptation methods have been given some thought considering the potential effects of global warming.[38] In addition, it is necessary to guarantee the water's safety for human consumption and environmental reuse. Climate change may lead to an increase in water scarcity, a decrease in water quality, a rise in soil salinity, a loss of biodiversity, a rise in the need for irrigation, or an increase in the expense of emergency and remediation measures. The number of studies looking at ways to utilize less water in irrigation procedures has increased because of these factors. A significant amount of freshwater is used in agriculture, especially for irrigation.

According to the Food and Agriculture Organization of the United Nations,[5] irrigation contributes to roughly 40% of the world's food output on 20% of the area used for agricultural production. In some locations today, water has become limited. Who obtains the water, and how much will they pay? A few of these issues may center on engineers and irrigation. Engineers will at the very least provide the data and methods that decision-makers will use to respond to those questions. Engineers must therefore be knowledgeable about the price of water and the methods used for its distribution. Virtual water is a crucial concept that has recently come to light. Water prices are influenced by demand, supply, and the value of developing water assets.[30]

The concept of "virtual water" argues that whenever a nation imports food, it also imports the water needed to grow that food. A metric ton of wheat is equivalent to importing 1000 m^2 of water, for instance, if producing wheat consumes 1000 m^3/ton of water.[19] Due to a lack of affordable, clever irrigation technology, developing countries require more water to produce the same amount of crops than industrialized countries. For instance, India's main agricultural products demand 2–4 times more water than those produced in other nations such as China and the USA, even though it only has 4% of the world's freshwater resources and 17% of the global population.[19] To effectively utilize fresh water, it is imperative to develop modern technology-based smart methods and systems.

There are practical uses for the IoT, which is no longer only a theory.[42] IoT platforms enable complex real-time control algorithms that include numerous layers of hardware, software, analytical methods, and application knowledge, posing significant technical and application challenges.[3] Since IoT makes it possible for items to be networked as services and allows for service mashups, realizing the anticipated effects of IoT on platforms is one of the major technical challenges. System development will thereafter be subsumed by service orchestration, and system logic will be substituted by dynamic plug-and-play compatible service composition. IoT provides adaptable solutions to address the requirements of certain applications in general.

IoT-based solutions have shown to be quite beneficial in several areas of the agricultural environment.[28] These clever techniques might also work well for irrigation that is economical with water. When creating a smart irrigation system, the primary elements to consider are soil moisture, precipitation, and evaporation. Agriculture relies heavily on irrigation to maintain a sufficient crop output by avoiding both under- and over-watering. In addition to the high cost of the energy required to move water and power irrigation systems, some regions also have high water costs.

This chapter focuses on applications of the "Internet of Agribot Things (IoAT)" for smart-based irrigation systems.

12.2 SUSTAINABILITY AND IRRIGATION SYSTEMS

One important factor pertaining to irrigation systems might be sustainability.[5] The three pillars of sustainability must be balanced to preserve sustainability within any system. Economic, social, and environmental sustainability are the three pillars of sustainability. The potential influence of irrigation systems on the economy, ecology, and society is depicted in Figure 12.1. The organizations engaged in the specific sector and related operations must take these factors into account because the components of sustainability could be evaluated in various contexts and media. Making sure that irrigation activities don't have a detrimental influence on the environment is one of the characteristics of sustainability.[10]

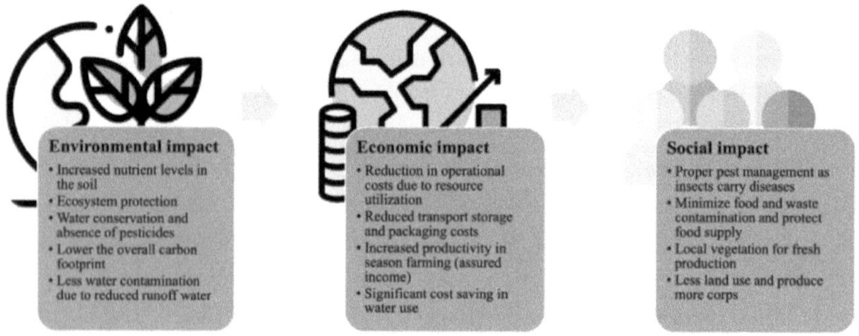

FIGURE 12.1 Environmental, economic, and social impacts of irrigation

Source: Reprinted from Ref. [44]. https://creativecommons.org/licenses/by/4.0/

Irrigation activity has been linked to waterlogging, an increase in waterborne infections, salinization of soils, relocation issues, and numerous other detrimental environmental effects. Given the significance of irrigation to the agricultural industry, it is essential that any activities related to irrigation do not jeopardize human or animal health. An element of sustainable irrigation systems may also involve water management. Water is a resource that is required for irrigation, so it is essential to effectively manage and control it to prevent water waste.[8] As a result, the development and management of irrigation are becoming subject to new demands because of the push for healthier and more sustainable food systems. Every irrigation system, from local to national, can increase agricultural output, increase water security, and promote inclusive growth and promote progress toward the SDGs.

Pumping power is required to run drip irrigation systems. Different energy sources are used in the process to provide pumping power, which also has an adverse effect on the environment. It is crucial to use green operating techniques to reduce energy use and carbon emissions impact when considering sustainable irrigation. Organizations involved in irrigation activities need to focus more on strategies that cut down on issues such as disease, pollution, and cost. When irrigation doesn't deplete natural or human resources, high irrigational sustainability can be achieved. Economic and environmental issues are the main definitions of sustainability in this context. To maintain economic sustainability, it is crucial to make sure that the cost of irrigation does not exceed the worth of its marginal productivity. If the irrigation cost process is too high, the operations might not be viewed as sustainable, in which case it might be required to alter the strategy for related activities.[48]

Water depletion is another idea that might be seen as being extremely important in irrigation and sustainability. As a finite resource, water has become increasingly problematic. For the agricultural industry, the scarcity of fresh water has proven to be a major problem, making sustainability even more crucial. In this context, limiting water waste and putting into practice efficient irrigation techniques may help save water that may then be used in the long run for the necessary activities. In many countries, the need for irrigation in agricultural cropping systems is one of the primary factors contributing to a sharp rise in water scarcity. Smart irrigation is essential for water conservation and is critical to giving each crop with the necessary amount of water.[40] If irrigation is applied too late, crops may dry out. The greatest solution for this problem is a drip irrigation system-based automatic controller. Indeed, combining modern technologies with irrigation can enhance many regions' usage of irrigation water. To help with the attainment of SDG 6, the IoT platform can provide unbiased information about water resources, its use, and management. ICT and IoT technologies were used to create the smart system of irrigation for South Algeria.[27,44] As a result, irrigation systems are now designed to be as economical, water- and labor-efficient as feasible.

12.2.1 CONTRIBUTION OF IRRIGATION TO THE REALIZATION OF THE SUSTAINABLE DEVELOPMENT GOALS

By ensuring food security and lowering poverty, irrigation contributes to the accomplishment of multiple SDGs. Smart irrigation supports a wider range of SDGs, stimulates more conscientious production and consumption for food security, and fosters industrial innovation. The present rate of the SDGs' progress is directly impacted by it. To boost the production of food crops, irrigation development requires water resources. Agriculture is essential to civilization because it uses water, one of the most precious resources on the planet, to produce food. Climate change and the quickening of population growth have a significant influence on agriculture and the water resources necessary for sustainable development. Thus, it has been demonstrated that smart irrigation systems significantly increase crop output and agricultural profitability. Using this technique, the sector is progressing toward the SDGs while improving irrigation management to make it more egalitarian, sustainable, and effective.[44] The contribution of the smart system of irrigation to sustainable development is seen in Table 12.1.

TABLE 12.1 Irrigation's Contribution to the SDGs' Realization.[44]

SDGs	Contribution of smart irrigation to the SDGs
SDG 1: No Poverty	Aids the remote areas' people of emerging nations.
SDG 2: End Hunger	Ends famine and increases agricultural productivity.
SDG 3: Promote overall health and wellbeing	Lowers the danger of pesticides and other ailments that are soil caused.
SDG 6: Clean Water and Sanitation	Uses utilities and agriculture irrigation to make sanitation accessible.
SDG 7: Accessible, Reliable, and Clean Energy	When used in conjunction with a solar power system, helps to achieve a clean energy option for farms.
SDG 8: Decent Work and Economic Growth	Enhances other industries and quickens the growth of rural economies.
SDG 9: Infrastructure, Industry, and Innovation	Encourages innovation and supports sustainable industrialization.
SDG 11: Sustainable Cities and Communities	Using intelligent irrigation systems to build sustainable cities.
SDG 12: Responsibly Sourced Production and Consumption	Ensuring responsible resource management and a decrease in the production of waste.
SDG 13: Action on Climate	Increasing crop production and eventually altering rainfall patterns.
SDG 14: Sub-aquatic Life	Compared to conventional irrigation systems, just a moderate amount of water is used.
SDG 15: Land-Based Life	Increases farm output in both quality and quantity and create a steady supply of food.

12.3 IRRIGATION TECHNIQUES

A farm field can be irrigated using a variety of methods depending on the sort of crop being grown there. In general, farmers in underdeveloped nations employ these three techniques: sprinkler system, drip system, and channel system. The irrigation technique known as the channel system is ancient.

However, a new technology for autonomously irrigating farm fields is the smart irrigation system.

a. **Channel system:** In agriculture, this irrigation method is often employed. This technique can be used to water a large agricultural field because it is so affordable. A water pump in this setup connects the pipes. Water flows via the pipe from a lake, river, or bore well to an agricultural region when the pump is operating. The farmer then fully dedicates himself to irrigating the crop field with the help of several laborers. During watering, there is a significant quantity of water waste and labor activity.
b. **Sprinkler irrigation system:** If there is less water available, this system is more practical. The main pipe and the parallel pipes both fill with water as soon as the pump turns on. The top of the perpendicular pipe has a connected nozzle that rotates automatically on a predetermined schedule. On sandy soil, this method is extremely beneficial. Water waste is decreased, and fewer workers are needed.
c. **Drip irrigation system:** In this method, the water trickles down at the roots one drop at a time. The best irrigation technology for fruit trees, gardens, and trees is this one. Smaller pipes are used to divide water from a main pipe. These sub-pipes have specially designed nozzles attached to them. Water is wasted very seldom, and irrigation requires no labor. If the farmer is aware of the condition of the farm field, he can start the motor and select a nozzle's direction to automatically water the plants. The farmer then checks the condition of the field after a while. The farmer turns off the motor once the entire crop has been watered.
d. **Smart irrigation system:** The three systems listed above are generally operated by a user, but with smart irrigation, the entire system is managed by autonomous mechanisms that can do so even if the farmer is not in his or her agricultural field (12.2). It notifies the farmer via messages about updates to the agricultural field's operation and relevant information. Unlike the other three ways, the system doesn't need a worker to operate and uses less water.

12.4 TRADITIONAL AGRICULTURAL MONITORING

Tension meter and drip irrigation with an automated controller system are used in some traditional irrigation systems to schedule irrigation while simultaneously monitoring the soil and water status in sandy soil.

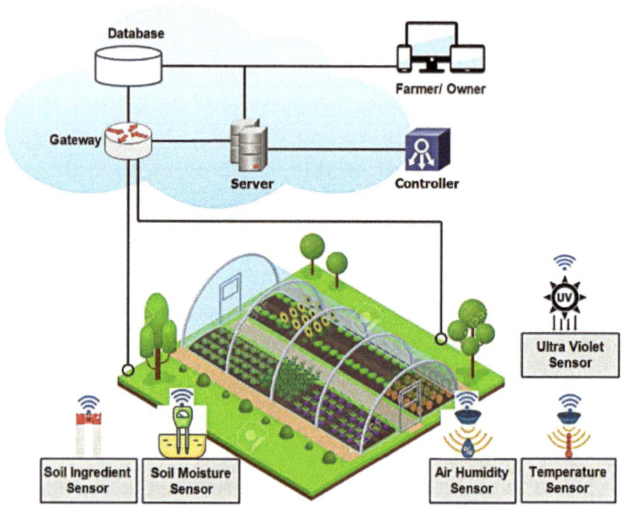

FIGURE 12.2 IoT-enabled smart agriculture.

Source: Reprinted from Ref. [61]. Copyright © 2022 by the authors. Licensee MDPI, Basel, Switzerland. https://creativecommons.org/licenses/by/4.0/

12.4.1 WEATHER AND SOIL MONITORING

To improve food production with the least amount of water loss, effective monitoring systems are essential to plant growth and development.[22] Monitoring in the specific context of precision irrigation encourages the collection of data that precisely portrays the condition of the plants, soil, as well as weather in irrigation zones by incorporating the Internet of Things (IoT) with wireless sensor network (WSN). IoT has enhanced the control and monitoring of the irrigation process, enabling the development of a real-time tracking system employing inexpensive technology. WSN also contributes significantly to the ongoing observation of precision farming. Using this technique, a connectivity of WSNs nodes may be built that can gather, analyze, and disseminate data on a variety of attributes. Soil moisture is one important element that the plant requires for growth. For the optimal irrigation strategy to be implemented, accurate soil moisture content monitoring may be deemed to be essential. The emphasis is on low-cost capacitance-based soil moisture monitoring, which mostly depends on a dielectric device's performance.

The SMART irrigation system's monitoring of the soil uses well-established technology to measure the quantity of soil moisture. Moisture sensors for soils are embedded in the rhizosphere of trees, shrubs, or turf to precisely detect the quantity of soil moisture and transmit the signal to the controller. Using this technique, it is possible to continue learning crucial details about how linked actions should be planned and carried out to get the best outcomes.[20] The two primary water level sensor-based irrigation techniques are moisture on-demand irrigation and paused cycle irrigation. A suspended cycle offers functions comparable to a regular timer controller, such as schedules for watering, length, start, and end timings. The key distinction here is that when the soil has enough moisture, the device will automatically cease the following watering session. Water on demand irrigation, however, doesn't need any programming or planning for the duration of the irrigation. In this technique, the user sets the threshold, and irrigation begins when the soil moisture falls below the necessary levels. A vast cropping area's weather is also evaluated through the process of weather monitoring, which takes place throughout the area and its surroundings. To detect hazards and create measures for mitigating adversity, the environmental circumstances of the work area are thus analyzed. When it comes to connecting various sensors to one another to monitor the physical and mental conditions of the environment, the deployment of WSN could once again be viewed as essential.[35] Through a feedback loop, mounted sensors' real-time monitoring and data processing are done. This further activates the control devices. The environment of the crop, including soil moisture control, air temperature, wind speed, humidity, and other parameters such as air temperature and temperature, has been tracked and assessed using a different IoT-based weather monitoring system. It uses weather-based sensors that are connected to wireless communication protocols in order to deliver real-time data. This approach makes it feasible to gather detailed meteorological data, which aids in the creation of strategies that could eventually enable irrigation.[23]

12.4.2 WATER MANAGEMENT

Regarding irrigation, water management could be seen as a critically important idea. Since there is a global worry about the lack of access to clean water, the agricultural and other businesses must pay close attention to and concentrate on this problem.[46] To ensure that the right amount and timing of water delivery, soil moisture could be thought of as part of water management. For the agricultural industry, efficient water management is essential since it can lower costs and increase crop yield. The ability of agriculture sector organizations to manage resources and complete necessary tasks in compliance is made possible by effective water management. Given the range of initiatives being undertaken at various scales, it is crucial to know whether they will be successful or not. Natural resource conservation is a growing area of concern for organizations in the modern world because of the limited availability of these resources. One of the most important and practical resources in this regard is water, which must be preserved and safeguarded at all costs.[41]

The institutions in and linked to these operations must be extremely vigilant and cautious about discovering tactics to use water in the most efficient manner possible because irrigation techniques use a lot of water. This means that the agriculture industry needs excellent water management technologies to reap a variety of benefits. Because of the external environment's high degree of uncertainty, agricultural activities may be significantly impacted. For instance, rising fuel costs may result in higher irrigation water pumping costs. If fuel prices rise, the cost of pumping irrigation water would rise, this could have an impact on how effectively the project is managed overall. By adopting water management, businesses may be able to construct larger reservoirs and use creative tactics, lowering the risks and adverse impacts. One of the most crucial components of effective irrigation water management is an understanding of the basic interactions between soil, crops, and water.

To successfully carry out the required agricultural operations, sufficient knowledge of the processes and final products must be acquired (irrigation). Only if such information is acquired may the needed activities be carried out in a sufficient manner. It would be impossible to analyze how irrigation tasks must be supervised and monitored in challenging situations without the right knowledge and information, which could ultimately lead to a decline in overall performance. Making sure that workers are operating at their highest levels of productivity is one of the many reasons why water management is important. It might be able to guarantee that agricultural crops receive the right amount of water in arid areas and during periods of low rainfall by using water management.[49]

Water management must be prioritized to distribute and use water effectively, given that many projects are being undertaken in locations where there is a water shortage. Additionally, rainfall is notably low in many regions, necessitating the storage of a lot of water to ensure that the lack of rainfall could be remedied. To meet future water needs and ensure uninterrupted operations, water management is even more essential considering the unpredictability and fragility. In the modern world, a sizable amount of water is being squandered through various uses, like irrigation.[54]

The need for better management is exacerbated by the massive amounts of water that are being wasted without being used in any way. It is necessary to establish strategies and procedures that would reduce resource waste and result in efficient resource use. The necessity for water management is also exacerbated by the possibility that there won't be enough access to clean water in the future. In the agricultural sector, various approaches, and techniques for managing water are being applied, and each of these techniques has advantages and constraints of its own.[4] Some of these include measuring, metering, managing; employing water-wise landscaping and irrigation; controlling reverse osmosis; collecting rainfall; and building reservoirs. These can be considered a few of the agricultural sectors' most efficient water management practices. Although these techniques might lead to effective water management, how well the results can be predicted depends greatly on how effectively they are implemented and executed.

12.5 SMART IRRIGATION SYSTEM FOR ACHIEVING SDGS

Smart irrigation uses data-intensive techniques to boost agricultural yield while minimizing environmental effect. The information that intensive farming operations collect from a range of sensors enables a higher understanding of the operational area and the operation actions.[33] Decision-making becomes more precise and effective as a result. Additionally, this enables resource optimization and accomplishing the sector's desired goals. When adopting these technologies in irrigation systems, the water is utilized in such a way that it significantly advances the Sustainable Development Goals (SDGs), particularly Goal 6 and Target 6.4. Smart irrigation systems can be used to achieve the SDGs for water and the environment and improve everyone's quality of life.[5]

The several aims and indicators of SDG 6 are connected to all water-related tasks and services, including irrigation. SDG 6 focuses on providing everyone with access to sanitary facilities. Water consumption potency and H_2O stress are two important measures included in the goal 6.4, which addresses water restrictions.[25] The quality of the water information that are easily accessible heavily influences how accurately this indicator is assessed. The demand for healthier and more environmentally friendly food systems has led to new specifications for how irrigation is built and handled. But the improper usage of agricultural components can have severe effects and pollute the environment.[13,31,51]

Two of the primary factors could be a lack of field expertise and a lack of land reservoirs. Water levels have fallen because of the regular withdrawal of water from the earth, which has aided in the expansion of regions of unirrigated land. As a result, governments are now striving to put in place efficient frameworks where systems could be operated effectively. Improving agriculture systems is therefore necessary. Modern agriculture and sustainable farming practices are using artificial intelligence (AI) technologies and solutions.[6]

Academics from a wide range of fields have recently focused a lot of emphasis on the complicated topic of sustainability. Sustainability encompasses a wide range of topics due to its multifaceted character, including weather patterns, ecology, the reduction of greenhouse gas emissions, food hygiene, sustainable agriculture, green energy, and so on. In view of this, interest in using AI in agriculture has lately grown. A recently created technique that optimizes irrigation systems and uses less water is called the SMART irrigation system. It raises productivity. The method helps farmers to meet their needs with a recently approved technology that conserves water for the irrigation process by adjusting irrigation to actual soil and weather conditions. IoT devices can make use of each of these parts. Thanks to the Internet of Things (IoT), smartphone apps, plus sensors, farmers are now even better equipped to assess the precise quality of their crops, including soil temperature, the quantity of water necessary, weather conditions, and much more.[2]

The Internet of Things (IoT) can be considered of as an augmentation of the edge network to any technological objects that can speak with one another and are linked to the internet, making technology more reachable and easier to use. As a result, IoT is linked to the digitization of all aspects of farming and agricultural tasks, enhancing output and efficiency all around.[5,42] Using sensors can help farmers better understand their crops, have less of an impact on the environment, and conserve resources. To boost yields while using less resources like seeds, water, and fertilizer, farmers can use SMART agriculture.[57]

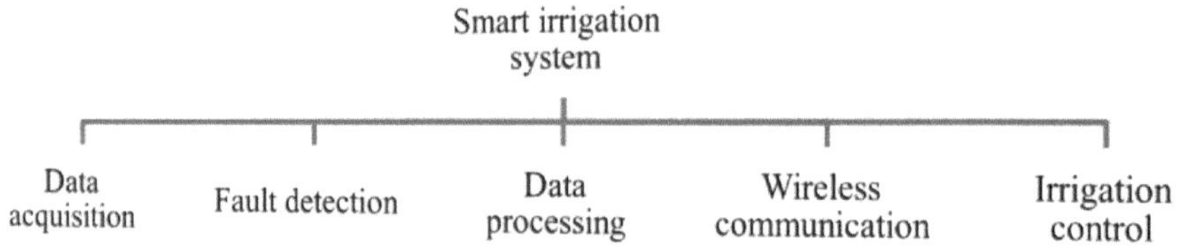

FIGURE 12.3 SMART irrigation system.

12.5.1 THE NECESSITY OF AUTOMATIC IRRIGATION

- By avoiding irrigation at the bad time of day, you can decrease yield losses and drainage from moisture stress damp soils.
- Conserving energy and resources to ensure that they are used appropriately and in sufficient amounts.

- If watering for fields or nurseries is computerized, producers may be able to administer the optimal volume of water at the right time.
- In an automatic watering system, valves control the ON and OFF states of the motor. Using controllers, it is simple to automate the ON and OFF of motors without the need for manual work.
- Installation and configuration are quick and simple.
- It is a precision irrigation technique and a helpful instrument for precise soil moisture control in the highly specialized greenhouse crop production.
- It is quicker to change the available soil moisture levels when human error is eliminated.

12.5.2 IoT SYSTEM BENEFITS FOR IRRIGATION

IoT systems for irrigation provide several advantages, some of which might be summed up as decreased overall water usage, high cost and performance efficiency, lower energy consumption, reduced crop loss, and more.[18] The advantages of IoT use in irrigation systems are depicted in Figure 12.4.

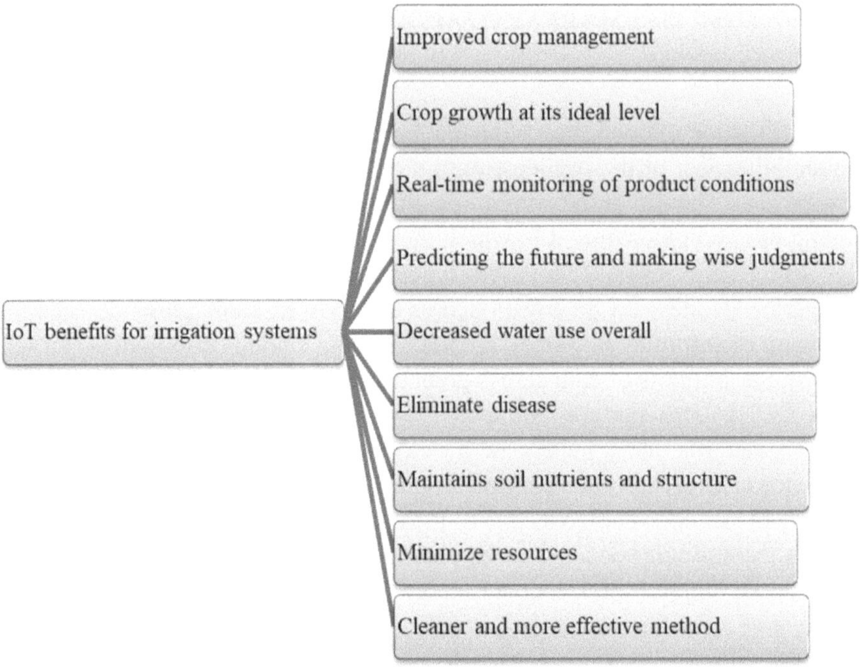

FIGURE 12.4 IoT advantages for irrigation systems.

The fact that IoT irrigation systems consume less water is one of its key advantages. An approach like this also reduces waste, automates most irrigation-related processes, and uses only the water that is required for irrigation. When human interaction was required in the irrigation process using traditional irrigation techniques, where most of the handling and operations were done manually, a substantial amount of water was wasted. Smart irrigation uses only the amount of water that is necessary and involves no or little human intervention. Since less water is needed and the operation is carried out with better precision, it also has the advantage of being more cost-effective.[55]

The strategy also considerably lowers energy use because machines run for shorter periods of time and controlled breaks occur throughout the process, which reduces overall energy usage. The need to control expenses and conserve resources is further necessitated by the fact that resources are few and firms must keep costs to a minimum. Smart irrigation considers the economic factor, making it feasible to successfully complete connected activities while reducing costs. Another advantage is that with better water management and irrigation efficiency, crops and plants only receive the water they need, which reduces crop loss due to insufficient or excessive water delivery.[9]

12.5.3 USE OF SMART SYSTEMS AND IOT IN IRRIGATION

12.5.3.1 TECHNOLOGY-BASED COMMUNICATION

The utilization of communication technologies could be seen as a crucial and essential aspect of the implementation of IoT devices to achieve successful operations. The use of the communication technologies could also be seen as being based on the context in which they would be employed.[14]

Two categories could be used to group the primary IoT technologies for irrigation. Nodes can be envisioned as low-power, short-range devices that transmit or communicate small amounts of data. Other technologies can send enormous amounts of data across great distances with little energy consumption. IoT devices can communicate via a variety of wireless protocols, and these standards can typically be divided into those that are used for long-distance or short-distance communication.[34]

Wi-Fi has been named as one of the most popular and efficient communication technologies because of its potential accessibility. Wi-Fi is supported by the vast majority of low-cost IoT devices today, and despite its drawbacks (sphere of influence and range), is still seen as a successful overall strategy.[21]

The Global System for Mobile networking, also known as a widely utilized wireless technology that enables long-distance communication, can be used with just a mobile plan from the network operator that functions and performs in that location. Other noteworthy recent advancements include the Message Queuing Telemetry Transit (MQTT) and Long Range (LoRa). Because LoRa has such extensive ranges, it is currently very useful and practical for outlying areas without service. However, despite MQTT's cheap overhead and power consumption, which have also made it a popular protocol, it is not yet widely employed for irrigation systems.[5]

12.5.3.2 CLOUD COMPUTING

The traditional database and the cloud are two of the most important and widely utilized storage technologies. These storage systems are crucial for numerous firms operating in numerous industries and sectors because they enable fast access to crucial information when needed. The idea of "big data," which describes vast quantities of datasets utilized by businesses for a variety of purposes, has emerged via the utilization of such storage systems.[24]

Middleware must be used to provide the necessary services that the IoT requires. It is now possible to connect programs that were not originally built to one another thanks to middleware. IoT middleware is further categorized depending on various capabilities and support for interface protocols. In the agricultural sector, sensors are used to gather and evaluate data, notably in irrigation-based systems. Numerous studies claim that the processing of the data takes place in the cloud itself, and that users can connect to the cloud to get the data and examine it. Keeping track of data and retrieving it as needed is how cloud computing is mostly employed in irrigation.[29]

Cloud technology gives clients both for-profit and nonprofit options for data storing, processing, and presenting on several devices and platforms. The usage of this technology to store work-related information greatly contributes to improving overall performance effectiveness. Among the various occasions and purposes for which the saved data is accessible, research and development may be one of the most significant ones. Many organizations in the agricultural industry are now able to store and access information that helps to increase productivity and effectiveness at work thanks to cloud computing.

Cloud technology has also been used to develop algorithms that produce alerts for the irrigation operation. Numerous risks and hazards that could have occurred have been reduced because to these alerts. Using these signals makes it easier to modify work processes and implement the appropriate safety measures, which could help to lessen the impact of any difficulties.[47]

Even though every software is unique and serves a different purpose, they are all connected to cloud technology and were all created to improve job performance. However, the way that these programs are used depends on a variety of criteria, including cost, applicability, services, and other considerations. Regarding the hazards, harms, and complexity involved in maintaining the irrigation system, it may be said that it is a complicated task that must be completed. However, people who run irrigation systems use cloud technology to minimize risks as well as improve the outcomes of their work to achieve their goals.[15]

12.5.4 IRRIGATION SYSTEM OPTIMIZATION USING IOT AND BIG DATA

IoT irrigation devices are like other IoT networks in that they generate enormous amounts of information when they continuously monitor a variety of indicators. Due to the existence of enormous amounts of information, it has become vital and imperative to develop systems that can efficiently access and manage the data.[26] Given the prospect that doing so might be a difficult task overall and could waste natural resources, it has been suggested that there is a pressing need to concentrate more on sustainable big data use.[58]

It has been suggested that blockchain technology be used, that useless data be removed, and only useful information be kept, that solar energy be used to power devices, that clustering techniques be used to reduce the overall volume of information, that efficient algorithms be used, and that sustainable resources be used.[37] Even though large data may be quite useful for the whole irrigation process, it is crucial to maintain appropriate administration and control of the data. Further research has shown that while data gathering from the sensors provides a plethora of data that may be used, data processing is necessary to improve irrigation in context of plant and weather conditions. While many organizations working in and related to irrigation can gather the necessary data, they fall short of correctly evaluating the material and drawing valuable conclusions from it. When it comes to increasing job efficiency and reducing associated hazards to the activities, this incapacity to analyze is a significant roadblock.[60]

The hurdles to smart irrigation are depicted in Figure 12.5. Furthermore, artificial intelligence (AI) is recognized as the technology that most organizations use for a variety of applications. The use of AI makes resource optimization and data collation about crops, such as crop diseases or rectified plant growth, more practical. To evaluate the information obtained from the sensors and perform out the irrigation-related tasks, a method approximating fuzzy logic is used. By employing this method, irrigation system scheduling will be enhanced. Machine learning is another method for making predictions in irrigation systems.[37]

The amount of water that is available for irrigation is determined using prediction algorithms. By anticipating potential difficulties that can arise and determining how the risks should be controlled to achieve maximum work efficiency, this enables enhancing the irrigation process. Therefore, reduced water use, more earnings, and improved crop yields could be seen as some advantages that could be associated with machine learning. By utilizing machine learning to reduce irrigation-related hazards, it may be easier to achieve effective performance while also generating financial gains.[11] Crop diseases, inadequate irrigation, poor water management, pesticide control, weed management, and poor storage management are just a few of the problems facing the agriculture sector. All these problems can be managed utilizing a variety of artificial intelligence techniques. Through algorithms, machine learning enhances all irrigation-related activities and processes and makes it possible to meet performance goals. Additionally, irrigation pattern estimations that heavily rely on forecasts of the weather and crop conditions are supported by machine learning. Because it encourages humans to respond and adopt strategies by considering anticipated future actions, machine learning may be strongly related to predictions. Therefore, these forecasts finally permit the adoption of appropriate actions that may assist irrigation in the long run.

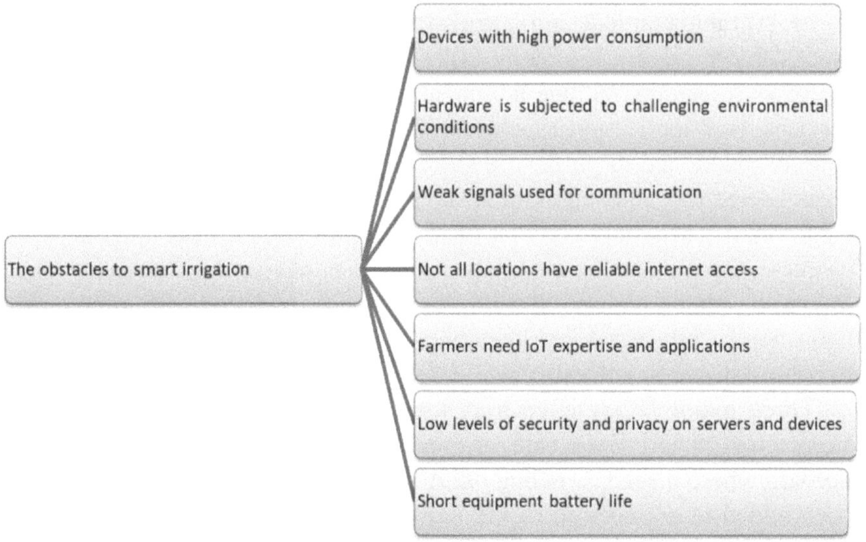

FIGURE 12.5 Smart irrigation system obstacles.

An overview of the main AI techniques and smart irrigation was presented by Froiz-Míguez et al.,[16] who also categorized AI advancements. Loua et al.[37] have developed a machine learning optimization method with an adaptable framework for data-driven decision making. It has also been shown that long-lasting precision irrigation systems can be created by efficiently managing sensed data about the weather, plants, and soil. To identify the best irrigation decision management, Adamides et al.[1] looked at the fusion of various machine learning models. The adaptive control of the changing environmental conditions could be accomplished by using precision irrigation systems. Numerous machine learning applications, including agricultural management, animal management, water management, and soil management, have been researched in the literature.

12.5.5 GENERAL ARCHITECTURE AND LAYOUT OF INTERNET OF THINGS-BASED SENSORY IRRIGATION SYSTEMS

Multi-agent architectures have been found to be rather popular and obvious in irrigation management and related IoT solutions. By distinguishing the various components that make up these specific architecture types, they enable development. The layer of architectural features is typically taken into consideration when developing the architectural differentiation. As an illustration, nodes that are now higher up in the hierarchy may end up working as a middleman for nodes that are further down.[10]

Most architectural designs are also thought to be broken up into functional blocks that represent the many tasks and actions that must be performed. These architectures include administration, devices, connectivity, security, services, and application as some of its key components. Multiple devices that are positioned to perform a variety of duties, including control, monitoring, detection, and action, make up the Internet of Things (IoT) systems.[56] Additionally, it is believed that these specific devices have interfaces that enable connections to be established with other devices and the transmission of crucial data. Likewise, the information obtained from different sensors will often be processed, and the conclusions drawn from them will be used to control different actuators.[12]

Three major layers have traditionally been regarded to make up IoT architecture. These three categories viz. perception, network, and application-are where these layers belong. A new layer called the service layer has been added in connection to the network and application levels. Fog and cloud processing are used to preserve and analyze the information on this specialized layer. One of the most prominent researchers who has helped to the design and dissemination of a variety of original design suggestions is Fernández-Pastor, who created four layered structures.[5]

Things, edge, networking, and cloud are the four layers mentioned above. The edge tier has been selected in this method to locate critical applications and perform essential control actions. Different tiered approaches have been used and deployed in relation to IoT systems for irrigation, with varying degrees of success. The bottom layer often consists of actuators and sensor nodes, while the intermediate layer frequently includes a gateway and facilitates data flow. Finally, cloud services, apps, or databases make up the third tier of the architecture. Although these three layers are the most widely recognized, they can also be distinct and diversified, with various properties. The general layout of the smart irrigation system is shown in Figure 12.6.

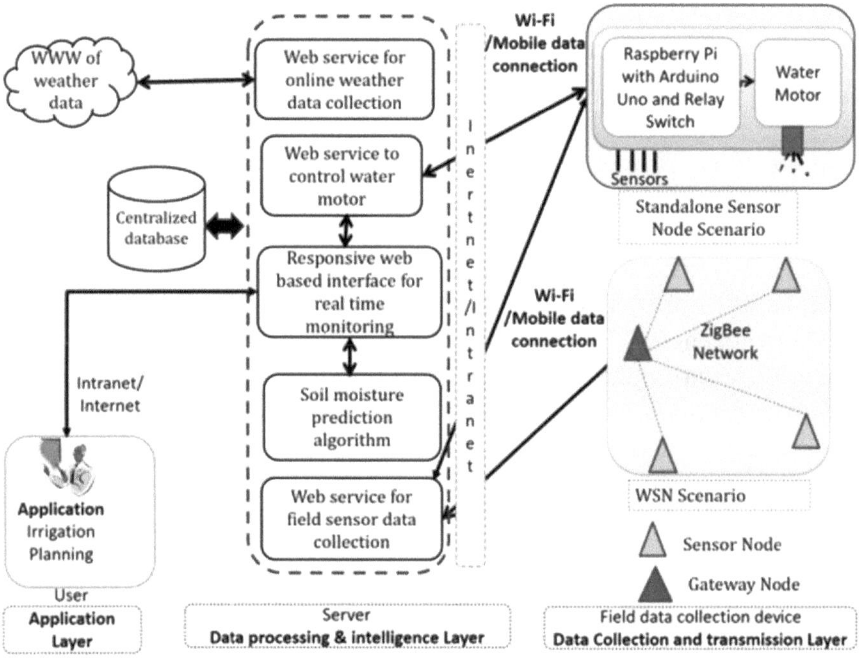

FIGURE 12.6 Smart irrigation system structure.

12.5.6 ACQUISITION AND SECURITY

The development of novel methods for gathering data from the field-deployed sensors has been facilitated by the advancements and improvements in technology. Drones are a successful way to collect data from sensor nodes despite the numerous innovations that have been achieved and put into use. Drone technology has also made it possible to collect new data, such as aerial photos of fields that could not otherwise be obtained.[39] One such technique for gathering data involves the employment of robots that integrate actuators and sensors to carry out a variety of tasks, including watering plants, monitoring soil moisture, scaring off animals, and weeding. Due to its ability to move to the needed sites, this robot technology could be employed for irrigation of land.

The robots also come with sensors that reduce the chance of collision and can gauge the soil's moisture content. Many businesses are increasingly utilizing robots because it has been proven that using them is one of the best efficient ways to complete activities related to irrigation. However, developments in autonomous robots are occurring, as are their uses and results. The wireless robot can go across a farm and sprinkle water while using soil and environmental monitoring to carry out several tasks. The coverage path planning process, which makes use of an environmental data set and a map of static images, could help irrigation robots navigate more effectively.

The three essential layers of the robot monitoring system could be further enhanced using the robot operating system. The first layer is used to read sensor data, the second layer is used for communication, and the third layer is used for course planning and decision-making. The robots can also be self-sufficient and powered by solar panels to ensure sustainability and effectiveness.[53] Because there could be a variety of dangers, ensuring security when implementing IoT devices could be challenging. Cloning, faulty software, information leakage, firmware assaults, routing intrusions, denial of service intrusions, eavesdropping threats, and a plethora of other security issues may arise and have detrimental impacts. It has become crucial to create security solutions that can lessen overall difficulties in light of all of these risks.[32]

Data protection becomes particularly important since it is essential to the operations and efficiency of organizations in the specific sector. Even though other ways have been created to address the associated problems, the blockchain technology is one of the more successful ones. With the help of this technology, IoT systems can be secured, enabling safer data storage and communication.[45]

Technologies like blockchain are primarily employed to safeguard the supply chain in the agricultural sector. The future smart irrigation system in connection to IoT irrigation schemes for agriculture uses blockchain to track and trace information exchange. There is a need for improved security, and as technology has advanced, security issues have been more over time. Through improved research and development, organizations have focused on minimizing risks and ensuring that hostile behaviors may be considerably decreased. Additionally, as agricultural practices become more technologically advanced and automated, the demand for a safe storage system has increased with time. At the same time, the likelihood and frequency of attacks are also significantly rising. A continuous improvement strategy through R&D may still help to attain the required results even though many firms have been successful in minimizing the threats.[59] The various sensor types that can be utilized in smart irrigation systems are shown in Figure 12.7.

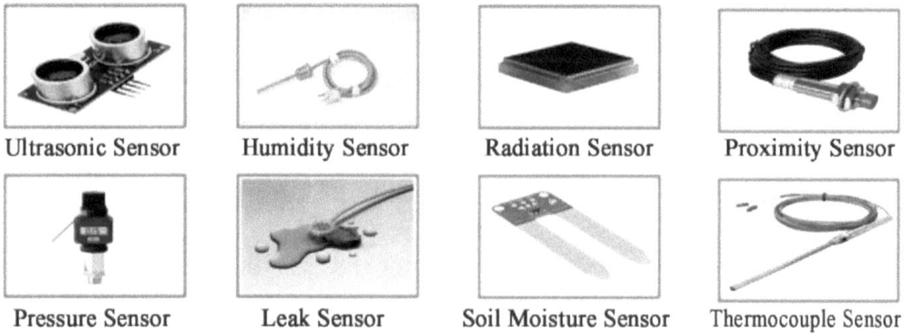

FIGURE 12.7 Smart irrigation system sensors.

Source: Reprinted from Ref. [44]. https://creativecommons.org/licenses/by/4.0/

12.6 APPLICATIONS OF SMART-BASED IRRIGATION

It is possible to consider several examples of SMART irrigation system components that have been successfully included and applied. Commercial enterprises in the farming sector and other sectors of the economy that have been involved in irrigation practices globally have a passion for applying intelligent irrigation technologies to reduce costs and increase output.

One of the scenarios that may be addressed involves the industrial company WaterBit. The innovative company has partnered with AT&T, one of the largest telecommunications companies in the world, to enable the administration and operation of local irrigation while securing Wi-Fi connections to its driverless irrigation system. As a result, the farmers were able to significantly reduce their resource usage while also increasing overall yield.[36] Through wireless technology, the WaterBit gateway securely and dependably transmits soil moisture data collected in the field to the cloud. Users have the option to view and control this specific data via a mobile-friendly application, which is routinely updated every few minutes.

The second example comes from Ipswich, a city council in Australia, which showed how irrigation can be controlled by an automated soil-moisture monitoring system while also significantly conserving water and lowering costs. Comparing the automated web-based technique to a method of allocation based on rainfall, it has proven to be significantly more effective. The second example comes from Ipswich, a city council in Australia, which showed how irrigation can be controlled by an automated soil-moisture monitoring system while also significantly conserving water and lowering costs. In addition, less water is being used and wasted thanks to the installation, which has improved overall results while preserving the resource.

At Maejo University of Thailand, a smart farm powered by solar energy was built as a source of backup power for an intelligent mushroom farm. To boost efficiency and automate watering, IoT technology was deployed. The Internet of Things regulates the essential airflow, temperature, humidity, and lighting conditions for mushrooms (IoT). Users were able to reduce energy use and save 20–29% of the energy utilized as a result of network sectoring.[17,44] Water for irrigation will be applied to the balance plants during the day. Utilizing the water that has built up in the soil, or if necessary, extending the time between watering operations. In light of this, farmers' energy expenditures may be reduced or eliminated if solar energy is used to power irrigation equipment. The use of clean energies will lower greenhouse gas emissions when compared to traditional electrical energy or diesel—powered used in irrigation systems. A smart irrigation strategic plan combined with solar photovoltaic feeds irrigation water to the network directly without the use of transitional storage elements, such as storage tanks or batteries.[43] A list of the cases that are presented is shown in Table 12.2.

12.7 PRESENT CHALLENGES AND FUTURE OPPORTUNITIES

Developing intelligent irrigation systems to manage various crops and advance sustainable agriculture involves several difficulties that must be overcome. These involve creating digital software and machine learning algorithms. To address the food shortages, the total food output must be increased. More income crops like cotton and latex should be developed to

TABLE 12.2 Examples of Smart-Based Applied Irrigation System.

Year	Technology	Remarks
2017	MATLAB with the Neural Network Toolbox	To guarantee optimal water resource management, the automated irrigation system for smart farms includes water usage optimization.
2018	Generic IoT architecture for enhancing irrigation in agricultural	An easy-to-use smartphone application has been created to deliver this information to farmers in their language. The general structure was validated using a chili crop irrigation system.
2018	Raspberry Pi and Arduino Uno	The design of the smart system of irrigation included both photovoltaic panels and control devices. The method is dependable, effective, and long-lasting.
2019	RBFN or Radial Basis Function Network	The IoE environment was used in the creation of the photovoltaic smart irrigation system. Prognostications for water level readings, weather factors including temperature and humidity, and irrigation data are provided by the irrigation system.
2019	Microcontroller by Arduino	A low-cost automated irrigation method has been devised for green walls. The system increases watering effectiveness while saving time and energy.
2020	The IRIS motes and TelosB	In a Greek strawberry greenhouse, a massive intelligent irrigation system was created. Smart irrigation is the main focus of the reference network architecture.
2020	AgriSens	Flexible irrigation scheduling system based on IoT (farmer-friendly user interface). Based on the requirements of the farmer, an algorithm for unsupervised dynamic-cum-manual irrigation is developed.

meet industrial demand, especially when combined with recyclable sources to halt soil erosion.[44] Additional difficulties brought on by these problems include the loss of agricultural labor, the amount of productive land being utilized for farming, the scarcity of water provisions, the consequences of climate change, etc. Given the rapid urbanization of the world, rural areas are rapidly aging and losing population. There are several potential applications for IoT technologies in agriculture and food production. Only a few of the numerous aspects of IoT in smart irrigation that need more investigation are its cost, unsupervised operation, flexibility, low maintenance, effectiveness, strong design, and reliability.

Once intuitive interfaces are aware of the potential of artificial intelligence and big data, it is predicted that agriculture will transform into a dynamic industry. The numerous agricultural tools, management strategies, and equipment that can be utilized for everything from planting to production estimations will be incorporated into these integrated systems. A new era of IoT in agriculture may be ushered in by high-tech tools like agricultural robotics, cloud technology, artificial intelligence, and big data. These devices are essential for sustaining sustainable agriculture. Farmers and other stakeholders have a lot of options when they combine deployable software applications with machine learning forecasts. Improved irrigation demand forecasts, volume and timing adjustments to meet plant water requirements, and adaptive water loss compensation can all help to increase water use efficiency. Due to this, irrigation water consumption will be reduced while output increases. A better-trained model will be used as the system grows more sophisticated and cleverer to make better irrigation decisions. For farmers and irrigation users, this can significantly lessen the stress and burden connected with irrigation.

12.8 SUMMARY

Today's business environment depends heavily on technological advancements, and businesses across all sectors are enhancing their operations to prosper and grow. In this context, irrigation and its implementations can be enhanced to offer the highest level of operational effectiveness while reaching the required performance outcomes. Water stress, water shortages or surpluses, and water crises are all aspects of the serious problem of water scarcity. Water management has become a concept, and businesses are now interested in finding ways to preserve the resource while simultaneously increasing the effectiveness of their operations. When corporations use technology to attain their performance goals, SMART irrigation systems have become necessary. Both IoT and sensor systems have had tremendously important impacts. IoT brings down technology's overall cost, opening the possibility of managing the irrigation process monitoring system. Wireless networks of sensors also make it easier to monitor irrigation and precision farming activities in real time.

KEYWORDS

- **Agribot**
- **agriculture**
- **IoT**
- **irrigation**
- **SDGs**
- **smart technologies**

REFERENCES

1. Adamides, G.; Kalatzis, N.; Stylianou, A.; Marianos, N.; Chatzipapadopoulos, F.; Giannakopoulou, M.; Neocleous, D. Smart Farming Techniques for Climate Change Adaptation in Cyprus. *Atmosphere* **2020,** *11* (6), 557–562.
2. Ahmed, I.; Akash, M. S. IoT Based Smart Irrigation System at University of Chittagong, Bangladesh. *Int. J. Comput. Appl.* **2020,** *176,* 39–45.
3. AlZu'bi, S.; Hawashin, B.; Mujahed, M.; Jararweh, Y.; Gupta, B. B. An Efficient Employment of Internet of Multimedia Things in Smart and Future Agriculture. *Multim. Tools Appl.* **2019,** *78* (20), 29581–29605.
4. Babaa, S. E.; Ahmed, M.; Ogunleye, B. S.; Khan, S. A.; Al-Jahdhami, S. A.; Pillai, J. R. Smart Irrigation System Using Arduino with Solar Power. *Int. J. Eng. Res. Technol.* **2020,** *9* (5), 7–15.
5. Badreldeen, M.M.; Ragab, M.A.; Sedhom, A.; Mamdouh, W.M. IOT Based Smart Irrigation System. *Int. J. Indus. Sustain. Dev.* **2022,** *3* (1), 76–86.
6. Ben Ayed, R.; Hanana, M. Artificial Intelligence to Improve the Food and Agriculture Sector. *J. Food Qual.* **2021,** *2021.*
7. Benyezza, H.; Bouhedda, M.; Rebouh, S. Zoning Irrigation Smart System Based on Fuzzy Control Technology and IoT for Water and Energy Saving. *J. Clean. Prod.* **2021,** *302,* 127001.

8. Bhanu, K. N.; Mahadevaswamy, H. S.; Jasmine, H. J. In *Iot Based Smart System for Enhanced Irrigation in Agriculture*, 2020 International Conference on Electronics and Sustainable Communication Systems (ICESC), IEEE, 2020; pp 760–765.
9. Biswas, S.; Sharma, L. K.; Ranjan, R.; Saha, S.; Chakraborty, A.; Banerjee, J. S. Smart Farming and Water Saving-based Intelligent Irrigation System Implementation Using the Internet of Things. In *Recent Trends in Computational Intelligence Enabled Research*; Academic Press: San Diego, 2021; pp 339–354.
10. Borrero, J. D.; Zabalo, A. An Autonomous Wireless Device for Real-Time Monitoring of Water Needs. *Sensors* **2020,** *20* (7), 2078.
11. Boursianis, A. D.; Papadopoulou, M. S.; Gotsis, A.; Wan, S.; Sarigiannidis, P.; Nikolaidis, S.; Goudos, S. K. Smart Irrigation System for Precision Agriculture—The AREThOU5A IoT Platform. *IEEE Sens. J.* **2020,** *21* (16), 17539–17547.
12. Campoverde, L. M. S.; Tropea, M.; De Rango, F. In *An IoT Based Smart Irrigation Management System Using Reinforcement Learning Modeled Through a Markov Decision Process*, 2021 IEEE/ACM 25th International Symposium on Distributed Simulation and Real Time Applications (DS-RT), IEEE, 2021; pp 1–4.
13. Elbasiouny, H.; Elbanna, B. A.; Al-Najoli, E.; Alsherief, A.; Negm, S.; El-Nour, A.; ... Sharabash, S. Agricultural Waste Management for Climate Change Mitigation: Some Implications to Egypt. In: *Waste Management in MENA Regions*; Springer: Cham, 2020; pp 149–169.
14. Farooq, M.; Hussain, A.; Hashim, S.; Yang, L.; Ali, M. In *Automated Irrigation System based on irrigation gates using fuzzy logic*, 2020 International Conference on Internet of Things and Intelligent Applications (ITIA). IEEE; 2020; pp 1–5.
15. Fernández-Ahumada, L. M.; Ramírez-Faz, J.; Torres-Romero, M.; López-Luque, R. Proposal for the Design of Monitoring and Operating Irrigation Networks Based on IoT, Cloud Computing and Free Hardware Technologies. *Sensors* **2019,** *19* (10), 2318–2322.
16. Froiz-Míguez, I.; Lopez-Iturri, P.; Fraga-Lamas, P.; Celaya-Echarri, M.; Blanco-Novoa, Ó.; Azpilicueta, L.; ... Fernández-Caramés, T. M. Design, Implementation, and Empirical Validation of an IoT Smart Irrigation System for fog Computing Applications Based on Lora and Lorawan Sensor Nodes. *Sensors* **2020,** *20* (23), 6865–6875.
17. García, A. M.; García, I. F.; Poyato, E. C.; Barrios, P. M.; Díaz, J. R. Coupling Irrigation Scheduling with Solar Energy Production in a Smart Irrigation Management System. *J. Clean. Prod.* **2018,** *175,* 670–682.
18. García, L.; Parra, L.; Jimenez, J. M.; Lloret, J.; Lorenz, P. IoT-based Smart Irrigation Systems: An Overview on the Recent Trends on Sensors and IoT Systems for Irrigation in Precision Agriculture. *Sensors* **2020,** *20* (4), 1042–1047.
19. Goap, A.; Sharma, D.; Shukla, A. K.; Krishna, C. R. An IoT Based Smart Irrigation Management System Using Machine Learning and Open Source Technologies. *Comput. Electron. Agric.* **2018,** *155,* 41–49.
20. Hasan, M. Real-time and Low-Cost IoT Based Farming Using Raspberry Pi. *Indones. J. Electr. Eng. Comput. Sci.* **2020,** *17* (1), 197–204.
21. Jain, R. K. In *Design and Development of IoT Based Smart Irrigation Sensing, Control and Monitoring System for Agricultural Applications*, E-business Technologies Conference Proceedings, 2021; Vol. 1 (1), pp 162–166.
22. Jain, R. K.; Gupta, B.; Ansari, M.; Ray, P. P. In *IOT Enabled Smart Drip Irrigation System Using Web/Android Applications*, 2020 11th International Conference on Computing, Communication and Networking Technologies (ICCCNT); IEEE, 2020; pp 1–6.

23. Kanchi, R. R.; Sreeramula, V. P.; Palle, D. V. In *Implementation of Smart Agriculture Using Cloudiot and Its Geotagging on Android Platform*, International Conference on Intelligent Computing and Communication Technologies; Springer: Singapore, 2019; pp 520–528.
24. Keerthana, P.; Pooja, M.; Raksha, M.; Smitha, M. IOT Based Smart Irrigation System. *Int. Res. J. Eng. Technol.* **2021,** *8* (7), 2464–2469.
25. Keesstra, S.; Sannigrahi, S.; López-Vicente, M.; Pulido, M.; Novara, A.; Visser, S.; Kalantari, Z. The Role of Soils in Regulation and Provision of Blue and Green Water. *Philos. Trans. R. Soc. B* **2021,** *376* (1834), 175–184.
26. Keswani, B.; Mohapatra, A. G.; Keswani, P.; Khanna, A.; Gupta, D.; Rodrigues, J. Improving Weather Dependent Zone Specific Irrigation Control Scheme in IoT and big Data Enabled Self-Driven Precision Agriculture Mechanism. *Enterprise Inform. Syst.* **2020,** *14* (9–10), 1494–1515.
27. Khelifa, B.; Amel, D.; Amel, B.; Mohamed, C.; Tarek, B. In *Smart Irrigation Using Internet of Things*, 2015 Fourth International Conference on Future Generation Communication Technology (FGCT), IEEE, 2015; pp 1–6.
28. Krishnan, R. S.; Julie, E. G.; Robinson, Y. H.; Raja, S.; Kumar, R.; Thong, P. H. Fuzzy Logic Based Smart Irrigation System Using Internet of Things. *J. Clean. Prod.* **2020,** *252,* 119902.
29. Kumar, A.; Ranjan, P.; Saini, V. Smart Irrigation System Using IoT. In *Agri-Food 4.0*; Emerald Publishing Limited, 2022.
30. Kumar, K. A.; Aju, D. An Internet of Thing based Agribot (IOT-Agribot) for Precision Agriculture and Farm Monitoring. *Int. J. Educ. Manag. Eng.* **2020,** *10* (4), 33–39.
31. Kumar, M.; Xiong, X.; He, M.; Tsang, D. C.; Gupta, J.; Khan, E.; Bolan, N. S. Microplastics as Pollutants in Agricultural Soils. *Environ. Pollut.* **2020,** *265,* 114980.
32. Kumar, R.; Kumar, D.; Kumar, D. Miltilevel Smart Irrigation System for In-House Farming. *Think India J.* **2019,** *22* (34), 1017–1021.
33. Kumari, A.; Sahu, P. K. Internet of Things-Based Smart Drip Irrigation Using Arduino. *J. Comput. Theor. Nanosci.* **2020,** *17* (9–10), 4598–4603.
34. Laabidi, K.; Khayyat, M.; Almohamadi, T. Smart Grid Irrigation. In *Innovative and Intelligent Technology-Based Services for Smart Environments–Smart Sensing and Artificial Intelligence*; CRC Press, 2021; pp 217–222.
35. Laksiri, H. G. C. R.; Dharmagunawardhana, H. A. C.; Wijayakulasooriya, J. V. In *Design and Optimization of IOT Based Smart Irrigation System in Sri Lanka*, 2019 14th Conference on Industrial and Information Systems (ICIIS), IEEE, 2019; pp 198–202.
36. Liu, Y.; Yu, Q.; Luo, X.; Yang, L.; Cui, Y. Continuous Monitoring of Diabetes with an Integrated Microneedle Biosensing Device Through 3D Printing. *Microsyst. Nanoeng.* **2021,** *7* (1), 1–12.
37. Loua, L. R.; Budihardjo, M. A.; Sudarno, S. In *A Review on a Machine Learning Approach of an Intelligent Irrigation Monitoring System with Edge Computing and the Internet of Things*, IOP Conference Series: Earth and Environmental Science; IOP Publishing, 2021; Vol. 896 (1), p 012029.
38. Lova Raju, K.; Vijayaraghavan, V. IoT Technologies in Agricultural Environment: A Survey. *Wirel. Pers. Commun.* **2020,** *113* (4), 2415–2446.
39. Mousavi, S. K.; Ghaffari, A.; Besharat, S.; Afshari, H. Improving the Security of Internet of Things Using Cryptographic Algorithms: A Case of Smart Irrigation Systems. *J. Ambient Intell. Humaniz. Comput.* **2021,** *12* (2), 2033–2051.
40. Munir, M. S.; Bajwa, I. S.; Ashraf, A.; Anwar, W.; Rashid, R. Intelligent and Smart Irrigation System Using Edge Computing and IoT. *Complexity* **2021,** *2021.*

41. Munir, M. S.; Bajwa, I. S.; Naeem, M. A.; Ramzan, B. Design and Implementation of an IoT System for Smart Energy Consumption and Smart Irrigation in Tunnel Farming. *Energies* **2018,** *11* (12), 3427–3432.
42. Nawandar, N. K.; Satpute, V. R. IoT Based Low Cost and Intelligent Module for Smart Irrigation System. *Comput. Electron. Agric.* **2019,** *162,* 979–990.
43. Obaideen, K.; AlMallahi, M. N.; Alami, A. H.; Ramadan, M.; Abdelkareem, M. A.; Shehata, N.; Olabi, A. G. On the Contribution of Solar Energy to Sustainable Developments Goals: Case Study on Mohammed Bin Rashid Al Maktoum Solar Park. *Int. J. Thermofluids* **2021,** *12,* 100123.
44. Obaideen, K.; Yousef, B. A.; AlMallahi, M. N.; Tan, Y. C.; Mahmoud, M.; Jaber, H.; Ramadan, M. An Overview of Smart Irrigation Systems Using IoT. *Energy Nexus* **2022,** 100124.
45. Okandeji, A. A.; Onaifo, F.; Kabir, M. T.; Yakubu, K. Design and Implementation of Internet of Things Based Irrigation System. *Arid Zone J. Eng. Technol. Environ.* **2020,** *16* (4), 663–674.
46. Pandey, S.; Shrivastava, A.; Vijay, R.; Bhandari, S. In *A Review on Smart Irrigation and Crop Prediction System,* Proceedings of International Conference on Sustainable Computing in Science, Technology and Management (SUSCOM), Amity University Rajasthan: Jaipur-India, 2019.
47. Patel, M.; Patel, P.; Jeya, M. R. IoT Based Approach to Automation of Farming Irrigation System. *Anna. Rom. Soc. Cell Biol.* **2021,** 5598–5609.
48. Podder, A. K.; Al Bukhari, A.; Islam, S.; Mia, S.; Mohammed, M. A.; Kumar, N. M.; ... Abdulkareem, K. H. IoT Based Smart Agrotech System for Verification of Urban Farming Parameters. *Microprocess. Microsyst.* **2021,** *82,* 104025.
49. Reghukumar, A.; Vijayakumar, V. Smart Plant Watering System with Cloud Analysis and Plant Health Prediction. *Procedia Comput. Sci.* **2019,** *165,* 126–135.
50. Sahu, C. K.; Behera, P. In *A Low-Cost Smart Irrigation Control System,* 2015 2nd International Conference on Electronics and Communication Systems (ICECS), IEEE, 2015; pp 1146–1152.
51. Shao, Y.; Wang, Y.; Yuan, Y.; Xie, Y. A Systematic Review on Antibiotics Misuse in Livestock and Aquaculture and Regulation Implications in China. *Sci. Total Environ.* **2021,** *798,* 149205.
52. Shilpa, A.; Muneeswaran, V.; Rathinam, D. K. D. In *A Precise and Autonomous Irrigation System for Agriculture: IoT Based Self Propelled Center Pivot Irrigation System,* 2019 5th International Conference on Advanced Computing & Communication Systems (ICACCS), IEEE, 2019; pp 533–538.
53. Singh, M.; Ahmed, S. IoT Based Smart Water Management Systems: A Systematic Review. *Mater. Today: Proc.* **2021,** *46,* 5211–5218.
54. Singh, S.; Haneef, F.; Kumar, S.; Ongsakul, V. A Framework for Successful IoT Adoption in Agriculture Sector: A Total Interpretive Structural Modelling Approach. *J. Glob. Bus. Adv.* **2020,** *13* (3), 382–403.
55. Sinha, B. B.; Dhanalakshmi, R. Recent Advancements and Challenges of Internet of Things in Smart Agriculture: A Survey. *Future Gener. Comput. Syst.* **2022,** *126,* 169–184.
56. Sobin, C. C. A Survey on Architecture, Protocols and Challenges in IoT. *Wirel. Pers. Commun.* **2020,** *112* (3), 1383–1429.
57. Suma, V. Internet-of-Things (IoT) Based Smart Agriculture in India—An Overview. *J. ISMAC* **2021,** *3* (01), 1–15.

58. Velliangiri, S.; Sekar, R.; Anbhazhagan, P. Using MLPA for Smart Mushroom Farm Monitoring System based on IoT. *Int. J. Network. Virtual Organi.* **2020,** *22* (4), 334–346.
59. Yamini, R.; Reddy, K. V. C. Smart Irrigation Using IoT. *Int. J. Adv. Res. Ideas Innov. Technol.* **2019,** *5,* 471–478.
60. Zinkernagel, J.; Maestre-Valero, J. F.; Seresti, S. Y.; Intrigliolo, D. S. New Technologies and Practical Approaches to Improve Irrigation Management of Open Field Vegetable Crops. *Agri. Water Manag.* **2020,** *242,* 106404.
61. Quy, V.K.; Hau, N.V.; Anh, D.V.; Quy, N.M.; Ban, N.T.; Lanza, S.; Randazzo, G.; Muzirafuti, A. IoT-Enabled Smart Agriculture: Architecture, Applications, and Challenges. Appl. Sci. 2022, 12, 3396. https://doi.org/10.3390/app12073396

CHAPTER 13

STATUS AND REMEDIAL MEASURES OF RECYCLED WATER IN NIGERIA

MOHAMMED RABIU MUSA, MUHAMMAD ISA ABDURRAHMAN, AMRENDRA KUMAR SINGH, and GAURAV SAINI

ABSTRACT

Nigeria is very low on the economic indices in the world. Agriculture is one of the key occupations and a significant contributor to the country's GDP. Unfortunately, the small-scale farmers dominate the agricultural scenario of the country and have low access to irrigation facilities, unlike the large-scale farmers. Despite the steps taken by the government, most of the farmers are still dependent on age-old water-lifting devices like *Shadouf*, rope-and-bucket, etc. for irrigation requirements. These systems are not only inefficient, have low water lift, are heavily manual in nature, and are non-ergonomic, leading to chances of injury in the longer run. Low-cost water-lifting devices can offer a solution. Similarly, the use of municipal wastewater can be a significant solution as well. This chapter presents details on the current irrigation situation in Nigeria, including its geography, major river systems, status of irrigation (including large- and small-scale farmers), and potential solutions such as the use of wastewater for irrigation. This work is not only significant for Nigeria but also for other developing and underdeveloped countries as well, where small-scale farmers face the same situations.

13.1 INTRODUCTION

Nigeria in West Africa lies between latitudes 4° and 14° North and longitudes 2° and 15° East. In the far north, there are areas that fall within the Sahara

Sustainable Use of Wastewater in Agriculture: Applications, Methods, and Technology. Megh R. Goyal & Gaurav Saini (Eds.)

Desert and there is evidence that the desert is encroaching on land that at one time was fertile; in the south of the country, there are tropical rain-fed areas where soil erosion problems are encountered.

Kano State, within which the study area lies, is in the northern part of the country and the climate and rainfall are such that even in the so-called rainy season it is usually necessary to irrigate crops if they are to reach their full potential.

The study area comprises the catchment along the River Kano below the Tiga Dam to the point where it meets River Challawa (Figure 13.1). River Kano flows in a northerly direction between the Tiga Dam and the Challawa; the distance between the two is 42.5 km. The catchment, which is some 13 km wide in the northerly section and widens out to approximately 34 km near the dam, covers some 79,000 ha. Within this area, there are essentially two types of farmers, those who are fortunate enough to be able to use water supplied by gravity from canals fed by the dam and those who have to lift water, if they are to irrigate.

As in some other parts of the world where irrigation is essential if crops are to flourish, several large dams have been built and water distribution systems based on canals have been established; it is usual to transfer water from these canals to the fields by siphoning. The Tiga Dam supplies the Kadawa Irrigation Scheme, which covers about 22,000 ha and is the largest scheme in Nigeria; about one-third of the scheme is within the River Kano catchment area, covering only about 9.3%. Wide range of crops are produced, some of these are being designed for the purpose of selling products rather than to feed the local population.

At the other end of the investment scale, there are a large number of irrigators, who have to lift water by hand to produce crops to feed themselves, with perhaps limited surpluses being available for sale. Most of these riparian farmers (defined in the Oxford Dictionary as the farmers of or inhabiting the banks of a river, lake, etc.) are situated along the banks of rivers and at times it will be necessary for them to dig into the dried-up bed of their river to reach water.

Some spectacular results have followed the construction of large-scale irrigation (LSI) schemes, but they have not been entirely without problems. Even if many more dams are constructed, the area that they can irrigate will remain relatively small when compared to the total cultivable land area, and for this reason alone there are many workers in this field who advocate the attraction of investing in programs that encourage small-scale irrigation (SSI) developments. Indeed, the term micro-scale irrigation is now being promoted by some interested parties.

In areas without electricity and where the capital and support services required to run solar or engine-driven pumps are absent, hand, wind, water, or animal-powered lifting systems are all that can be used. Along the Kano River, this usually means that the Shadouf is employed by many farmers, and this then leads to low efficiency, reduced agricultural productivity, and economic loss not only for the farmers but also for the nation as a whole.

A detailed description of the prevailing conditions (including meteorological, climatic, geological, and hydrological) is presented for Nigeria, with specific focus on Kano State. The development of various irrigation schemes, with focus on farmer-operated devices is presented and the drawbacks of such systems are highlighted. A number of solutions in terms of developments in such hand-operated systems and the application of wastewater for irrigation are presented.

It is hoped that the outcome of this work will eventually enable the small-scale riparian farmers to use the water more efficiently and perhaps in consequence increase his farm size and output.

13.2 FACTORS INFLUENCING THE SUPPLY OF WATER AND ITS UTILIZATION IN KANO STATE OF NIGERIA

Water supply in Kano State is influenced by both natural factors (such as soil and climatic characteristics) and artificial factors (such as ecological to human activities).

13.2.1 LOCATION AND TOPOGRAPHY OF NIGERIA

Nigeria is one of the largest countries in Africa. It comprises an area of about 923,768 Km[269] and is situated in the tropics between latitudes 4° and 14° north of the equator. Nigeria's neighbors are, in the north, the Republic of Niger, in the north-east, the Chad Republic, in the east, the Republic of Cameroon, and the Peoples Republic of Benin in the west (Figure 13.1).

The country is divided into three unequal parts by the rivers Benue and Niger, the Benue flowing from the east and the Niger from the northwest. The rivers meet at Lokoja near the new Federal capital of Nigeria, Abuja, and then flow south into the Atlantic Ocean (Figure 13.1).

Until 1996, Nigeria had 16 states in the north and 14 states in the south, with Abuja being situated more-or-less in the center of the country. During 1996, the number of states was increased to 36.

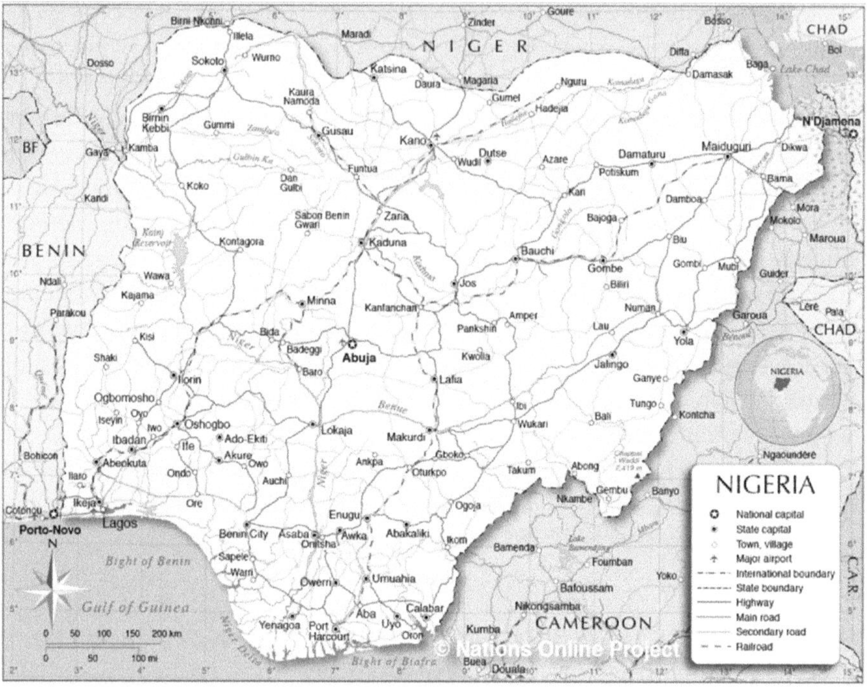

FIGURE 13.1 Political map of Nigeria.

Source: Nations Online Project. Based on a UN Map. https://www.nationsonline.org/oneworld/map/nigeria-political-map.htm. Copyright Nations Online Project. Free to use for educational purposes.

13.2.2 CLIMATE AND RAINFALL

Nigeria has a typical tropical climate which is characterized by dry and wet seasons. The dominant factor influencing the climate is the presence of two air masses—the tropical continental air mass and the equatorial maritime air mass. The latter is characterized by south-westerly winds associated with cool and moist monsoon winds from the Atlantic.[26] The tropical continental air mass, on the other hand, produces dry north-easterly winds and these are associated with the dusty harmattan from the Sahara Desert.[22]

The boundary between these two air masses is known as the inter-tropical front or inter-tropical convergence zone (ITCZ) and it runs approximately from east to west. The retreat of the ITCZ to the north during the summer months, under the influence of a low-pressure zone over the Sahara Desert, allows the maritime air mass to enter the area and this marks the beginning of the rainy season. At the end of this season, the southward movement of the boundary occurs and once more the continental air mass from the Sahara is brought back into the area; this marks the onset of hotter, less humid conditions.[26]

The tropical continental air mass, on the other hand, produces dry north-easterly winds and these are associated with the dusty harmattan from the Sahara Desert. The boundary between these two air masses is known as the inter-tropical front or ITCZ and it runs approximately from east to west. The retreat of the ITCZ to the north during the summer months, under the influence of a low-pressure zone over the Sahara Desert, allows the maritime air mass to enter the area and this marks the beginning of the rainy season. At the end of this season, the southward movement of the boundary occurs and once more the continental air mass from the Sahara is brought back into the area; this marks the onset of hotter, less humid conditions.[26]

As a result of climatic differences between the north and south, there is a large corresponding variation in vegetation in the country.[22] The tropical, humid south, with its natural forests, supports trees such as cocoa, rubber, coffee, and a variety of long-growing seasonal plants like cocoyam and cowpea. In contrast, parts of the extreme north are desert but the bulk is savanna of one type or another.

Northern Nigeria has high temperatures, low humidity, and long periods of uninterrupted sunshine, reduced only by the cloud cover (normally from mid-July to September) and the dust from the Sahara during the harmattan which reduces the duration of the sunlight and its brightness of the sunshine to 75–80% of the maximum possible.[57] The combined effect of these factors is that water evaporation rates are particularly high throughout.

Annual rainfall increases from 500 mm in the far north (concentrated in the period mid-May to September) to 2500 mm in the south, as shown in Figure 13.2.[55] The study area and the catchment area of the Kano River between the Tiga Dam and the Challawa River lie between Kano and Jos, being nearer to Kano than Jos. Long-term rainfall data for Kano and Jos was recorded by the Water Resources and Engineering Construction Agency (WRECA) from 1914 to 2013 and the monthly averages are shown in Table 13.1. The average annual rainfalls for these periods are 830 mm and 1400 mm, respectively.[74]

Rainfall data for Kano in 1994 and 1995 was 726.1 and 549.8 mm, respectively. Therefore, it can be seen that there is considerable year-to-year variation.[40,41] Another relevant fact concerning the growth of crops is that the rainy season is short: in 1995, for example, over 99% of the annual rainfall fell within a 13-week period beginning in early June. In general, Northern

Nigeria has high temperatures; in Kano, the range of a mean decade maximum temperatures is 28.94–38.61°C and mean a minimum of 12.28–24.39°C.[40]

TABLE 13.1 The Average Annual Rainfall (mm) for Kano and Jos During 1914–2013.

Month	Rainfall (mm)		Month	Rainfall (mm)	
	Jos	Kano		Jos	Kano
January	0	0	July	303	232
February	4	4	August	287	308
March	16	7	September	201	142
April	91	17	October	58	20
May	158	63	November	1	0
June	197	129	December	1	0

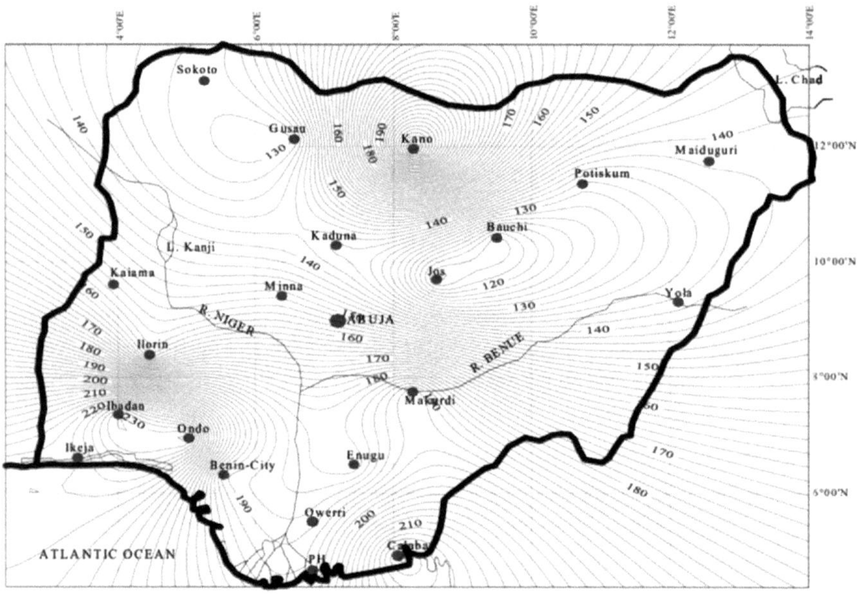

FIGURE 13.2 Peak daily rainfall (mm) Isohyet map of the 100-year return period.

Source: Reprinted from Ref. [76].

13.2.3 EVAPORATION

Northern Nigeria has high temperatures, low humidity, and long periods of uninterrupted sunshine, reduced only by the cloud cover (normally from mid-July to September) and the dust from the Sahara during the harmattan which reduces the duration of the sunlight and its brightness of the sunshine to 75–80% of the maximum possible.[57] The combined effect of these factors is that water evaporation rates are particularly high throughout the region.

The magnitude of the problem is shown by the fact that at Samaru, Northern Nigeria, annual potential evaporation is 57 inches (1448 mm), this being between 65 and 85% of the annual rainfall in that area.[34] Evaporation isograms (EI) determined using class A evaporation pans, covering the whole of the Kano River basin shows that at Jos, near the escarpment of the river, the El is some (3080 mm) and increases (to about 3600 mm) near the end of the river at Kano, well above the annual rainfall.[59]

In the northeast of the Kano River basin, less than 10% of the total surface runoff from the Yobe River system reaches Lake Chad.[20] Most of the water losses are due to evapotranspiration; groundwater percolation and infiltration are thought to be much less important sources of loss.

13.2.4 GEOLOGY, SOILS, AND VEGETATION

The Kano River basin, like most areas in Northern Nigeria, lies on the Precambrian basement complex generally known as the basement complex.[2] To the northeast of the old Kano State, now Kano and Jigawa States is a continuous hydrological boundary separating this formation from the Chad Formation, the latter being covered by a thick succession of sedimentary rocks. The northwest part of this boundary in the state runs from longitude 8° 30 E to longitude 10° 00 E at its south-eastern extremity.[48]

Muslim reported that the basement complex consists of fractured and jointed rocks and this forms a good aquifer. Great variations in permanent water tables exist in the area covered by the basement complex, but in most places, it is between 7.62 and 22.86 m deep. However, in places with rocky terrain, the water table may rise to ground level to form springs. It was noted that around Kadawa (southwest of Kano) the groundwater table was extremely high, in some places being only some 40 cm below the soil surface.

Nigeria is fortunate in that most of its soils can be cultivated; some in the far north crops are not grown, but the reason for this is primarily because of the water shortage: most of these have been classified as sandy.[59] The three other major soil types in their classification are:

- The lateritic soils of the savanna areas;

- The rainforest soils; and
- The alluvial and swampy soils of the coastal region.

The Kano River basin falls within the savanna region where the soils according to this classification are generally lateritic. Singh and Chari also concluded that the soils are fairly rich in potassium content but are low in organic matter, magnesium, molybdenum, nitrogen, and sulfur and that they are acidic.

An attempt has been made to map Nigeria's soils according to their potential productivity.[54,56] The most striking feature of these works is the way that the country's two major rivers qualify for the designation "high productivity"; unfortunately, the Kano River is not shown on this map.

The tropical, humid south, with its natural forests, supports trees such as cocoa, rubber, coffee, and a variety of long-growing season plants like cocoyam and cowpea. In contrast, parts of the extreme north are deserts but the bulk is savanna of one type or another.

The Nigerian government recognizes the importance of maintaining high levels of vegetative cover—for example, it supplies trees free of charge for planting; in some cases, fencing material to protect the trees from animals is provided and every year it organizes ceremonies for the tree planting campaign. Forest assistants are also employed to monitor the situation in villages. Unfortunately, the general population tends to appreciate trees only for the shade, fruit, and firewood they provide—the fact that they influence climate is difficult to demonstrate on a wide scale. The four main direct causes of desertification are over-cultivation, overgrazing, deforestation, and mismanagement of irrigated cropland.[25]

The density of the vegetation is crucial in maintaining the momentum of the hydrological cycle, which in turn influences the rate of desert encroachment and probably global warming.

The rate of desert encroachment is such that the region is now steadily moving toward a major ecological disaster.[26] Different parts of Northern Nigeria are classified in terms of desertification as either; severe, moderate, or slight;[45] and the River Kano falls within the moderately and severely affected zones.

Unfortunately, there has been a considerable reduction in the tree population in the study area. Mass destruction of vegetation cover for agricultural purposes has taken place over the last few years, particularly where LSI schemes have been established, and the population explosion has resulted in a high firewood demand. If the people were well informed on the environmental implications of their gathering of firewoods they might reduce their use of it, and it is now much cheaper to use either kerosene or gas than the firewood. Kerosene currently sells at N40.00 per gallon and a large-sized cooking gas cylinder (25 kg) costs N800.00 (where £1.00 = N125.00).

TABLE 13.2 Projected Agricultural and Non-Agricultural Population in Nigeria.

	1961/62	**1963/64**	**1967/68**	**1973/74**	**1979/80**
	Population in millions				
	Northern Nigeria				
Total population	28.4	29.8	32.9	38.7	46.2
Agricultural population	–	20.8	22.7	25.9	30.0
Non-agricultural population	–	9.0	10.2	12.8	16.2
	Eastern Nigeria				
Total population	9.8	10.3	11.3	13.3	15.9
Agricultural population	–	7.2	7.8	8.9	10.3
Non-agricultural population	–	3.1	3.5	4.4	5.6
	Mid-Western Nigeria				
Total population	24	2.5	2.8	3.3	3.9
Agricultural population	–	1.8	1.9	2.2	2.5
Non-agricultural population		0.7	0.9	1.1	1.4
Sub-total	52.4	55.0,	60.7	71.4	85.2
Agricultural population	–	38.5	41.9	47.8	55.3
Non-agricultural population	–	16.5	18.8	23.6	29.9
Lagos	0.6	0.6	0.6	0.8	1.0
Total population					
	Grand Total				
Total population	53.0	55.6	61.4	72.2	86.2
Agricultural population	–	38.5	41.9	47.8	55.3
Non-agricultural population	–	17.1	19.5	24.4	30.9

While Kano is by no means the most densely populated state in the country (Kano probably has between 300 and 450 people per 2.59 km^2 (per mile^2)) (Figure 13.3). It is obviously important to make the best possible use of whatever water is available for crop production so that the state's population can be adequately fed. This is more so since Kano has been shown to have a low rainfall, high evaporation rate and not to have the most productive soils.

13.2.5 POPULATION

Several attempts have been made to estimate the human population in Nigeria, but they have met with difficulties.[67] The first attempt was in 1911 and the last in 2006. The problem in the early attempts was that people were reluctant to give the actual figures for fear of taxation; there was also the problem of accessibility, hence the general belief was that the figures were underestimated. On the other hand, the most recent censuses (1963, 1972, 1991, and 2006) might have given inflated results because of political and budget allocation implications, so the 1972 and 1973 figures were not published. A census conducted in 2006 gave a population of 140 million. Table 13.2 shows the importance of the agricultural population (based on the 1963 census) and Figure 13.3 shows the population density.[67]

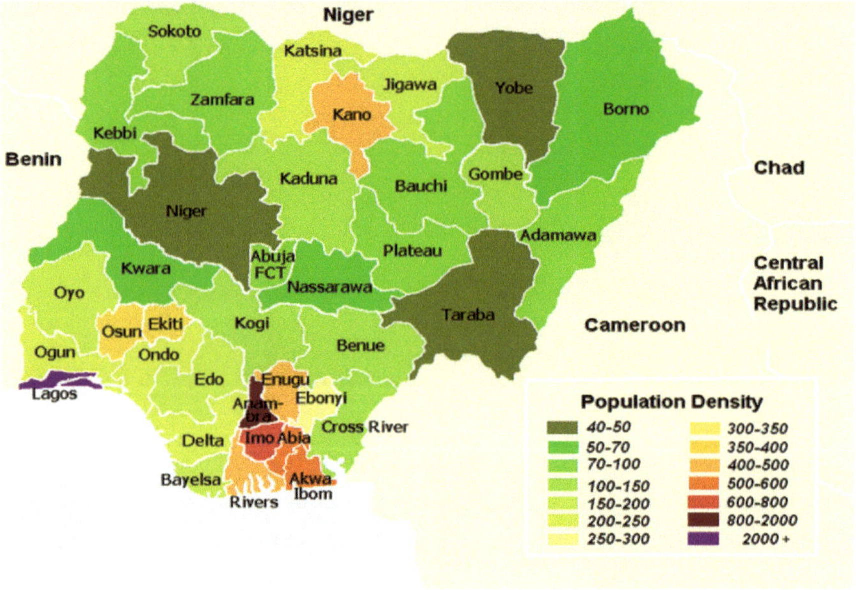

FIGURE 13.3 Nigeria's population density

Image credit: Marcel Krüger. https://creativecommons.org/licenses/by-sa/3.0/

13.2.6 AVAILABLE SURFACE WATER AND IRRIGATED LAND

The predicted long-term decrease in rainfall in Kano State will inevitably result in the depletion of both surface water and groundwater availability.[41] It is also expected that there will be an increase in population, that higher per capita demand for water will arise, and that agricultural, commercial, and industrial water requirements will increase.

Table 13.3 indicates estimated potential areas for formal irrigation schemes and their estimated water requirement: it will be seen that the latter far exceeds the storage capacity of the two dams. Since 1985, there have been no significant changes in the area developed for formal irrigation, so although the potential for additional change remains there are many financial, political, and environmental factors that will have to be taken into account in any future developments.

TABLE 13.3 Areas Developed for Irrigation by Various Agencies.

Agency	Developed area (ha)	Water requirement (Mm^3)	Potential area (ha)	Potential water requirement (Mm^3/year)
Khrb and Rda	23,000	346.00	87,000	1305
KNARDA	5170	77.55	20,000	300
Manr, Kano	1131	16.965	5840	87.60
Cbda	12,000	180.00	157,500	2362.50
Manr, borno	1200	18.00	N/A	N/A
Total	42,501	638.515	270,340	4055.10

There are no estimates published of the demands on the study area's water supply made by riparian farmers, but in comparison with the formal schemes their requirements must be considered to be small.

The irrigable land area of both the developed and potential categories was estimated to be 312,841 ha, and the annual requirement for irrigation per hectare is estimated to be 1.5×10^4 m^3; it should be noted that these figures are in addition to about 7.3×107 m^3 of surface water extracted by Water Board for Kano City's water supply. Based on the above estimate, it can be seen that if the potential agricultural areas are developed the water requirement will be much higher than the available storage, even with the commissioning of Challawa Dam.

The downstream implications of exhausting the water resources of the Kano, Challawa, and Hadejia rivers indicate that the full implementation of all planned water-demanding projects to be supplied by these rivers would result in the reduction of the frequency and extent of flooding within the Hadejia-Nguru wetlands; this would harm both the active agricultural activities now going on there and the ecosystem of the area.[3] It will also result in further falls in the groundwater level.

The combined water resources available in the Tiga Dam and Challawa Gorge (commissioned in 1992) will be barely sufficient for the ultimate water requirements of Kano City's water supply, the Kano River (phase 1) project, and the Hadejia irrigation project. All these show the need for efficient management along with an effective system of water conveyance and distribution in the Kano River basin.

13.3 ORIGIN AND GEOGRAPHY OF THE KANO RIVER

The terrain through which the Kano River flows is generally higher than that found in the south of the country, with the altitude varying from 300 to 1500 m above sea level. It rises in the Jos plateau extensive plains and broad valleys, which is centrally situated in Nigeria.[51] This plateau and the high ground around its northwest edge form the escarpment of the Kano River. The basin is roughly rectangular in shape and it lies in the western part of the Lake Chad basin. The basin covers 7770 km^2 (152 km in length and 56 km in width).[6] The Kano River joins the Challawa River to form the Hadejia River near Tamburawa village, which is about 20 km from Kano City.

13.3.1 KANO RIVER DISCHARGE AND CHANNEL RESPONSE

With the completion of the construction of Tiga (1974) and Bagauda (1970) and Ruwan Kanya (1977), there was a considerable change in the water flow in the Kano River. Measurements taken at the Chiromawa gauging stations during two five-year periods before and after the construction of the dams are shown in Table 13.4;[54] and it can be seen that the overall mean discharge was reduced by 74% by the dams. Before construction, there were wide variations in the seasonal flows: there were floods and loss of potentially valuable water. After construction, more water was available to the users from the river due to deliberate discharge from the dams, during the months of November through April.

13.3.2 RIVER FLOW ALONG THE KANO AND HADEJIA RIVERS

It is usual for workers studying river systems to estimate flows and then to refer to changes as gains or losses, the latter is perhaps misleading as it includes water that has been extracted from the river and presumably put to good use. As can be seen from Table 13.5, the losses along the River Kano, as it flows through the area of interest for riparian farmers, amount to 39%. This figure includes extraction by the water authority (WRECA) and by riparian farmers, using *Shadoufs* and pumps, the provision of the latter being assisted by Kano State Agricultural Development Authority (KNARDA).

TABLE 13.4 Mean Discharge of the River Kano at Chiromawa (m^3/s) Before and After the Construction of Tiga and Bagauda Dams.

Month	1965–1970	(before)	1974–1979	(after)	Change
	x	s.d.	x	s.d.	
April	0.05	0.1	6.52	6.5	+6.47
May	10.27	8.7	8.05	6.2	–2.22
June	32.14	21.8	7.93	7.4	–24.21
July	64.92	37.5	8.34	6.1	–56.58
August	146.25	27.2	13.97	12.5	–132.28
September	164.41	80.5	27.11	34.7	–137.30

TABLE 13.5 Losses from River Kano.

Date	Site of measurement of discharge	Discharge, cumecs	Loss between stations, %	Cumulative loss from Tiga/ Wudil, %
18-03-85	Tiga	6.96		From Tiga
	Chiromawa	6.75	3	3
	Tamburawa	4.04	40.4	39
19-03-85	Wudil (with Shimar)			
	Flow of 2.05	5.12	23.6	From Wudil
20-03-85	Joda	2.85	44	44
20-03-85	Dabi	2.6	8.7	49
	Suntalmawa (2 sections: 1.51 + 0.32)	1.83	29.6	64
20-03-85	Sarawuya	1.04	43.2	80
	Hadejia	0.81	22	84
	Hadejia	0.21	74	96
	Marma	Impounded with a gauge height of 0.4 m		

Note: Local intensified irrigation activities in the downstream reach have also been included, and the flow of 2.05 from Shimar has also been taken into account.

Beyond Kano, the river becomes the Hadejia, having been joined by the Challawa at Tamburawa. The WRECA estimate is that 74% of the flow in the Hadejia is "lost" as a result of evapotranspiration in the swamps, percolation into the ground, and legitimate extraction. Confirmation that the "losses" are estimated to be 70%.[11] It was concluded that the flow in the river channel was not efficient, and Shultz also pointed out that considerable losses occurred in irrigated areas due to unwise irrigation practices and seepage into the ground from irrigation canals.

13.3.3 CHANGES IN THE KANO RIVER MORPHOLOGY AND ITS EFFECTS ON RIPARIAN FARMERS

The overall reduction and control of flow of the river after the dams' construction means that the cross-section of the channel required is now much smaller than hitherto. The measurement of the geometric variables of the River Kano before the dam construction and of the incised channel, after it, are shown in Table 13.6.[54]

TABLE 13.6 Geometric Variables of Kano River: Before and After the Construction of Dams (The Data in Brackets are After the Construction).

Profile	Top width, m		Mean depth, m		Cross-sectional area, m^2	
	before	after	before	after	before	after
			Kano			
Kl	270	(30)	1.70	(1.05)	459	31.5
K2	250	(40)	1.70	(1.00)	425	40.0
K3	360	(47)	1.67	(1.00)	601	47.0
K4	300	(40)	1.75	(0.90)	525	36.0
K5	280	(40)	1.77	(0.97)	496	38.8
K6	300	(33)	1.71	(1.00)	513	33.0
K7	130	(30)	2.45	(1.05)	319	31.5
K8	240	(28)	2.20	(1.36)	528	38.0
K9	120	(42)	2.76	(1.00)	331	42.0
K10	150	(15)	2.53	(1.40)	380	21.0
Mean	240	(34.5)	2.02	(1.07)	458	(35.9)
s.d.	77	(9.24)	0.40	(0.17)	88	(7.14)

The net effect of this change is that a large proportion of the former Kano River channel, which was infertile and could not be used to produce crops, has now been transformed into a potentially fertile soil where crops such as sugarcane, vegetables, wheat, rice, fruit, and forage for grazing animals are grown. These crops are produced during both the rainy and dry seasons. The proportion of the original channel transformed into an area that can be cropped is about 83%.[28] Although the width of the land created is not uniform, he estimated that it averaged 200 m: if a conservative estimate of 150 m is accepted, over a distance of 60 km taken downstream from the dam to the Kano Challawa confluence, the area of land created by the change is some 9000 ha.

The creation of such a large area capable of being cultivated within the former riverbed has produced important benefits both for farmers and the government and, in addition, the distance through which water now has to be lifted to irrigate this land is relatively low. In some cases, the incised flow of the new river is just by the banks of the old one; this can cause the sides to collapse and thus reduce the area of agricultural land at those places. This phenomenon has resulted in border conflicts not only between individuals but also between local governments where the river was a boundary.

The desire to raise revenue and to settle border and individual conflicts involve local government in the distribution and ownership of the newly formed land. Hence, in most of the places where land was created, it now belongs to the local government and is normally let out to riparian farmers who pay rent for it.

This general increase in land availability along the Kano River, coupled with price increases for farm products, has improved the livelihood of the riparian farmers to some extent. Some of these profits have been spent on the purchase of irrigation pumps, fertilizer, and other inputs to raise production levels. Unfortunately, the operation and maintenance of these pumps encountered serious problems, as discussed later.

The beneficial effects experienced along the River Kano following the construction of the dam have not always been experienced in other locations, as was shown when a study similar to that done by James on the Kano River was carried out to evaluate the economic activities below the Kainji Dam after its construction.[7] This dam was completed in 1968; before this river flows were lowest between February and July, then they rose in August and reached their peak in September. Around March/April most of the areas downstream from Jebba had to rely for their fishing and irrigation activities on water trapped in lakes and pools in the river channel. It was originally thought that after the dam construction, those tributaries downstream of it would provide an adequate supply of water to support crop growth in the area during dry season cultivation. However, they proved to have an insufficient flow and in consequence, thousands of acres that were previously cultivated could not be cropped after the dam was made, and even those that were cultivated were adversely affected in terms of yield.

Even in areas not served by main rivers or their tributaries, irrigation can be practiced by drawing water from below ground level. If the water table is deep, it will be necessary to install a borehole and a suitable pump. However, where the water table is high (e.g. in a depression) a shallow excavation may be all that is required to give access to the water. The term used to describe the latter situation, Fadama, is of Nigerian origin and has become widely used.

Table 13.7 shows the decrease in the number of fadamas cultivated following the change in the Kainji River regime. One can argue that the percentage of acreage reported "lost" was misleading: before the dam construction water was simply flooding the farms whereas afterward, the land was still there; efforts now have to be made to pump and convey water to the farms. Once this is done the water supply will be controllable and it follows that better water management will be possible and higher crop yields should follow.

TABLE 13.7 Decrease in Fadama Cultivation Following the Change in the Kainji River Regime.

Village	Population 1963	Fadamas cultivated 1967–68	Approx. acreage	Fadamas Cultivated 1968–69	Approx. acreage	Acreage Lost	% Acreage Lost
Rabba	2139	210	500	86	150	350	70
Muregi	3053	1230	2700	852	1500	1200	44.4
Katcha	6939	720	1600	326	750	850	53.13

13.4 IRRIGATION STATUS AND PRACTICES IN KANO STATE

To supplement food supplies many farmers in Kano State resort to irrigation, especially during the dry season. Most of them use inter-cropping on their farms even though sometimes this can be a disadvantage. Some cultivate large areas while the majority are small-scale farmers.

13.4.1 WATER SUPPLY

The water released from the Tiga Dam has to supply a number of needs: water for drinking and other human requirements, water for industrial and commercial purposes, and water for irrigation. The rate of discharge through the dam can be controlled and a balance has to be struck between too high a rate, which would deplete the resources too quickly, and one that is too low and would restrict some of the users and eventually lead to a general lowering of the water table downstream.

The first major recorded extraction, which predates the dam's construction, started in 1938 when 13,000 m^3 d^{-1} was taken by the first Challawa Water Works. From this treatment plant, the water was pumped to a storage reservoir at Goron Dutse in Kano City from whence it was allowed to flow by gravity into the distribution network.

After the dam was completed in 1975, the Challawa system's capacity was later increased by building a Waterworks, serving over 1.5 million people. Then, in 1983, there was a shortage of water in Kano City due to the increased demand of a growing population and expansion of commercial and industrial activity. The two treatment plants could only be upgraded to supply 135,000 m^3 d^{-1}.[66] To meet this shortage, the government awarded a contract to build a further 90,000 m^3 d^{-1} conventional treatment works which was commissioned in 1991.

Despite the existing issues, Kano City's water supply is generally regarded as being the most dependable and adequate system of water supply in Nigeria. Some of the difficulties that have been met include pump breakdowns, power failures, and periodic shortages of chemicals and spare parts.

The quality of water in the Kano Rivet is generally good, a typical chemical analysis of it is shown in Table 13.8. Water of this type is considered suitable for distribution after direct chlorination, that is, it would not have to pass through a pre-treatment plant first. It is also completely satisfactory for agricultural irrigation use as it is well within the limits of soluble salt content.[10]

At the moment, the demand for water is being largely met; when interruptions in supply occur these are due to breakdowns rather than an overall shortage of water from the source. As time passes, however, demand will grow until the catchment area may not yield sufficient water to meet with present attitudes to water use.

One aspect that is particularly evident is that of water wastage, caused by reported and unreported leakages. At certain times, the former cannot be repaired due to a lack of fittings; some of the latter are a result of public apathy whereas others are caused by poorly laid or aging pipes, sometimes buried in gutters where leaks are difficult to detect. Yet another source of wastage can be regarded as the fault of consumers, who are simply careless and/or extravagant with water, probably because their use of it is not metered.

TABLE 13.8 Chemical Analysis of Water Samples.

Water Resources and Engineering Construction Agency (WRECA)	
Water Quality Laboratory Challawa	
Kano State, P.M.B. 3165, Kano – Nigeria	
Name and Address of Sender: Water Lab	Sender's Ref No: xxx
	Lab. Ref. No: 1102
Nature of Samples: Raw water	Date of collection: 01-03-18
Where Collected: Raw water chamber	Date of test: 01-03-18
Appearance: 58 NTU	
Color: 10 Hazen	
Odor: Odorless	
Temperature:	
	pH: 7.3
Phenolphthalein Alkalinity (as $CaCO_3$)	0.00 mg/L
Methyl Orange Alkalinity (as $CaCO_3$)	55.00 mg/L
Total Alkalinity (as $CaCO_3$)	55.00 mg/L
Total Hardness (as $CaCO_3$)	40.40 mg/L
Calcium (Ca^{2+})	10.80 mg/L
Magnesium (Mg^{2+})	4.38 mg/L
Chloride (Cl^-)	0.00 mg/L
Nitrite Nitrogen (as N)	0.00 mg/L
Nitrate Nitrogen (as N)	0.00 mg/L

13.4.2 INTER-CROPPING AND LAND USE

Most farmers in Northern Nigeria believe that inter-cropping, i.e. growing two or more crops in rows or clusters, is more profitable than single cropping. Despite all the efforts by extension workers to persuade farmers to move to single cropping, they were not convinced.[38]

In a cropping investigation, published as the report on the Kano River Project phase 11 (KRP 11), four blocks of 45 ha along the Kano/Hadejia River were investigated. Ground survey and aerial photographic techniques were used in August/September 1976 and it was found that 94% of the land was inter-cropped with two or more crops; even three and four-crop mixes were common, covering 39% and 37% of the land, respectively. Although there have been changes in attitude in the 41 years since this survey, they have been relatively minor. On a few large farms, changes have been made to accommodate some mechanization, but traditional farmers have been reluctant to move away from established methods of crop production (See Appendix A).

The farmers of the region practice multi-crop systems with the assumption that in this way the total yields are generally higher. It has also been pointed out that the risk of losing all crops to pests and diseases appears to be greater when single cropping is practiced. The farmers also cite the example of inter-cropping plants such as sorghum and maize; although similar crops in many ways they have different water requirements and different lengths of time to reach maturity (See Appendix A). The hope, of course, is that both crops will yield well but in reality some seasons will suit one of the crops rather than the other and so, while inter-cropping has an in-built insurance component, the amount of water applied may not be optimum for either crop.

Riparian farmers also practice inter-cropping rather than single cropping. They follow no definite system and what they do varies from locality to locality, season to season, farm to farm, and farmer to farmer. Within the study area, it was found that farmers using basins basically inter-cropped in five ways as follows:

- crops inter-planted within each basin,
- crops inter-planted on the edges of the furrows between the basins,
- crops inter-planted on the soil bunds of the basin,
- plants inter-planted with fruits, and
- when small basins are used (say of 2 m × 2 m area) and each one of these is planted with a different crop the overall effect is similar to inter-cropping.

Near the WRECA, well No. 11 on the Kano River, 104 plots of land were visited at intervals over two seasons. These will in the future be referred to as the sample area. It was observed that, as a result of pressure on the farms to produce crops, only one farm was allowed to follow in the second season and all the farms were invariably inter-cropped. This supports the view that until it has been proved by experimental work and widely demonstrated that single cropping is superior to inter-cropping there seems little likelihood that farmers will change their ways, so perhaps in the meantime, some effort should be made to improve the standard of the inter-cropping method currently in use.

Inter-cropping relies on considerable inputs of labor; when mechanization is introduced, even at low level, single cropping is much more convenient. The traditional method of irrigation in basins also has a high labor content and it could be difficult if not impossible to mechanize, so in future perhaps a change to growing crops on or in furrows or in some other way that facilitates mechanization may have to be introduced.

13.4.3 LARGE-SCALE IRRIGATION SYSTEMS

To feed the large and increasing population, Nigeria will have to utilize both its water resources and agricultural land effectively. Although peasant farmers have irrigated their crops for centuries, early this century it was decided that efforts should be directed mainly toward tire needs of LSI schemes, however, in more recent years the contribution that small-scale irrigators can make to overall production levels has become widely recognized.

One of the first steps in establishing LSI schemes was taken in 1918 when the first survey on irrigation in Northern Nigeria was carried out by Collins (a military engineer) in the Sokoto and Rima valleys. As a result of this survey, the Sokoto and Rima rivers were impounded and water was used for irrigation projects,[53] (see Table 13.10). It can be seen from Table 13.9 that in 1956, after 1965 pumps had been installed on the Hadejia River near Kano, 473 ha of rice and wheat was irrigated. This project was subsequently enlarged when the Tiga Dam was constructed.

LSI is a scheme, which is "*typically fairly large by African standards (20–100,000 ha) and is developed and managed by a government or other body external to the community.*"[18, 19] The Kano River irrigation project at Kadawa is the only scheme developed and managed by the government on the Kano River; there are other schemes but these are managed by bodies with commercial interests (SSI schemes, in contrast, have been defined as being usually on small plots, in which small farmers have the major controlling influence, and using a level of technology which the farmers can effectively operate and maintain).

The Kano River scheme draws its water supply from the Tiga Dam by gravity, through an 18 km long main canal. It is currently one of the largest schemes not only in Nigeria but also in the whole of West Africa.[40] An area covering about 15,000 ha was under irrigation at that time while an additional

TABLE 13.9 Crop Production and Value in the Wet and Dry Seasons of 1983–1987 of the Kano River Project (Phase-1), Kadawa Farmers.

Crop	Area (Ha)	Yield (tons)	Value (=N=)	Area (Ha)	Yield (tons)	Value (=N=)
		1983			**1984**	
Wheat	778.69	1903.06	761,244.00	5000.00	12,500.00	5,625,000.00
Rice	980.40	2844.10	1,422,050.00	3717.85	3717.85	2,044,817.50
Tomatoes	806.12	20,153.00	201,530.00	2000.00	30,000.00	300,000.00
	1985			**1986**		
Wheat	7190.06	14,380.12	6,471,054.00	4674.12	9348.42	4,674,210.00
Rice	156.87	470.61	343,427.00	1735.43	6941.72	6,941,720.00
Tomatoes	867.60	17,352.00	2,776,320.00	1,670.26	33,405.20	5,344,832.00
Wheat	4789.18	11,972.95	9,578,660.00			
Rice	4824.05	19,286.00	19,296,200.00			
Tomatoes	1680.34	33,606.80	6,721,360.00			

TABLE 13.10 Chronology of Irrigation Development in Northern Nigeria.

Date	Consultant	Water Source	Construction	Project location	Ha irrigation	Crop grown	Current status
1918–25	Col. Collins	Sokoto and Rima River valleys	Impound water behind the dike	Sokoto area	Small plots	Sugarcane Cassava Sweet potatoes and maize	Destroyed by Flood in 1922
1925–40	Col.Collins	Sokoto, Rima and Zamfara River valley	Canals and a bound for water impoundment	K-ware Project	243	Sugarcane Cassava Sweet potatoes and maize and wheat	Destroyed by Flood in 1940
1949	irrigation Division I8RD and NEDECO	Niger River Valley Streams	Impound water behind the dike	Niger valley Project (Mokwa) and parts of Sokoto Province	405	Rice	Parts still operational
1950–59	Irrigation Division	Niger River Valley streams	Impound water Behind the dike	Niger Valley Projects (Bedeggi and Edozhtght)	405–810	Rice	Still operational
1956–65	USA ID USA ID NEDECO FAO	Yobe River	Installed pump near river	Daya and Abadan	384–425	Rice and wheat	Still operational
		Hadejia River		Kano	–73		
		Ebeji River		Gamboru	400		
		Rima River		Nurno	162		

7000 ha was at various stages of development.[75] Some of the crop production figures and the value of crops produced in this area between 1983 and 1987, in both wet and dry seasons, are shown in Table 13.9.

The land covered by the Kano River scheme is owned by individuals, whose farms vary in size from a fraction of a ha to several ha. The government is charged with the responsibility of organizing and supervising enlightenment programs, marketing arrangements, training courses, the formation of cooperative societies, and the maintenance of an adequate water supply. It also must supply and distribute agricultural inputs through its agents, namely the Kano State Agricultural Supply Company (KASCO).

On the question of water management, the general belief by farmers, despite many years of experience and the provision of information, is that the more water applied to plants the more they produce. That they hold this view is extremely unfortunate. Farmers upstream who have the advantage of having adequate water supplies, apply as much irrigation as possible; unfortunately, this results in waterlogging, especially when crops like rice, which can tolerate this condition, are inter-cropped with maize. This mix of crops is relatively common in the area and one can often observe the inter-cropped maize under stress—it will look partially brown and will produce only very low yields.

The other consequence of overwatering can mean that little or no water is available for farmers downstream; this is especially so during the daylight hours when irrigation is at its peak. The problem is so acute in some downstream locations that farmers will sometimes divert any night flow of water into their land and this frequently leads to severe flooding because there is no one present to control the process. This not only results in the wastage of valuable water but also in land and crop damage.

Another issue raised by imperfect management of large schemes is that some parts of them can become neglected; the most common problem is that channels are filled with sediment or vegetation and this reduces water flow and thus restricts the output of crops from the scheme.

Reporting on problems with LSI schemes in Africa in general, researchers identified the lack of practical and management skills and the necessary attitudes on the part of the national staff. It is recognized that the Kano River Project has some of these characteristics which result in poor maintenance of irrigation structures and hence inadequate distribution of water. Four other common problems with large schemes[18] are:

- Their extremely high capital cost ($ 10,000–20,000 per ha being common);
- The over-optimistic benefit projections and development rates quoted;

- That insufficient account may be taken of indigenous farming systems, skills, and incentives.
- The under-provision of funds to meet operation and maintenance costs by government, lending agencies and donors: this obviously has an influence on the situation described above.

The main achievement in all large-scale schemes, as has been pointed out, is the provision of a reliable and readily available irrigation water supply for farmers to use. The fine-tuning of the management of any scheme comes later: once the supply is there it is mainly a question of organization and training to exploit it to everyone's advantage.[19]

The early (1970s) development of LSI projects "in Nigeria in particular, where investment was huge, the predicted rate of growth seems bizarre when set against achievement. One reason for this was lack of funds, however, more serious was the growing understanding by both aid donors and governments of the huge battery of problems with large-scale government irrigation in the African context."[4]

However, not all LSI schemes have bad disappointing outcomes, for example, several large non-governmental schemes have sprung up along the Kano River and its associated irrigation channels following the strict control by the Nigerian government on the importation of rice and wheat and the introduction of massive subsidies on farm inputs. The size of these developments ranges from a few ha to several hundred ha; they are managed and owned by companies, entrepreneurs, civil servants and military officers and in general it seems that they operate effectively.

13.4.4 SMALL-SCALE IRRIGATION SYSTEMS

In recent years, there has been a considerable surge of interest in the development of SSI in Nigeria and many other parts of Africa. This is largely due to various national and international assessments of the performance of SSI projects. Some of these assessments are detailed in publications by the World Bank (2020). In its assessment, the World Bank concluded that the money Nigeria had spent on LSI projects—estimated to be about one billion US dollars—would have been better spent on rain-fed agriculture and SSI. The capital and political goodwill put into Nigeria's LSI schemes encountered some of the problems mentioned earlier. The country's SSI sector has had little or no assistance from anywhere, nevertheless, it has apparently been a tremendous success. This has been recognized by the fact that the government now assists small-scale farmers through the Agricultural Development

Projects (ADP), now sponsored by the World Bank, the Federal government, and the State governments. The ADP have gone a long way to encourage progress by drilling tube wells as well as by making improved seeds and extension services available and providing pumps at subsidized prices.

In particular, the success of SSI is evident when one visits riparian farms within and along the Kano River; such schemes are invariably small (0.93 ha on average) (see Appendix B) but are generally well managed. A farmer interviewed in the course of this study showed the writer an area of one hectare close to the Kano River which produced wheat in the 2017/2018 season valued at N300,000.00; the estimated inputs (labor, seeds, fertilizer) were put at N131,000.00, leaving a profit of N169,000.00. To put this in context, this sum would buy 2 number 75 mm engine-driven irrigation pump sets and perhaps 100 m of irrigation pipe.

Despite the huge success achieved by the riparian farmers along the Kano River, they also have problems to overcome, four of which are discussed below. Perhaps the biggest difficulty is that of a general shortage of fertilizer; despite all the efforts of the authorities concerned, supplies to peasant farmers are very restricted. Unfortunately, practically all of these materials have to be imported and as yet the correct balance between supply and demand has not been achieved.

The second difficulty arises when farmers, having produced their crops, attempt to sell their produce at the nearest market. If it is a perishable crop like tomatoes, the numerous small farmers all tend to produce ripe crops at the same time. This causes a glut, so prices collapse and the way around this problem is not easy. One solution with this crop, but not for many others, might be to start a canning operation; for others the operation of market forces could eventually produce a change toward extending the season and diversification. There is obviously long-term potential for development in the area—there is labor, excellent light conditions, and some water— but such developments require capital injection if marketing on a wider scale is to be contemplated.

Now most of the farmers under the riparian irrigation scheme are relatively inaccessible, a consequence of this is that sophisticated methods of both pest and disease control are not available: this again is a situation that will presumably be resolved in time.

It is perhaps appropriate at this point to comment on the difficulty of obtaining reliable information on the numbers of people, farms, and irrigation schemes in a situation such as is found in the study area. Information on LSI project has been provided by the HJRBD while on smaller schemes KNARDA is the best source of information. Other sources used include Land and Regional Planning Directorate, Bayero and Ahmadu Bello Universities,

Diyam consultants, Kano State Water Board, Ministry of Water Resources, WRECA, and the International Crop Research Institute in Kano.

All these organizations will admit to difficulties in collecting reliable statistics in a region with a high proportion of underdeveloped rural communities, so where local information was required, the writer approached the traditional rulers and individual farmers directly.

In recent years one aspect of the relationship between the people who farm on a large scale and the riparian farmers on the Kano River has been one of competition for land ownership. It is evident there has recently been an inflow of wealthy individuals and long-term assets. Yet another is that farming has now become a respectable and prestigious business and retired Nigerian leaders with both civilian and military backgrounds have become successful farmers. The former head of state and later president in Nigeria, General Obasanjo, for example, said that after leaving the public office he decided to go into farming to dignify it by example.[8] He thought that many Nigerians did not believe that he would go into farming because it was regarded by some as “a mean job.” Other reasons for his decision include his determination to practice what he preached while in government and to increase Nigerian awareness of the overall importance of farming to the nation.

Although the reasons given by General Obasanjo could be regarded as in the best interest of farming in Nigeria, yet unfortunately such moves have an adverse effect on riparian farmers in terms of both land ownership and, of course, water supply. If wealthy farmers use large pumps to irrigate hundreds of ha, they will tend to exhaust the available surface water and lower the groundwater table. Major developments in farming of this type were not foreseen when the Kano River Project was originally planned and they may influence the way the scheme progresses.

There is no data available on the number of riparian irrigators in the River Kano catchment between the Tiga Dam and Tamburawa. Most of them are located along both banks of the River Kano and its several tributaries, but throughout the catchment, there are also small groups of farmers cultivating land around fadamas and the many small areas of open water that form in slightly low-lying areas of this generally level tract of land. Even if there is no water visible it can sometimes be found just by digging a hole.

Similarly, there is no official estimate of the relative number of farmers (a) without motorized pumps or (b) either owning or having access to such equipment. On the other hand, the introduction of pumps and facilities to repair them are improving, but it is a painfully slow process (see Appendix C).

13.5 WATER-LIFTING TECHNIQUES FOR WASTEWATER AND IRRIGATION

Water-lifting devices have been used throughout history to supply water for both human consumption and agricultural purposes. The earliest example might have appeared 5,000 years ago, the numbers involved today are enormous,[69] indeed in China alone there were over one million water-lifting devices of various types in use.[20] Without these water lifters irrigation would be confined to areas that can be supplied with water by gravity, the consequences of this are that many areas of the world, including some parts of Northern Nigeria, would remain barren.

In the developed world, pumping is probably a more appropriate term than water lifting, which tends to be reserved for systems using animals and manpower, although wind and solar energy-powered systems are also found.

When they are available and can be afforded, pumps are used in preference to be lifting devices. Pumping usually employs either internal combustion engines or electric motors: in Northern Nigeria, both are to be found, but there are limitations to their use among riparian farmers. Nevertheless, an outline of the role they play and the way in which they are being introduced has been included.

13.5.1 MOTORIZED PUMPS

To encourage farmers to purchase motorized pumps to enable them to increase crop yields, the Nigerian government has made subsidies available. These pumps have been sold, both in the marketplace and by government agencies in the states. For example, in Kano State, Kano Agricultural Supply Company Limited (KASCO) is responsible for the sale of both pumps and their spare parts. These pumps are mostly 50 mm or 75 mm self-priming centrifugal units fitted with 2 or 4-stroke Honda petrol engines.

One such pump at the time the water was being used to irrigate tomatoes, onions, and cabbages. The cloth wrapped around the delivery pipe was being used in an attempt to seal a leak that could not be repaired with the facilities available.

The advantage of these pumps when compared with diesel engine units is that they are relatively light; in fact, they are light enough for a farmer to carry home on a bicycle or donkey after use—a procedure that is essential for security reasons. The increase in the number of pumps being imported has been accompanied by the appearance in most towns of mechanics that repair them and carry stocks of spare parts.

Pump outputs are such that the discharge can usually be accommodated in existing channels, used to distribute water lifted by *Shadoufs*, so no major changes to existing layouts and techniques are required.

The price of these pumps has increased considerably in recent years; for example, a 50 mm unit was N650.00 in 1987 (Kano State Agricultural and Rural Development Agency (KNARDA)),[30] and the current price of a similar pump (Honda) in 2018 is N50,000.00. The writer enquired users about pump life and most of the farmers asked said that between three and five years of useful work is commonplace (Appendix C). This relatively short life span is probably due to the operator's lack of mechanical knowledge and poor maintenance.

In an effort to develop alluvial groundwater for the use of riparian farmers, the KNARDA installed many jetted boreholes for farmers who were prepared to pay for their service and purchase a pump.[18] The method of constructing the borehole is shown in Figure 13.4.

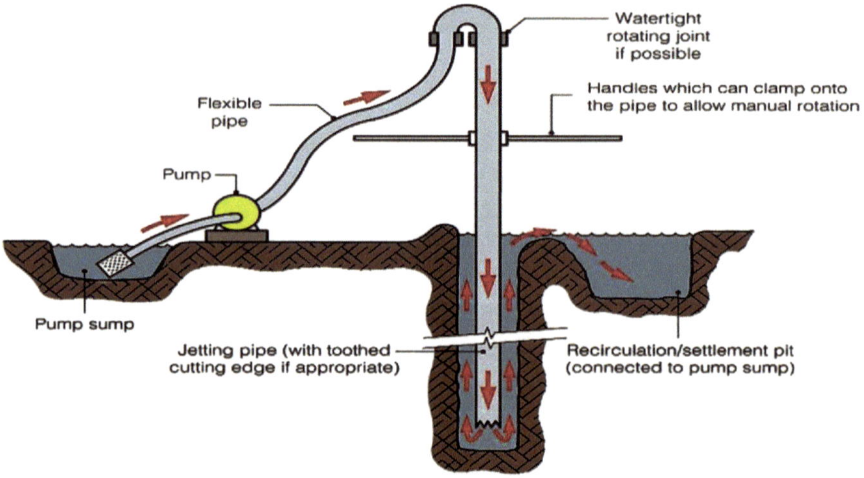

FIGURE 13.4 Major components of a simple jetting rig.

13.5.2 NON-MOTORIZED PUMPS AND LIFTING DEVICES

Designs have evolved in several parts of the world, some using animal power to operate them, others being designed with man's capabilities in mind.

In Northern Nigeria it is manpower, rather than animal power, that lifts water when motorized or gravity-fed systems are not available. This approach is still widely used in many other developing countries and essential features of all such equipment are that they need to be cheap and easy to produce and maintain locally. Ideally, they should use local materials. Unfortunately, devices which have such characteristics and constraints are normally associated with low lifts, low efficiencies, and—because of the limited power available from man—small discharge rates.

A general classification of manual units is shown in Table 13.11, and it is based on whether a rotary or reciprocating motion is used to propel the lifting element and on the sets of muscles mainly employed.[74]

In any one country, it seems that only a limited number of types of water lifters are found. Reasons for this are not at all clear, but in some cases, it may be that essential materials are not readily available, whereas in others, the tradition may be that individuals look after their own requirements rather than work to produce a large system capable of supplying several farms.

In the study area, as in other parts of Northern Nigeria, there is a tendency for individuals to make their own arrangements within the extended family group. They are not in a position to either take water by gravity from the Tiga Dam or to purchase a pump, they will almost certainly use a Shadouf, a simple bucket, a rope, or a hand pump, although the latter is primarily used for domestic water supply and perhaps to irrigate small "Micro plots" which happen to be located nearby.

13.5.2.1 HAND PUMPS

Thousands of imported hand-operated pumps were introduced into the study area by KNARDA, Hadejia Jama'ara River Basin Development Authority (HJRBDA), and WRECA—the three authorities responsible for the supply of water in Kano State. It has been reported the importation of 1000 Zwaans and Bumi diaphragm pumps, which are made of cast iron and are extremely strong (see Figure 13.5).[31] One attraction of these pumps is their high discharge rate (about 5 m^3/h), which is higher than that expected from a *Shadouf* and twice that of the Rower pump which is described below. One drawback of the pump is that the forces required to operate it are higher than some people can exert continuously, although for short periods it can even be operated by children provided that there are two or three of them prepared to work together. This underlines the importance of ergonomics in any human operated pump.

TABLE 13.11 Manual Power Applications.

	A. Reciprocating motions	
Performed with	**Transmission**	**Water Lifter**
arms and/or upper torso	rope (may also include pulleys, etc.)	Bucket
	Rope	swing basket
	Rope	water balance
	rope and pole	counterpoise Lift
	rope and pole	shovel
	Handle	gutter
	Handle	diaphragm pump
	handle and connecting rod	piston pump
	crank and rocker mechanism, gear train	centrifugal pump
Legs and lower torso	connecting rod, foot pedal, and return device (e.g. spring or counterweight)	piston pump
	pedal and spring return	diaphragm pump
	Pedal	air displacement pump
	rocking footrests	bellows pump
	Direct	doons
	Walkway	picottah
	Walkway	gutters
	Walkway	picottah-style doon
	B. Circular motions	
Performed with	Transmission	Water Lifter
Arms and /or upper torso	Crank	screw
	Crank	water ladder
	crank and small gear train	rotary
	Lever	semi-rotary
	crank and windlass	bucket
	crank and shaft	chain pump
	crank and cam	piston pump
Legs and upper torso	horizontal axis treadmill or	paddle wheel noria
	pedal (on lifter or separate with connecting shaft)	Persian wheel tympanum
	horizontal axis treadmill or pedals connected via a shaft	screw water ladder chain pump
	horizontal axis treadmill (or pedal) and cam	piston pump
	horizontal axis pedals and gear train	centrifugal pump

FIGURE 13.5 A typical hand pump.

13.5.2.2 ROWER PUMP

The pump is a reciprocating piston type utilizing a 50 mm plastic pipe as its cylinder. This is inclined at 30^0 to the horizontal and the operator stands on a platform to push and pull the T-shaped handle, attached directly to the piston as shown in Figure 13.6.

This action, which gives the pump its name, produces discharges of 1.56 $m^3 h^{-1}$ according to tests done in Northern Nigeria, though no pumping head is given.[12] This discharge is less than that of the Shadouf, as shown later in this chapter. One other problem with this pump is its leather seals, which last for about half a season when fully in use and then require replacement. In general, this pump requires more frequent repairs than the Shadouf and has appealed to farmers in Kano.

13.5.2.3 ARCHIMEDEAN SCREW

This is one of the oldest types of lifting device, the traditional version being built before Roman times, and it is still in operation in a similar form in Egypt today.[69] Early models were made entirely of wood whereas modern ones employ galvanized steel for both the flight and the cylinder.[62]

The cylinder is fixed in an inclined position, usually 33–40° to the horizontal, depending on the pitch and diameter of the internal helix; its lower end is in the water and the M top end has a cranked handle by which it is rotated. The screw operates only on low heads, of between 0.25 and 1.30 m, and when fitted with good bearings it is one of the most efficient devices in terms of its specific output.[62] Sometimes two operators are necessary if the discharge head is high.

The screw can be fitted with gears and can also be operated by animals. Modern versions of this ancient device include screw pumps[62] which operate in close-fitting (but not touching) open inclined channels; the latter are normally made of accurately formed concrete and the rotors are motorized. The only screw pump found in Kano by the writer is near the airport at the sewerage and drainage section of the Urban Development Board.

The advantages of the Archimedean screw are its relatively high discharge rate and its high efficiency, reported to be up to 80%.[69]

The low groundwater table in much of the study area means that there are very few places where an Archimedean screw can be used, however, where the water table is high, for example, in the newly formed channels in the Kano River systems (discussed in Chapter 2), this device could be very useful.

13.5.2.4 TREADLE PUMP

In the late 1970s, a manually powered twin-cylinder pump, operated by a treadle mechanism was introduced by the Ranjpur Dinajar Rural Service in Bangladesh (Figure 13.7). This pump has been remarkably successful and sales were estimated to be some 65,000 units per year by 1991.[55] In their report, Orr and his colleagues estimated that 83% of the farmers in Bangladesh would still be under 1 hectare in area by the year 2000, so the importance of SSI in the country cannot be underestimated.

Information on the performance of the pump is given and the initial publication has been followed up[16] in which the number of units in use was put at 500 000. The claimed capacity was 5000–7000 liter/h—sufficient to irrigate between 0.25 and 0.5 ha with family labor, but these figures are quoted without reference to the lift achieved. The 1995 publication gives a detailed description, extending to 44 pages, of how to manufacture the pump. It is mainly made of steel and involves cutting, drilling, and welding, plus the use of a lathe for turning wooden components.

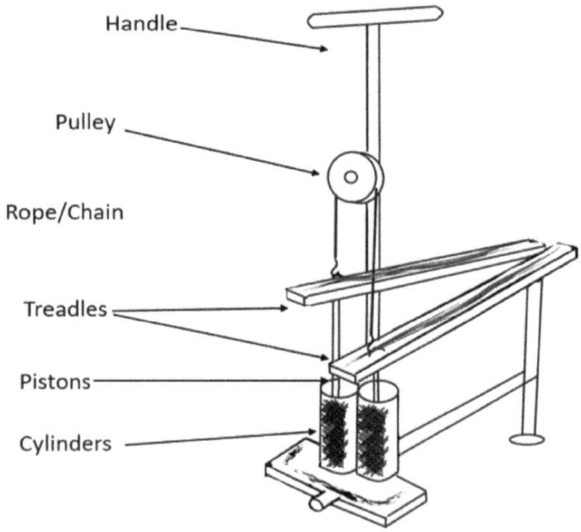

FIGURE 13.7 Complete treadle pump.

13.5.2.5 WIND-POWERED PUMPS

Wind-powered pumps have been in use in the study area for many decades, although they have not been used for irrigation. They are mainly used for domestic water supply, the provision of water to institutions such as schools, and the watering of livestock (Figure 13.8). There are about 64 of these pumps in Kano State, out of which only half were operational due to damage or lack of maintenance.[31]

FIGURE 13.8 A wind-powered pump.

Note: The concrete storage tank is necessary because of the intermittent wind in the area.

One major reason why these pumps are not more popular is because of the light winds in the area; their output has been estimated[29] and is given in Table 13.12. This table shows the average wind speed to be lowest during the dry season when the water demand for irrigation is at its peak, it is generally believed that in places where the wind's average speed is less than 8–11 km h^{-1} wind pumps are unlikely to be an attractive means of pumping water.

13.5.2.6 SOLAR-POWERED PUMPS

In recent years, considerable interest has developed throughout the world in the application of solar-powered pumps. Research and development projects on them have been numerous with a major initiative in the form of sponsorship coming from the United Nations Development Program and the World Bank.[69]

There is an abundance of energy arising from the sun which the Earth's atmosphere receives at the rate of 151×10^{18} kWh per year. Considerably less than this amount of energy reaches the Earth's surface and since the present technology involved in capturing this is inefficient, only an insignificant portion of the total energy available can be converted to a useful form.

Nevertheless, progress is being made and solar energy applications are thought to have enormous potential. The method used for converting solar energy to pump water can be divided into two: thermodynamic and photovoltaic.

Of the two the former has been investigated for by far the longest time, but with little, practical success. It generally involves the use of mirrors to concentrate solar energy on a boiler which raises steam to pass through a turbine or a steam engine. Such systems are inevitably large, expensive, and of low efficiency and they were more-or-less discarded during the era of cheap oil.

They are now considered to be less attractive than the equipment based on photovoltaic devices, which convert solar radiation directly to electricity. Many materials can be used to make photovoltaic cells but the only one used commercially now is silicon in the form of thin wafers of highly purified crystals. Though silicon is a cheap and abundant element, its purification, crystallization, and slicing is a high-technology process, which makes the cells expensive.[25]

TABLE 13.12 Output of Wind-Powered Pumps in Kano.

Parameter	Jan	Feb	Mar	Apr	May	Jun	Jul	Aug	Sep	Oct	Nov	Dec
	Km/day											
Average daily: Windrun[1]	116	116	122	153	199	195	169	127	115	93	98	108
Min average: Windrun[2]	101	88	94	112	177	119	129	95	83	72	67	82
Max average: Windrun[3]	139	135	157	179	232	231	205	178	232	128	123	127
No. of years of data[4]	11	11	11	11	12	12	12	12	12	12	12	9
Kijito Type Wind Pump: Tower height 6 m and diameter 6 m												
Average windspeed at 6 m height[5] (Kmh^{-1})	5.8	5.8	6.1	7.7	10	9.8	8.5	6.4	5.8	4.7	6.2	6.4
Estimated power output[6] (watts)	31	31	36	72	158	149	97	41	31	16	38	42
Estimated water output[7] (m^3h^{-1})	1.4	1.4	1.58	3.17	6.9	6.6	4.3	1.8	1.4	0.7	1.7	1.8
Climax Type Windmill: Tower Height 9 m and rotor diameter 6 m												
Average windspeed at 9 m height[5] (Kmh^{-1})	6.3	6.3	6.6	8.3	10.8	10.6	9.1	6.9	6.2	5.0	5.3	5.8
Estimated power output[6] (watts)	40	40	46	90	200	189	119	52	38	20	24	31
Estimated water output[7] (m^3h^{-1})	1.7	1.7	2.0	4.0	8.8	8.3	5.2	2.3	1.7	0.9	1.0	1.3

Notes (super-script) in column one for Table 13.12:

1) As recorded at Kano Meteorological Station at 2 m height
2) Lowest monthly average
3) Highest monthly average
4) Between 1970 and 1981
5) **Calculated from: $H/2 = 0.17 (V_H)/V_2$;** where: H = height of windmill tower (m), V_2 = Wind velocity at 2 m height, V_H = Wind velocity at height H, 0.17 = approximate exponent for an average ground surface roughness.
6) **Calculated from: p = KAV3 × 0.5926 × h;** where: p = power output in kW, K = 0.0000137, A = area of rotor in m^2, v = wind velocity in Km/h, 0.5926 = maximum possible efficiency of wind-powered device, h = efficiency of wind pump being considered as a fraction of maximum possible (= 70% for Kijito and Climax).
7) **Calculated from:** Q = (p × h × 102 × 3.6)/H; where: Q = discharge in $m^3 h^{-1}$, p = power input in kW, h = pump efficiency (assumed to be 60%), and H = pumping head (assumed as 5 m).

13.5.2.7 ROPE AND CONTAINER "GUGA"

By far, the most widely used method of lifting water for domestic water supply in Northern Nigeria is the use of a rope and container. The container is often a metal can but sometimes leather vessels are used. The size of the container will vary with the depth of the water table and the capability of the person who has to draw the water, but a typical capacity is 4 liters,

In some cases, small-scale farmers, especially those in upland situations, will use this method of lifting water when raising plants for transplanting or sale near the well. The method has the advantage of being very efficient in terms of water application because water can be applied to the crops by hand, reasonably uniformly, and without waste. This method is obviously not suitable when larger areas are to be irrigated.

The introduction of a windlass into this class of lifting devices has not happened on any scale in the study area. It is not clear why this is so; pulleys are sometimes installed so the rope can be pulled horizontally. Windlasses are only used to remove spoil when wells are being excavated.

13.5.2.8 SHADOUF

One of the simplest devices in use for lifting water for irrigation is the *Shadouf* and this is by far the most common method throughout the Kano area. There were between 150,000 and 200,000 in use in Kano State. They are mostly owned by individuals, but in some instances, they are shared between two or more people, often from the same extended family. The fact that it is so popular does not mean that it is without drawbacks, and it is not necessarily the most appropriate device available.[69] Its use is traditional and, in this area, traditions change slowly.

The *Shadouf* is a simple, almost primitive, manually operated device which is used to lift water from streams, rivers, ponds, and shallow wells.[63] It is not suitable for use in boreholes or when the water is found more than 3 m (in exceptional cases 4 m) below ground level. In Northern Nigeria, it is most frequently used by riparian farmers to irrigate vegetables during the dry season (October to March). The irrigation using the Shadouf has been practiced for many centuries in India and Egypt;[51] it apparently originated along the river Nile and was first recorded in tomb drawings at Thebes, Egypt, which have been dated to 1250 B.C.[29]

The date of its introduction into Northern Nigeria is uncertain, but it has been suggested that it was brought in either by Nigerian Muslims who were going on a pilgrimage to Mecca by road, or by the Arab traders who have had links with Northern Nigeria for many centuries.[50] Some of these devices, for example, the sweep, differ considerably from the conventional Shadouf, but they are nevertheless normally put in this category.

In its commonest form in Northern Nigeria (Figure 13.9), the Shadouf consists of a pole, mounted on a pivot, which carries a weight at one end and a water container at the other. The container is normally made from a calabash—the one shown in Figure 13.9 is roughly hemispherical and is carried in a bamboo cradle. The calabash will vary in size and metal buckets, lined baskets, and petrol cans are all used for this purpose.[43]

The calabash is attached to a light vertical pole by means of a rope, and this is also tied to the pivoting pole: the use of ropes in two places gives the operator a chance to both swing the bucket to collect water from the source and also to empty it easily.

The main pole is usually mounted on a cross pole in a manner that allows it to pivot (Figure 13.9) and the counterweight in the photograph is typical of that used in the study area in that it is made out of dried mud reinforced with grass.

The size of the counterweight and the moment arms of the vertical pole are adjusted so that when empty the calabash stays in the raised position (Figure 13.9), but when filled it has to be lifted. In practice, the weight is approximately half that of the full calabash. The actual dimensions and weights are adjusted to suit the situation and the individual operator and it has been found that farmers are very skilled at "fine tuning" their Shadouf to their own liking.

FIGURE 13.9 Typical traditional *Shadouf* in the study area.

In some cases, it was found that counterweights were heavy enough to balance the whole weight of the bucket while full, which meant that the operator had to use his weight to return the bucket to the water source rather than use his lifting muscles to pull it up.

The height and stability of the bank, where the Shadouf is located, will determine where the operator will have to stand. Sometimes he will be able to stand on the bank, while in another place the bank will have to be excavated to make a low-level standing area. Another possibility is that a platform may be constructed adjacent to the water source; in this case, it would be surprising if the output was high because of the unsuitability of the operator's location.

One of the attractions of the traditional *Shadouf* is that most of the materials necessary for its construction and maintenance are usually available in the nearest village market. Sometimes the farmer does not even have to buy components—they may simply have to be cut from trees on his farm. Other materials such as the straw mat traditionally used for the *Shadouf* head channel may now be replaced by old mats from the house or by a piece of polythene sheeting.

The performance of different types of human-powered water lifters in various developing countries has been evaluated including the traditional Shadouf.[33] The output of most of the devices was within the 10- and 20-Watts range with the Shadouf being 24 Watts. The output values were 16.0, 17.4, 20.2, and 27.5 Watts,[74] and the best performance was on a design that lifted water at a rate equivalent to 26 Watts.[51]

13.5.3 THE COST OF ACQUIRING AND OPERATING PUMPS AND LIFTERS

The main source of information on the cost of owning and operating various water-lifting devices in the study area is the KNARDA report.[31] It is seen that costs for both family and hired labor are considered for human and animal-powered devices as well as the costs of motorized and solar-powered pumps.

The Shadouf using family labor is the cheapest method in an area such as that occupied by a typical riparian farmer (0.1 ha). On larger farms, other methods have to be employed. This estimate was made with 1983 prices and, as has been shown earlier, there have been significant price changes since then and labor costs have also increased.

There has not been any further published costing by KNARDA since their 1995 report, but any current study in this area would have to take into account of the major change in fuel costs that occurred between 1995 and 2020. In 1995, petrol and diesel oil cost approximately N3.25 per liter, but there were severe shortages due to considerable illegal exporting operations to neighboring countries where prices were substantially higher. A black market developed, and many essential services were curtailed by this shortage of fuel, however, the government then imposed increases in fuel costs (N145/liter of petrol and N200/liter of diesel, in 2020) and the problem disappeared.

It is acknowledged that fuel is only one of the costs of operating a motorized pump, but this large change in fuel cost must have changed the relationship between motorized pump costs and all the other devices.

It is difficult to judge the difference in the cost of operating manual water lifting equipment between hired and family labor. Assumptions have to be made about the true cost of family labor and every family would probably put. a different cost per hour on their time; much would depend on individual circumstances—the number of people available, their age and health, and their opportunity to contribute to the family's well-being by earning money outside or doing other family-based work.

A further point can be made namely the attempt to indicate a break-even point. This obviously depends to a great extent on the topography, the soil type, the types of crops grown and the skill (or lack of it) in growing and marketing produce.

It has been suggested that below 0.5 ha only the Shadouf, or possibly a small petrol-powered pump would be used and that animal-powered pumps were in use. Both suggestions are queried. Now, animals are seldom, if ever, used and it is doubted whether this situation has changed in the 35 years since the report was published. During this time, there has been a slow increase in the number of pumps in the area, and in consequence, there is a limited amount of pump sharing and pump hiring taking place.[29]

13.6 THE FACTORS INFLUENCING THE IRRIGATION EFFICIENCY OF UTILIZATION

The efficiency with which water is used to produce crops has been extensively studied; it is normally expressed as the ratio between the quantity of irrigation water effectively used by the crop and total quantity of water supplied from the source.[72]

of irrigation. Technically, overhead irrigation would be better in such an environment, but they lack the equipment to follow such a route.

Furrows can be used to convey water within crops and both Kano River Project farmers and riparian farmers use them in appropriate circumstances. There was some indication in the survey carried out among the Garu famers that they are most used in crops such as okra and peppers where the water logging sometimes associated with basin irrigation might be a problem.

The furrow lengths found on riparian farms during this survey ranged from 10 to 30 m and seemed to depend more on the size and shape of the "farm" rather than on the soil texture. The width of the furrows at the top varied from 15 to 30 cm and they were 10–25 cm deep and mostly V-shaped. The furrows would normally be straight and run down a slope parallel to one another, all joining a header furrow into which the water flow was directed. Each furrow would be opened to the header in turn and then closed by the movement of soil with a hoe when thoroughly wetted. Considerably longer furrows are used by those farmers supplied with water from the dam.

FIGURE 13.10 The area in the foreground is being prepared for cropping; the irrigation water will be supplied through the channel that is being formed.

Having briefly described the situation in the study area, it is now appropriate to review some of the research work on water use efficiency associated with the various systems. It will be noticed that the amount of such work with basins is much less than that with furrows—one explanation for this is

FIGURE 13.11 The basins on the right have been irrigated prior to a crop being transplanted into them (the water supply has been fed in from the right-hand side).

that much of it has been done with large-scale mechanized farming in mind and with irrigation on such a scale it is impractical to think of using basins.

FIGURE 13.12 A sand bund in the process of being broken to allow water to enter the basin.

FIGURE 13.13 Water advancing along an unlined channel in a basin irrigated area.

13.6.3 BASIN IRRIGATION

Basins are most suited to cereals, orchards. and rice crops. This method of irrigation is described as the simplest and oldest of all irrigation systems and it involves dividing the area or farm into rectangular or square units and then flooding them; this is most satisfactorily done in areas where the topography is relatively flat.[32]

The dimensions of the units are adjusted to the type of soil and the rate at which water can be supplied. Soil bunds are made to form the edges of these units and they retain the water which flows into the basin through a man-made break in the bund.

In some cases, the water will remain in the basin for several hours or even days, the time is influenced by the type of soil and the amount of water applied.[32] It follows that the system is only ideal for crops such as rice and orchards that can withstand short-term water logging; other crops, such as vegetables, maize, and cereals, can also be grown in basins only when the soils are such that water infiltrates more quickly.

It is important to note that soil type has relatively little effect on intermittently flooded basins and according to the authors, this can be explained by the fact that the basins are small, so they flood quickly and can usually be made level. If, however, the surface of the basin slopes or undulates this will influence distribution efficiency adversely. It is worth noting that on the lighter soils, there is approximately 10% difference in water utilization efficiency between sprinklers and basins and between basins and the furrow methods used.

13.6.4 BORDER IRRIGATION

This method of irrigation uses parallel earth bunds or borders which are used to control a sheet of water flowing across the field. The area enclosed between two borders is known as a strip and its width will vary from 3 to 30 meters and length from 100 to 800 meters.[28]

It is essential that there are no cross slopes within the strips otherwise the flow will be mainly adjacent to the lower border and they have to be either flat or graded (3% maximum). If flat they are provided with built-up dikes at the downstream end to control runoff.[17] The supply ditch from which water is siphoned is located along the highest contours and for effective distribution, which is by gravity, minor undulations within the strip must be removed.

This method is suitable for relatively large fields (4 ha or more) and it requires relatively large flows of water because the whole area has to be covered with water during irrigation. It is best suited for deep rooting crops (such as orchards, grass, Lucerne, and grain) growing on deep, medium-textured soils.[57]

This system is practiced in the study area, mostly on the LSI scheme at Kadawa where the rate at which water can be applied is adequate. Even when the soil and leveling are ideal, the efficiency of water utilization is lower than that in basins;[17] in areas with variable soil efficiency will be adversely affected, but the attraction of the system is that both capital investment and labor requirements are low.

13.6.5 FURROW IRRIGATION

Furrow irrigation is accomplished by running water in small channels that are constructed either down the slope of a field or across it in such a way that water will infiltrate into the soil from the bottom and sides of the furrows. It moves laterally and downwards to wet the soil, and in doing so it transports soluble salts, fertilizers, and herbicides with it.[1] For this system to function well, it requires a slope of between 0.5 and 1.5%; if slopes exceed 2% erosion will become a problem;[14] if they are too small movement along the furrows will be too slow.

Several workers have examined the method of improving water application with furrows, the aim usually being to cut water use while sustaining little or no reduction in crop yield.[1,37,47,56] The methods investigated include the use of more widely spaced furrows than normally practiced, only irrigating alternate furrows in row crops, alternating the furrows each time irrigation occurs, and furrow compaction.

Furrow irrigation is again an ancient method of distribution, predating the introduction of water pumps and all the systems of irrigation that depend on them. Its main drawbacks are that careful land preparation is required and that there is inevitably uneven application of water along the furrow, with the amount applied falling as the distance from the feed-in point increases.

Despite these drawbacks, furrow irrigation is still widely practiced even in areas where pumps and overhead irrigation equipment are available; one reason for this is that the capital and energy requirements of the furrow approach are very low.

Furrow irrigation also has attractions because it only wets part of the soil surface, thereby reducing the losses through direct evaporation; alternate or "skip" furrow irrigation takes the advantage further than the irrigation of every furrow.

All the work carried out by the above workers was designed with large irrigation schemes in mind, with plant row lengths of up to 800 m and in situations where sophisticated irrigation facilities were available.[61] This does not mean that this work is of little relevance to the situation faced by the small-scale riparian farmer, for the basic information gained on economical water use and its relationship to crop yield will presumably still be valid.

The high evaporation rates around Kano could make the alternate furrow technique potentially attractive so this aspect of the body of research work is presented first. One report on research carried out in Texas High Plains, using three crops (sorghum, sugar beet, and potatoes) for five seasons. It was shown that irrigating alternate furrows reduced water intake by 13–33% without significant loss in yield when compared with the normal system of irrigating every furrow. Two furrow spacings were used in the work (1.5 and 2.0 m) and the soil was a slowly permeable clay loam.

Researchers have also found that irrigating alternate furrows was detrimental where the length of the furrow was greater than 350–550 m. Most of the yield loss with these very long rows occurred toward the lower end of the length of the row.[46] It must be assumed that the water supply had to be turned off to minimize tail runoff losses before it had time to move sufficiently sideways in this part of the field. On large fields, there is obviously a minimum length of the furrow that it is practical to use, although this is difficult to define; "short" furrows will require extra cross ditches to be made to collect tail-water and this will increase labor demand and involve some loss of land that might be cropped.

The tail-water problem is more serious in large irrigation schemes than in small-scale riparian farms, partly because of the scale of the operation and partly because the person lifting the water will ensure that the results of his work will not be wasted by running it out of the far end of the scheme.

Four-year research program was conducted on alternate furrow irrigation.[22] The soil was a silty clay loam and two varieties of soybean were used in the work. The average yield from alternate furrow watering over the period showed a statistically significant reduction in yield compared to every furrow irrigation, nevertheless, acceptable yields were still produced with 40 – 50% saving in water. This level of water saving was so substantial that despite the lower yields it was an acceptable trade-off in an area where water is particularly scarce.

Work on widely spaced furrow irrigation was carried out at Goodwell—USA.[61] The experiments were conducted on 60 m and 800 m long plots and the beds were 1.4 m and 1.0 m wide, respectively. The crop used was grain sorghum and it was planted on Richfield clay soils in both experiments. The results show that decreases in yields can be reduced if the practice of widely spaced furrow irrigation is abandoned at times when critical climatic conditions and stages of crop growth occur. A similar effect can be achieved by alternating the irrigated furrows in alternate furrow irrigation.

In both cases, the expected water saving was found to range from 20 to 50%, depending on (i) the quantity of water used prior to planting and (ii) when it became necessary to revert to every furrow irrigation to avoid crop stress.

In mechanized systems of row crop farming where wide machines are used, only some of the furrows will coincide with the tractor wheels if conventional spacing is employed. If all furrows are then irrigated, this will result in excessive ran off from furrows that have carried the tractor, because the compaction will increase the rate of water advancement.

Where widely spaced furrows are used and irrigation is restricted to the wheel tracks, compaction will tend to limit the amount of water that can be applied at any one time. This effect was investigated;[47] they found that the average water intake during one season's irrigation of wheel track was

82 mm, compared to every furrow irrigation. This was accompanied by a drop in the profile drainage losses from 29.4 to 9.1% of the water applied, more significantly the reduced water intake due to furrow compaction did not affect grain yields.

13.6.6 OTHER LOW-COST IRRIGATION SYSTEMS

There are relatively few documented works covering technical and management issues in African SSI.[13] This was still the case as confirmed in early 1990 by a computer search of the literature, however, since then a report of preliminary work carried out on low-cost, high efficiency irrigation systems at Chiredzi, south-east Zimbabwe has been produced.[15]

The work in Zimbabwe involves the use of small plots or gardens and the aim was to evaluate several cheap methods of irrigation that employ a level of technology that is suitable for use by farmers in rural areas. These approaches were all capable of operating at low water pressures and the view after the experiment had been set up was that they were all potentially more efficient than conventional methods of irrigation.

The investigations studied three irrigation methods—pitcher, subsurface, and low-head drip, and two variations of each of the three were compared. The crop used in the initial experiment was field beans and the soil, which was described in great detail, was of a "heavy" type.

FIGURE 13.14 A schematic of pitcher irrigation.

Equal amounts of water were applied to each treatment at every irrigation, and Figure 13.14 shows the pitcher irrigation arrangement. The yields produced ranged from 1.11 ha^{-1} (small pitchers) to 1.71 ha^{-1} (emitter).

Unfortunately, the treatments in this experiment were not replicated, it was not laid out as a randomized design and it did not have a control treatment such as basin irrigation, so the conclusions reached are of limited value. It was reported that among the objectives of the experiment were studies of soil moisture distribution and crop water use, and some progress.

In the final report on the work in Zimbabwe, Batchelor and his colleagues[39] include a further method of irrigating small plots, namely flood irrigation mulched ground. One of the major aims of this work was to improve irrigation efficiency by reducing non-productive soil evaporation which, in a flood-irrigated maize crop, has been shown to reach an amazing 54% of the amount of water applied.

Another low technology irrigation system known as the bubbler system has also been studied. The system was claimed to have advantages over the conventional trickle systems—it is cheaper, easier to install and there is minimum risk of pipe blockage.

The operating pressure of the system is l–5 m head (which suits the manually operated lifters or hand pumps) and in situations where the topography is suitable the system could be connected to wells using a rope-and-bucket lifter.

The use of a watering can to irrigate crops is perhaps the simplest method of overhead irrigation and does not require any investment in equipment other than the container. It is not widely used in the study area, but it can be found where individual trees have to be irrigated or where the irrigation water has to be lifted from a very deep ground; water table, as sometimes occurs in upland situations, when it is usual to employ a rope-and-bucket. The area of the "farm" that can be irrigated in this way depends on the time taken to fill the watering can/bucket and the distance from the water source to the crops. It was reported that it was realistic for a man with a plot adjacent to the water source to manage to irrigate about 500 m^2, whereas if the distance between the two was 100 m he could only manage 250 m^2.

The difficulties involved in carrying out censuses in Africa are well known, nevertheless, it is generally accepted that Nigeria is the continent's most populous country. As is the case in most developing countries, a considerable proportion of the population is involved in the agricultural sector; in 1977 NWA reported that 75–80% of Nigerians were thus employed, but since then there has been some decline in this proportion as other industries have been introduced and developed.

Agriculture is largely in the hands of small-scale farmers, as in many other African countries: Uganda, for example, has reported that their small-scale farmers constitute about 90% of the farming population.[9]

In addition to its enormous human resources, Nigeria is blessed with the basic essentials necessary for food production: there is a more than adequate supply of land that can be cultivated and a considerable proportion of the country enjoys both a climate that is suitable for crop growth and a reasonably adequate supply of water. What is sometimes lacking is effective management and utilization of these resources. Where water is scarce, management and skill in utilization are even more important and without these, the goal of self-sufficiency in food production is difficult to achieve. Researchers have spelled this out in a paper on manpower planning in irrigated agriculture when they said that in Nigeria it was considered that the most important general skills which are missing are those of management.[18]

Training is crucial in the overall performance of both small-scale and large-scale farmers. Contact with farmers during the experimental work in Nigeria over the last six years has tended to confirm these views; in particular, it seems that emphasis should be put on the training of small-scale farmers by local extension workers, especially on the importance of good water management. The impression gained is that this aspect is reasonably well understood on some large farms where the staff tend to have access to relatively well-qualified advisers and modern technology.

Apart from the few large irrigation schemes, such as the Kano River Project and those operated by other River Basin Development Authorities, the only areas that can be used for farming in Northern Nigeria during the dry season are the fadamas. These are either low-lying areas or are adjacent to supplies of water which allow them to be irrigated; their total area is limited to only 10% of the land surface.[64,65]

In a "normal" year, there are adequate supplies of food, some coming from the rain-fed agricultural areas and the remainder from irrigated areas. However, the country is occasionally subjected to droughts, the most recent one being in 1987. In that year thousands of peasant farmers and their families were reported to be facing starvation because of the combination of a drought and a locust invasion; the worst hit area was Kano State.[71] Not only did rain-fed crops fail but there was an inadequate supply of water in the reservoirs for all the irrigation needs to be met.

During the drought, people were forced to migrate, causing hardship and social problems in the cities, which they tended to move in the hope of finding sources of income and food to purchase. The response of the government was to provide some 8500 bags of assorted grains for sale to farmers at subsidized prices, but this was obviously not enough to keep starvation at bay.[71]

As well as having a good potential to produce food, Nigeria is also an oil-producing nation. Unfortunately, the known oil reserves are limited and currently production levels are declining. In such a situation, it is important for the nation's prosperity that the food-producing sector becomes more efficient—it could make the country a food exporter let alone make it self-sufficient in food. But to do this in the northern parts of the country means that water supplies must be adequate and reliable.

Some writers have stressed the importance of encouraging self-reliance in individuals, and it is agreed that this must be promoted by all reasonable means, but when there is a shortage of an essential input into a system (in this case water) there is little the individual can do: planning and implementation of actions to guarantee such fundamental inputs requires regional or government action.

Various suggestions have been made as to how best to ensure the security of adequate water supplies for both human and agricultural purposes in Northern Nigeria. Among other suggestions several writers—for example, researchers have advocated an inter-basin transfer of water, from areas where there is more than enough to areas of scarcity.[44] Unfortunately, this would involve major engineering work and the cost would be enormous. Technically such a project is possible, but it cannot proceed without careful analysis of hydrological and hydro-geological data, good planning and design, and a comprehensive cost-benefit analysis to determine its viability.

Another way in which it is thought that water supplies may be improved in Kano State is by conducting a detailed investigation of the Kano Hadejia River system to determine the areas of heavy losses.[75] Possible solutions to the problem of losses might emerge; for example, it might be possible to line the river in some areas, but this is unlikely to happen because of cost. A more realistic outcome might be to accept these seepage losses and tap them by pumping out from wells along the river system in the affected areas, rather than from the river itself.

The third important factor that will improve the water supplies in the area is the effective utilization of the resource by all parties involved: for example, the Water Board and its consumers should detect leakages in pipelines at early stages and effect repairs promptly. Metering of all the consumers of piped water and increasing the water rates would encourage everyone to attach a higher value to the resource. Whether this is done or not, is a political issue which will have to be given due consideration. In agriculture, the most obvious measure is that as far as possible both the large irrigators and small-scale farmers should be properly trained and monitored to ensure good water management.

Farmers receiving irrigation water from the canals originating from the dams are charged for it, but the level of charges is minimal, and they are levied on a "per hectare" basis. Again, more careful use of this resource could follow a move toward metering, although it is recognized that this would be difficult to enforce and probably have political consequences.

Finally, water reservoirs should be constructed where feasible alongside smaller water courses so that the water that would have otherwise run to "waste" can be conserved and utilized by the small-scale farmers.

There is some debate about the definition of the term SSI. Large-scale is relatively easy to visualize, but as the physical size of the individual unit decreases it is difficult to know where to draw a line. The definition is "those schemes which are under local responsibility, controlled and operated by the local people in response to their felt needs."[67] Formal irrigation schemes have been defined as being large (the suggestion was over 20 ha) and managed by a body other than the farmers; presumably, this classifies all the remaining irrigators as "small."[18] Small-scale in the context of irrigation is a management concept rather than a matter of size as such. The key factor in the small-scale philosophy is to keep management as far as possible in the hands of individual farmers or farmers' groups.

The term micro-scale irrigation can also be confused with the term micro-irrigation, which was adopted by the International Commission on Irrigation and Drainage as an alternative to trickle, drip, or localized irrigation.[18,19,36]

The most recent definition of SSI takes into account three important aspects: these are in the order of importance, farmer control, technology, and scale.[18] Hence SSI is defined as irrigation, usually on small plots, in which small farmers have the major controlling influence, and using levels of technology that the farmers can effectively operate and maintain. However, in most areas of the developing world, where SSI is practiced the amount of water available is often the limiting factor. It is suggested that if possible this aspect should be included in any definition.

Many reasons were given as to why irrigation planning at the national level tends to concentrate on large schemes; with these planning, design, appraisal, funding, construction, and finally operation and management are more efficiently done in terms of cost ha^{-1}.[67] He also maintained that large schemes have more psychological appeal and prestige value to donors and politicians than small ones.

Small irrigation schemes have been studied in Bauchi State, which borders Kano State (BSADP, 1984). This study highlighted the fact that negligible attention had been given to small-scale irrigators either by government agencies or educational institutions. The report found that there was very little, if any, good empirical data on field water management practices under traditional (Shadouf) and improved (motor pump) irrigation systems in Nigeria, despite the traditional farmer's long history and positive contributions to the agricultural sector. This situation on the ground has changed little since the publication of the BSADP report.

Despite the problems of SSI, the writer's impression is that it is presently expanding in the study area. The population in the area has been increasing, there is an increased demand for the produce grown and prices are good, however, this does not mean that this sector is not market sensitive. The basic research work has not been carried out to identify the limit of the growth in the SSI sector: there are presumably finite water reserves available, and other constraints operate.[5] Among the latter are water management, crop disease, marketing, manpower, land tenure, and public health: yet another feature of the whole scenario is the potential conflict between irrigators and the Fulani pastoralists who now are beginning to find that areas their animals were previously able to graze are now being used to grow irrigated crops.

A literature review on water management, particularly the aspect that deals with the task of determining the most appropriate amounts of water to be used on crops, yields an enormous amount of information. The justification for this mass of experimental work is that if it can be transferred into practice it should assist farmers who should be interested in minimizing the depletion of both surface and groundwater resources, minimizing pumping energy costs or labor costs in water lifting and keeping production costs low. The real aim in places where water resources are limited should be to use it efficiently; in practice, this usually means minimizing the amount of irrigation water applied while at the same time suffering acceptable reductions in yield.

The keywords in the preceding two sentences are "should be." It is probably true to say that individuals are not usually particularly concerned about using resources sparingly to conserve them, especially when they are seen as being "free." If they have to be paid for by metering or by the effort of manual lifting a different perspective applies, if the cost is simply one of providing fuel for a pump, then this is less likely to inhibit the tendency to over-irrigate.

The literature review revealed a mass of research work that has influenced the development of irrigation systems and practices, mainly applied in the developed world. The work covers the wide spacing of furrows, the compaction and smoothing of irrigated furrows (to decrease rates of infiltration), the discipline involved in frequent monitoring to minimize tail-water runoff, the use of surge flow techniques in graded furrows, and the use of subsurface trickle and overhead methods of irrigation. Other factors associated with effective water utilization are the avoidance of deep cultivations wherever possible (to minimize evaporation losses) the recognition that surface cultivations should only be as frequent as is consistent with the needs for weed control, moves toward crops that can withstand reduced water application, the minimization or elimination of pre-planting irrigation, the application of mulching operations, the lining of channels and the use of substances which either obstruct the stomata or interfere with the internal metabolism of the crop to reduce its rate of water loss. It may also be possible to shelter or shade the crop in some way to modify its environment to one where its evapotranspiration rate is curtailed.

In almost every part of the world where irrigation is practiced, there is ample scope for improvement but a realistic view of what might be possible in any location must be taken. It is difficult to see how some of the factors mentioned above have much direct relevance to the riparian farmers who live around the Kano River and its tributaries. The map of the study area shows the rivers and all the roads in the area, including earth roads and tracks. No part of the area is completely inaccessible, but if one takes 2 km from even an earth road as being reasonably accessible, then only 19.0% of the study area falls into this category. If this exercise is extended to the proximity of such roads and tracks to the river and its tributaries, that is, where the riparian farmers are mainly to be found, then only 35.8% of their length is near a basic road.

The official advisory services in developed countries have computerized lists of farmers, they have telephone contact with them, they have the local press to carry notices of meetings and demonstrations, newsletters and emails, local radio, and television.

There are also advisory services financed by the many commercial concerns, although these will usually be orientated toward product or service promotion. Contrast this situation with that in which the riparian farmer/irrigator finds himself isolated, few of the services mentioned above are available, lacking basic education, possibly superstitious, without electric light and often concentrating on survival rather than advancement—then ask why even the most basic concepts of irrigation technology and water management are slow to be transferred.

Although there is much work required when preparing land for cropping, establishing the crops and looking after them while they are growing, by

25. Fraenkel, P. *Water Pumping Devices: A handbook for Users and Choosers*; Intermediate Technology Publications: London, 1986.
26. Grainger, L. *The Threatening Desert: Controlling Desertification*; Earthen Publications Ltd.: London, 1990.
27. Haider, H. Climate Change in Nigeria: Impacts and Responses. In *Knowledge, Evidence and Learning for Development*; 2019, vol 38, 68975 [Online]. https://www.preventionweb.net/publications/view/68975; (accessed Mar 31, 2023).
28. James, O. O. *Dam-affected River Channels and the Green Revolution Programme in the Savanna zone of Nigeria. Geographical Journal Nigeria - Consortium of Geography Students*. (COGS): Nigeria, 1984.
29. Kano State Agricultural and Rural Development Authority (KNARDA), *KNARDA General Cost Survey*, Kano, 1983.
30. Kano State Agricultural and Rural Development Authority (KNARDA), *Report on the 1984 Large-scale Reconnaissance Survey,* Kano, 1987; p 147.
31. Kano State Agricultural and Rural Development Authority (KNARDA). *Kano State: Nigeria Impact Evaluation Report Kano Agricultural Development Project (Loan 1 982-UNI)*, Kano, 1995.
32. Kassa, M.; Andualem, T. G. Review of Irrigation Practice in Ethiopia: Lessons from Israel. In *Irrigation & Drainage Systems Engineering*; 2020; vol 9 (1), pp 1–6.
33. Kennedy, W. K.; Rogers, T. A. *Human and Animal-Powered Water-Lifting Devices*. Intermediate Technology Publ.: London, 1985.
34. Klassen, G. The Rower Pump. *Appropriate Technology* **1980,** *7* (3), 9–10.
35. Kowal, J. M.; Knabe, D. T. *An Agroclimatological Atlas of the Northern States of Nigeria*. Ahmadu Bello University: Zaria, 1972.
36. Lambert, R. A. In *Micro-scale Irrigation in Africa*. In *Inca, M. (ed), Rural Development in Africa*. Proceedings of 13th WEDC Conference, Malawi, 17 April, 1987.
37. Liu, Y.; Cui, E.; Neal, A. L.; Zhang, X.; Li, Z.; Xiao, Y.; Du, Z.; Gao, F.; Fan, X.; Hu, C. Reducing Water use by Alternate-furrow Irrigation with Livestock Wastewater Reduces Antibiotic Resistance Gene Abundance in the Rhizosphere but not in the Non-rhizosphere. *Sci. Total Environ.* **2019,** *648,* 12–24.
38. Longtau, S. R. A Review and Description of Rice Production Systems in Nigeria: Multi-agency Partnership in West African agriculture. *Eco-systems Development Organization*, JOS: Plateau State – Nigeria, 2013.
39. Lovell, C. J.; Batchelor, D. K.; Mugweni, O. *Small-scale Irrigation Using Collector Wells Pilot Project - Zimbabwe: Final Report*. IH Report ODA #95/14, Institute of Hydrology: Wailingford – UK, 1996.
40. Maina, M. A.; Wayayok, A.; Sani, I. Soil Salinity Assessment of Kadawa Irrigation of the Kano River Irrigation Project (KRIP). *J. Food, Agric. Environ.* **2012,** *7* (10), 1028–1034.
41. Mbanasor, J. A.; Nwachukwu, I. N.; Agwu, N. M.; Onwusiribe, N. C. Impact of Climate Change on the Productivity of Cassava in Nigeria. *J. Agric. Environ. Sci.* **2015,** *4* (1), 138–147.
42. Mohammed, M.; Abdulhamid, A.; Badamasi, M.; Ahmed, M. Rainfall Dynamics and Climate Change in Kano-Nigeria. *J. Sci. Res. Rep.* **2015,** *7* (5), 386–395.
43. Molenaar, A. *Water Lifting Devices for Irrigation FAO Agricultural Development Paper No.60*. FAO: Rome, 1956.

44. Musa, M. Harnessing Ground Water, Surface Runoff and Non-perennial Streams in Semi-arid areas of Nigeria. *Paper Submitted at the National Workshop on Ecological Disasters-Draught and Desertification*, Kano–Nigeria, 1985.
45. Musa, M. *Report on Desertification and Desert Encroachment in the Eleven Northern* States of the Federal Republic of Nigeria, 1988.
46. Musick, J. T.; Dusek, D. A. Alternate Furrow Irrigation of Fine-textured Soils. *Trans. ASAE* **1974,** 17 (2), 289–294.
47. Musick, J. T.; Pringle, F. B.; Johnson, P. N. Furrow Compaction for Controlling Excessive Irrigation Water Intake. *Trans. ASAE* **1985,** *28* (2), 502–506
48. Musick, J. T.; Walker, J. D. Irrigation Practices for Reduced Water Application -Texas High Plains. *Appl. Eng. Agric.* **1987,** *3* (2), 190–195.
49. Muslim, I. Report on geological Cross-Sections in Kano State. *WRECA* **1984,** *1,* 1–30.
50. Nwa, E. U. An Evaluation of Small Pump and Shadouf Systems in Northern Nigeria. *Samaru J. Agric. Res.* **1981,** *1* (2), 191–201.
51. kna, E. I. U. Ergonomic Study of *Shadouf* Irrigation in Northern Nigeria. *J. Agric. Eng. Res.* **1989,** *43,* 137–147.
52. Odunuga, S.; Badru, G. Landcover Change, Land Surface Temperature, Surface Albedo and Topography in the Plateau Region of North-Central Nigeria. *Land* **2015,** *4* (2), 300–324.
53. Oiganji, E.; Igbadun, H. E.; Ahmed, A.; Oliver, U. E. Implication of Bakolori Dam Irrigation Activities on its Physical Resources. *Afr. J. Agric. Res.* **2015,** *10,* 690–696.
54. Olofin, E. A. *Some Aspect of the Physicals Geography of Kano Region and Related Human Responses*. Debris Standard Printers, 734 Zaria Road, Gyadi – Gyadi, Kano - Nigeria, 1984.
55. Orr, A.; Nazrul, A. S. M.; Barnes, G. *Manual Irrigation for Small Farmers in Bangladesh*. Bulletin 102, College of Agriculture, Bangladesh Agric. University, 1991.
56. Putt, S. N. H.; Shaw, A. P. M. *The Socio-Economic Effects of the Control of Tsetse Transmitted Trypanosomiasis in Nigeria,* Veterinary Medicine Pub. Co.: Kano, 1982.
57. Sable, R.; Kolekar, S.; Gawde, A.; Takle, S.; Pednekar, A. A Review on Different Irrigation Methods. *Int. J. Appl. Agric. Res.* **2019,** *14* (1), 49–60.
58. Salau, O.; Fasuba, A.; Aduloju, K.; Adesakin, G.; Fatigun, A. Effects of Changes in ENSO on Seasonal Mean Temperature and Rainfall in Nigeria. *Climate* **2016,** *4* (1), 5–20.
59. Sangari, D. An Evaluation of Water and Land Uses in the Kano River Project, Phase I, Kano State. *J. Appl. Sci. Environ. Manag.* **2010,** *11* (2), 105–111.
60. Singh, A.; Chari, A. V. Certain Aspects of Soil and Water Management in Major River Basins of Northern Nigeria. *Paper Presented at the 12th International Congress of Soil Science*, New Delhi – India, 1982; p 8.
61. Stone, J. F.; Reeve, H. E.; Carton, J. E. Irrigation Water Conservation by using Wide-spaced Furrows. *Agric. Water Manag.* **1982,** *5,* 309–317.
62. Thepwong, Development of Archimedean's Pipe-screw for Thai Sail Windpump. *Am. J. Environ. Sci.* **2014,** *10,* 244–259.
63. Thorne, M.; Wallender, W. W.; Chancellor, W. J.; Grismer, M. E. Performance Characteristics of the Shadouf: A Manual Water-lifting Device. *Appl. Eng. Agric.* **1999,** *15* (3), 225–231.
64. Turner, B. *The Fadama Lands of Central Northern Nigeria: Their Classification Spatial Variation, Present and Potential Use*. PhD Thesis, University of London, 1977.
65. Turner, B. *The Classification and Distribution of Fadamas in Central Northern Nigeria. Z Geomorph.* N.F. Supplement Bd52 87–113. Berlin-Stuttgart, 1985.

66. Umar, A. Problem of Water Supply in Greater KANO WATER SUPPLY AREA. Paper Presented at the Regular Bimonthly Meeting of the Kano State branch of the Nigerian Society of Agricultural Engineers, 1983.
67. Underhill, H. W. *Small-scale Irrigation in Africa in the Context of Rural Development.* FAO: Rome, 1984.
68. Vollset, S. E.; Goren, E.; Yuan, C.-W.; Cao, J.; Smith, A. E.; Hsiao, T.; Bisignano, C.; Azhar, G. S.; Castro, E.; Chalek, J.; Dolgert, A. J.; Frank, T.; Fukutaki, K.; Hay, S. I.; Lozano, R.; Mokdad, A. H.; Nandakumar, V.; Pierce, M.; Pletcher, M.; Robalik, T.; Steuben, K. M.; Wunrow, H. Y.; Zlavog, B. S.; Murray, C. J. L. Fertility, Mortality, Migration, and Population Scenarios for 195 Countries and Territories from 2017 to 2100: A Forecasting Analysis for the Global Burden of Disease Study. *Lancet* **2020,** *396,* 1285–1306.
69. Yannopoulos, S. I.; Lyberatos, G.; Theodossiou, N.; Li, W.; Valipour, M.; Tamburrino, A.; Angelakis, A. N. Evolution of Water Lifting Devices (pumps) Over the Centuries Worldwide. *Water* **2015,** *7* (9), 5031–5060.
70. Yamusa, I. B.; Yamusa, Y. B.; Danbatta, U. A.; Najime, T. Geological and Structural Analysis using Remote Sensing for Lineament and Lithological Mapping. *IOP Conf. Ser. Earth Environ. Sci.* **2018,** *169*, 012082. https://doi.org/10.1088/1755-1315/169/1/012082; (accessed March 31, 2023).
71. West Africa. *Famine in the Far North?* **1988,** 85.
72. White, G. F. *Environmental Effects of and Land Irrigation in Developing Countries.* UNESCO: Geneva, 1978.
73. Withers, B.; Vipond, S. *Irrigation: Design and Practice*. Bats Ford Academic and Educational Ltd.: London, 1985.
74. Wood, A. D. *Water Lifters and Pumps for the Developing World.* M.Sc. Thesis, Colorado State University: Collins - Colorado, USA, 1976; p 125.
75. WRECA (Water Resources & Engineering Construction Agency). *Ground Water Monitoring and Imbalance in Kano State*; 1985.
76. Olofintoye, Oluwatosin & Sule, B. & Salami, Adebayo. Best-fit Probability Distribution model for peak daily rainfall of selected Cities in Nigeria. New York Science Journal. 2009. 2. 1–12.

APPENDIX A Intercrops Crops in the Sample Area: Months of Planting and Harvesting.

Farm #	Distance between water and crop, m	Crops grown within the sample area	Month of planting	Month of harvest
1.	4.5m	Maize and sugarcane	November	March/Sept.
2.	7.0	Sugarcane	November	September
3.	9.5	Sugarcane	November	September
4.	7.5	Vegetables and sugarcane	November	Jan/Sept
5.	10.0	Sugarcane	November	September
6.	20.0	Vegetables and sugarcane	November	Jan/Sept
7.	6.0	Vegetables and sugarcane	November	Jan/Sept
8.	8.0	Maize and sugarcane	November	March/Sept
9.	15.0	Vegetables and sugarcane	November	Jan/Sept
10.	12.0	Maize and sugarcane	November	March/Sept
11.	13.0	Maize and sugarcane	November	March/Sept
12.	15.0	Vegetables, maize, and sugarcane	November	Jan/March/Sept
13.	31.0	Maize and sugarcane	November	March/Sept
14.	10.0	Maize and sugarcane	November	March/Sept
15.	12.0	Vegetables, maize, and sugarcane	November	Jan/March/Sept
16.	16.5	Vegetables	November	January
17.	20.5	Vegetables	November	January
18.	16.51	Vegetables and sugarcane	November	Jan/Sept.
19.	6.0	Vegetables and sugarcane	November	Jan/Sept.
20.	12.0	Vegetables, maize, and sugarcane	November	Jan/March/Sept.
21.	15.0	Vegetables and sugarcane	November	Jan/Sept.
22.	15.0	Vegetables and sugarcane	November	Jan/Sept.
23.	6.5	Vegetables	November	January
24.	7.5	Vegetables and sugarcane	November	Jan/Sept.
25.	5.0	Vegetables and sugarcane	Jan/Sept.	Jan/Sept.
26.	16.0	Vegetables	November	January
27.	15.0	Vegetables	November	January
28.	20.0	Vegetables	November	January
29.	5.0	Vegetables and sugarcane	November	Jan/Sept.
30.	10.0	Vegetables	November	January
31.	5.0	Vegetables	November	January
Avg.	11.98			

APPENDIX B Demographic Information for the Farming Community.

Farmer	Age (yrs)	Weight (kg)	Marital status	Children (boys only)	Area of land farmed (ha)	No. of plots in the sample area	No. of plots outside the sample area
1	65	60	Married	2	0.60	1	1
2	60	52	Married	5	0.40	1	1
3	80	50	Married	4	1.60	1	2
4	65	55	Married	3	0.40	2	1
5	5	48	Married	5	2.00	2	3
6	55	60	Married	2	0.70	1	1
7	58	51	Married	1	0.80	1	1
8	45	54	Married	1	0.06	1	0
9	50	60	Married	1	1.20	1	2
10	45	59	Married	1	2.00	2	3
11	30	75	Married	3	1.80	1	3
12	67	55	Married	2	0.24	1	1
13	37	58	Married	5	0.80	1	1
14	30	60	Married	0	1.20	1	2
15	50	58	Married	3	0.04	1	0
16	50	60	Married	2	2.00	2	2
17	45	63	Married	3	1.60	1	2
18	42	66	Married	1	1.00	1	2
19	60	58	Married	3	1.00	1	1
20	45	62	Married	4	0.40	1	1
21	25	53	Married	1	1.00	1	2
22	32	49	Married	1	1.20	2	2
23	52	58	Married	1	0.40	1	1
24	27	53	Married	2	0.60	1	1
25	40	52	Married	4	1.60	1	2
26	40	61	Married	2	1.20	2	1
27	45	58	Married	2	0.20	1	1
28	44	66	Married	3	0.80	1	1
29	35	53	Married	2	0.44	2	1
30	32	53	Married	3	0.36	1	1
31	71	62	Married	4	1.20	2	1
Average	47.65	57.58		2.45	0.93	1.26	1.42

APPENDIX C Some Water Lifting Techniques.

Farmer	Method of lifting water	Pump owned/ shared	Age of pump (years)	Hours of operation per day	Pattern of irrigation
1	50 mm pump	Shared	8	2	8 am–10 am
2	50 mm pump	Shared	5	2	4 pm–6 pm
3	50 mm pump	Shared	5	2	4 pm–6 pn
4	50 mm pump	Shared	3	4	8 am–12 noon
5	50 mm pump	Shared	6	5	8 am–1 pm
6	50 mm pump	Shared	5	3	8 am–11 am
7	50 mm pump	Shared	8	2	4 pm–6 pm
8	50 mm pump	Shared	7	2	4 pm–6 pm
9	50 mm pump	Shared	6	2	8 am–10 am
10	75 mm pump	Owned	6	4	8 am–12 noon
11	75 mm pump	Shared	6	2	S am–10 am
12	50 mm pump	Shared	4	2	8 am–10 am
13	50 mm pump	Owned	3	2	4 pm–6 pm
14	50 mm pump	Owned	3	3	8 am–11 am
15	50 mm pump	Owned	7	2	4 pm–6 pm
16	50 mm pump	Owned	7	3	7 am–10 am
17	50 mm pump	Owned	7	2	4 pm–6 pm
18	50 mm pump	Owned	3	2	4 pm –6 pm
19	50 mm pump	Owned	5	2	10 am–12 noon
20	*Shadouf*	N/A	N/A	2	8 am–10 am
21	50 mm pump	8	8	2	8 am 10 am
22	50 mm pump	2	2	2	4 pm –6 pm
23	*Shadouf*	N/A	N/A	2	9 am–11 am
24	50 mm pump	Shared	7	3	9 am–12 noon
25	50 mm pump	Owned	7	2	4 pm–6 pm
26	50 mm pump	‘ ’	3	3	8 am–11 am
27	50 mm pump	‘ ’	5	2	4 pm–6 pm
28	50 mm pump	‘ ’	7	2	8 am–10 am
29	50 mm pump	Shared	5	2	8 am–10 am
30	*Shadouf*	N/A	N/A	2	4 pm–6 pm
31	50 mm pump	Owned	6	4	8 am–12 noon
Avg.	–	–	5.10	2.45	–

INDEX